ADVANCES IN
MOLECULAR AND
CELL BIOLOGY

Volume 12 • 1995

CYTOSKELETON

ADVANCES IN MOLECULAR AND CELL BIOLOGY

CYTOSKELETON

Series Editor: E. EDWARD BITTAR
Department of Physiology
University of Wisconsin
Madison, Wisconsin

Guest Editor: JOHN HESKETH
Rowett Research Institute
Aberdeen, Scotland

VOLUME 12 • 1995

JAI PRESS INC.

Greenwich, Connecticut *London, England*

55 Old Post Road, No. 2
Greenwich, Connecticut 06836

JAI PRESS LTD.
The Courtyard
28 High Street
Hampton Hill, Middlesex TW12 1PD
England

Transferred to digital printing in 2009

ISBN 9781559388450

CONTENTS

LIST OF CONTRIBUTORS

Kenneth A. Beck
Department of Molecular and Cellular Physiology
Stanford University School of Medicine
Stanford, California

Avri Ben-Ze'ev
Department of Molecular Genetics
Weizmann Institute of Science
Rehovot, Israel

Scott T. Brady
Department of Cell Biology
University of Texas Southwestern Medical Center
Dallas, Texas

Janet L. Cyr
Department of Cell Biology
University of Texas Southwestern Medical Center
Dallas, Texas

José Luis Rodríguez Fernández
Department of Molecular Genetics
Weizmann Institute of Science
Rehovot, Israel

Eric Fyrberg
Department of Biology
Johns Hopkins University
Baltimore, Maryland

Benjamin Geiger
Department of Chemical Immunology
Weizmann Institute of Science
Rehovot, Israel

Ursula Glück
Department of Molecular Genetics
Weizmann Institute of Science
Rehovot, Israel

Arild Johannessen
Department of Biochemistry and Molecular Biology
University of Bergen
Bergen, Norway

E. Birgitte Lane — Department of Anatomy and Physiology
University of Dundee
Dundee, Scotland

W. James Nelson — Department of Molecular and Cellular Physiology
Stanford University School of Medicine
Stanford, California

Ian F. Pryme — Department of Biochemistry and Molecular Biology
University of Bergen
Bergen, Norway

Beat M. Riederer — Institute of Anatomy
University of Lausanne
Lausanne, Switzerland

Daniela Salomon — Department of Chemical Immunology
Weizmann Institute of Science
Rehovot, Israel

Ann O. Sperry — Department of Cell Biology
University of Texas Southwestern Medical Center
Dallas, Texas

John Trinick — Department of Veterinary Medicine
Bristol University
Bristol, England

Anni Vedeler — Department of Biochemistry and Molecular Biology
University of Bergen
Bergen, Norway

Margot Zöller — Department of Radiology and Pathophysiology and Pathophysiology
German Cancer Research Center
Heidelberg, Germany

PREFACE

It is now clear from a wide range of research that cytoplasm is not merely a buffered solution of proteins and enzymes but contains a series of complex filamentous structures. The cytoskeleton is the collective term given to these filaments. There is a considerable amount of data available on the protein composition of the major filament systems (microfilaments, microtubules, and intermediate filaments) but we are still comparatively ignorant about the role of the cytoskeleton in cell physiology. However such major cytoplasmic components (actin and tubulin, the monomeric constituents of microfilaments and microtubules, are major cell proteins) must have important roles to play in cell function, and investigations into the functional role of the cytoskeleton currently represent a major area of cell biological research.

Much of our understanding of the actin microfilament system has come from studies of the specialized muscle cytoskeleton, the sarcomere. Recently, it has been found that the muscle sarcomere contains several very large proteins of unusual properties—titin and nebulin. The first chapter in this volume (by Trinick) describes the structural characteristics and possible functions of these proteins; if non-muscle counterparts of these proteins exist they could be very important in determining the organization of the cytoskeleton and thus imparting important effects on cell shape and motility.

It has long been surmized that the cytoskeleton may play a role in cell organization: for example, in determining the spatial arrangement of membrane receptors and thus in the development of cell polarity; and in providing a structural basis for the organization of glycolytic enzymes. In recent years it has become evident that a proportion of cell polyribosomes is associated with the cytoskeleton and that this may be an important factor in protein synthesis. This, together with the evidence for the role of the cytoskeleton in the transport of mRNAs from the nucleus, is discussed in the chapter by Pryme et al. It now appears that the cytoskeleton may be crucial in protein sorting, both by targeting of the mRNAs so that synthesis occurs in specific sites and by subsequent transport and targeting of the proteins. There is now considerable evidence that the latter, as discussed by Nelson and colleagues, represents an important mechanism in the development of cell polarity. These areas of investigation will become of increasing importance in cell biology as we struggle to understand factors governing the regulation of cell shape and form both during differentiation and during tissue formation.

There has been considerable interest in the cytoskeleton of neurones for a number of years, as this reflects the potential roles of the cytoskeleton in intracellular transport and cell shape as well as the significant amounts of cytoskeletal proteins found in neurones, in nerve endings, and in dendrites. The neurone is a cell where subcellular compartmentation and therefore targeting of components to specific regions appears to be of particular importance. It is of potentially great interest therefore that different regions of the neurone (e.g., axon and dendrite) differ in the properties and isoform composition of the cytoskeleton; aspects of such subcellular organization is discussed by Riederer.

It is becoming increasingly appreciated that the interactions of cells with each other and with the extracellular matrix is important in regulating a wide range of cellular functions. Since the cytoskeleton appears to be directly linked to membrane receptors involved in cell–cell or cell–matrix interactions, it is likely that it may have a crucial role to play in the regulation of cell function by extracellular interactions. This type of regulatory role is discussed in the review by Ben-Ze'ev et al., who have used molecular biological techniques to investigate the role of various cytoskeletal proteins in the mechanisms by which the extracellular matrix influences cell growth.

In recent years rapid advances in molecular biology have begun to influence research on the cytoskeleton. This trend is sure to continue and the techniques of molecular biology and genetics are set to make major contributions to our understanding of the cytoskeleton, as illustrated in this volume by several reviews; the use of transfection techniques by Ben-Ze'ev, the power of *Drosophila* genetics is described by Fyrberg and the major advances made in the kinesin field using molecular approaches as described by Cyr et al. The chapters by Fyrberg and Cyr et al. also illustrate two other areas where major advances in our understanding of the cytoskeleton is occurring; the great array of different motor proteins involved in intracellular movements and the study of the cytoskeleton in developmental biology.

Overall, Volume 12 in this series illuminates our increasing knowledge of the important roles of the cytoskeleton in cell function, particularly how it is central to metabolic organization, intracellular transport, interactions with matrix, and nerve function. Our knowledge of the cytoskeleton is now reaching a stage where it is clear that abnormalities in the organization of the cytoskeleton can lead to important clinical manifestations of disease; an example of how such research is now impinging on medical science is presented in the final chapter by Lane on keratin diseases.

John Hesketh
Guest Editor

ASPECTS OF THE MUSCLE CYTOSKELETON

John Trinick

Advances in Molecular and Cell Biology
Volume 12, pages 1–39.

ISBN: 1-55938-845-5

I. INTRODUCTION

The cytoskeleton is usually taken to mean the structural apparatus by which cells maintain and adapt their shape in response to different circumstances. The elements involved in this apparatus include microtubules, intermediate filaments, and actin filaments (see Small, 1987). In the particular case of muscle, the primary cell function is to produce axial shortening and tension, based on the force produced by interaction between the thick and thin filaments. Here, the term cytoskeleton is often used in a more restricted sense, to mean the protein structures that are responsible for assembling the contractile apparatus correctly, and for maintaining its correct spatial organization. Also relevant to this subject are the intermediate filament connections between myofibrils (e.g., Wang, 1984) and to the vinculin-containing costameres that connect myofibrils to the periphery of the cell (e.g., Pardo et al., 1983).

To adequately cover these diverse topics in detail here is not possible and this article concentrates on the mechanisms by which the internal structure of the mature myofibril is specified and maintained. One of the most remarkable features of muscle is the great precision with which the structure of the myobribril is specified. The past three years have seen a rapid increase in knowledge in this area and this review is mainly concerned with these developments. Much recent progress has centered on elucidation of the properties of the giant proteins titin and nebulin. Evidence now suggests that molecules of both these proteins are approximately 1 μm long and that they are arranged parallel to the thick and thin filaments *in situ*. Both molecules span approximately half the sarcomere and are therefore in a position to influence accurate assembly and maintenance of the myofibril structure over what are, in molecular terms, very large distances.

II. TITIN

A. Nomenclature

Titin is also known as connectin and there is unfortunately no agreement yet as to nomenclature. Some of the work on connectin predates that on titin (Maruyama et al., 1976), but it was only with the advent of large-pore SDS polyacrylamide gels (Wang et al., 1979) that it became clear the chain weights of the proteins being dealt with are extremely large. Both titin and connectin preparations could initially only be isolated in denatured form and both preparations tended to be contaminated by other proteins (titin by nebulin and connectin by actin and phosphorylase); however, used in the present context of well purified preparations titin and connectin are synonyms.

B. Amount of Titin Present in Muscle

Titin is present in many types of vertebrate striated muscle (Hu et al., 1986; Locker and Wild, 1986; Labeit et al., 1990), and in rabbit skeletal muscle forms roughly 8% of myofibrillar protein (Trinick et al., 1984). It is not present in smooth muscle but has recently been reported in intestinal brush border (Eilertson and Keller, 1992). In invertebrates there is a smaller related protein called twitchin (see below).

C. Molecular Weight

Since titin only just entered very weak (~3%) polyacrylamide gels, it was evident that its chain weight was likely to be in the megadalton range. It was not a trivial task to calibrate gels in this range or to determine the chain weight by classical analytical ultracentrifugation and several years elapsed before there was agreement on even the approximate chain molecular weight. A value of ~3 MDa for the chain weight was obtained, based on calibration of SDS polyacrylamide gels with cross-linked oligomers of the myosin heavy chain (Maruyama et al., 1984), and these workers also measured 2.8 MDa for the native protein, using sedimentation equilibrium under native conditions. Subsequently, sedimentation equilibrium in guanidine hydrochloride gave a value of 2.4 MDa for the chain weight of a titin fragment (Kurzban and Wang, 1988). The fragment used in this work migrates significantly more rapidly than the parent molecule on gels and it therefore seems likely that the chain weight of the parent molecule is approximately 3 MDa. Also consistent with a value in this range is the figure of 2.7 kDa/nm for the mass/unit length of the titin molecule, determined from scanning transmission electron microscopy (STEM) (Hainfeld et al., 1988). Since the titin molecule is now thought to be approximately 1 μm long, the STEM data also imply a molecular weight ~3 MDa. As such, titin is the largest polypeptide yet described.

D. Shape of the Titin Molecule

Initially, titin could only be purified in denatured form by gel filtration chromatography in SDS, taking advantage of its huge size to separate it from the other muscle proteins which are all substantially smaller (Wang et al., 1979). Subsequently, however, it was found possible to solubilize it from muscle in high salt and then to purify it by conventional methods (Maruyama et al., 1984; Trinick et al., 1984; Wang et al., 1984). These preparations were examined by electron microscopy to give the first information about the shape of the molecule. When visualized after rotary metal shadowing, titin molecules were seen to have a string-like appearance with a width of about 4 nm and with variable length up to approximately 1 μm long. The convoluted tracks followed by the strings suggested considerable flexibility.

These early preparations of titin were isolated by procedures that took several days. The heterogeneity of lengths of the strings seen by electron microscopy and increases in mobility on gels both suggested that during this time degradation was taking place, caused by endogenous proteinases. Recently improved methods have been developed that reduce or completely stop this degradation by accelerating purification and including proteinase inhibitors (Nave et al., 1989; Kimura et al., 1992; Soteriou et al., 1993b). A modified method, involving removal of buffer by

Figure 1. Electron microscopy of purified titin molecules. The molecules are visualized by rotary shadowing with platinum. Prior to drying and shadowing the molecules were straightened by a flow of solvent after being deposited on a mica substrate. Magnification ×53,000. (Reproduced from Soteriou et al., 1993b.)

centrifugation prior to drying *in vacuo*, has been used to prepare the molecules from these preparations for electron microscopy (Nave et al., 1989); when this was done the flow of buffer straightened the molecules, which considerably simplified their appearance (Figure 1). The straightened molecules were all approximately 900 nm long which is consistent with the molecular length inferred from antibody labelling studies of whole muscle and myofibrils (see below). A new feature visible in these micrographs was an enlarged head region at one end of the molecule and sometimes the molecules were seen to be dimerized or trimerized through interactions between these heads (Nave et al., 1989).

In addition to the ~3 MDa polypeptide visible on gels of titin, the preparations also showed evidence of smaller polypeptides. The presence of proteins with chain weights of approximately of 190 and 165 kDa was reported (Nave et al., 1989). Antibodies showed these to be immunologically distinct and to be both located in the head region of the titin molecule. *In situ* the antibodies bound to the M-line demonstrating that the titin head is part of the M-line. The 165-kDa protein is an M-line component which had already been described called myomesin (Grove et al., 1984), although its molecular weight was lower than previously estimated. In addition, the presence of a band of about 100 kDa in titin preparations was described (Kimura et al., 1992; Soteriou et al., 1993b). This had a mobility very similar to α-actinin, although peptide fingerprinting and antibody cross-reactivity did not confirm this identity. To date, it is not known whether these smaller polypeptides are subunits of the titin molecule, or whether they are separate proteins that co-purify, perhaps reflecting an interaction *in vivo*.

E. Location *In Situ*

Early work on titin and connectin led from the outset to suggestions that these preparations originated from a third type of filament in the sarcomere, additional to the thick and thin filaments. The model of sarcomere structure that has formed the basis for most modern work on muscle contraction includes only the thick and thin filaments; but there were consistent indications, in work dating back many years, that there is at least one more type of longitudinally orientated filament present. It was therefore not unnatural to suggest that the titin (and connectin) might form or be part of such a third set of filaments.

One of the most striking pieces of evidence for the presence of such extra filaments came from work on muscle stretched to the point where the thick and thin filaments no longer overlap (Sjostrand, 1962). Bridging the gap between the ends of the thick and thin filaments in thin sections of such muscles were thin fibrous connections termed gap filaments. Since this gap could be varied in size, the micrographs also suggested that gap filaments might be elastic. Gap filaments have also been extensively studied by Locker (e.g., Locker and Leet, 1976). Other electron microscope evidence for extra sarcomere filaments includes structures termed S-filaments (Huxley and Hanson, 1954), super-thin filaments (McNeill and

Hoyle, 1967; Walcott and Ridgeway, 1967), C-filaments (Auber and Couteaux, 1963), connecting filaments (Pringle, 1977) and connectin filaments (Dos Remedios and Gilmour, 1978).

The difficulty with most of this evidence was that it was based on thin sections of muscle embedded in plastic. Such sections do not normally retain details smaller than about 5 nm and this was the approximate size of the extra filaments. Additionally, it also now seems likely that the extra filaments tend to become cross-linked to thin filaments during chemical fixation for electron microscopy. It was therefore difficult to distinguish the extra filaments from thin filaments or to be sure about their structure. For several years it appeared that they were only present in certain subsets of muscles, such as those used by insects for flight (Pringle, 1977), but it is now believed that they are ubiquitous in striated muscles. A recent and direct demonstration of the presence of extra filaments in the I-band was revealed by electron microscopy of muscle fibers where the thin filaments were selectively removed by gelsolin treatment (Funatsu et al., 1990).

As yet there are few details of the structure of the connecting filaments available because of the technical problems referred to above. However, in view of the striations traversing the I-band in gelsolin-treated muscles, it seems unlikely that they are uniform throughout the I-band. In particular, the first ~100 nm of the connecting filaments emerging from thick filaments appears thicker and more rigid than the remainder. This region may correlate with structures termed end-filaments seen in separated thick filaments by negative staining (Figure 2) (Trinick, 1981). The somewhat better resolution of negative staining than embedding shows the end-filaments to be crossed by striations spaced ~4 nm apart. This periodicity, together with the location of the end-filaments, suggests they are composed of titin molecules.

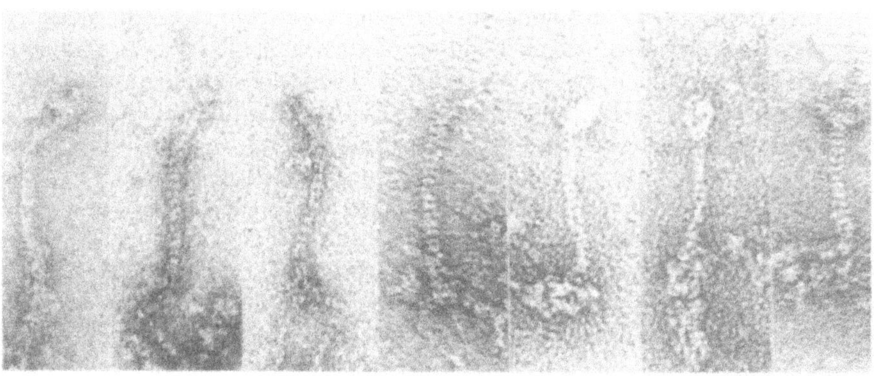

Figure 2. End-filaments found at the ends of negatively stained thick filaments from rabbit psoas muscle. The end-filaments are about 85 nm long by 5 nm wide and are crossed by striations approximately 4 nm apart. It is assumed that end-filaments are formed by association of sections of three titin molecules emerging from the end of the thick filament, however, it is not known why the titin strands sometimes appear separately (see Figure 5). Magnification ×350,000. (Reproduced from Trinick, 1981.)

Confirmation that the location and properties of titin are compatible with its forming the extra filaments has mainly been based on antibody labeling experiments. Early work was carried out at the light microscope level using polyclonal antibodies raised to denatured titin purified in SDS. Fluorescent antibodies mainly labeled the junction of the A- and I-bands, but the labeling moved to near the Z-line when myosin was dissolved, suggesting some type of elastic behavior (Wang et al., 1979). In some cases such antibodies appeared to label the whole sarcomere, suggesting titin is also present in the A-band. Since anti-nebulin (see below) antibodies labeled the I-band, it seemed possible that titin and nebulin would both be involved in the third filaments, with nebulin providing the linkage between titin and the Z-line (Wang and Williamson, 1980). However, it now appears likely that nebulin is a thin filament protein and that titin alone forms the extra filaments.

Recent titin localization data have mainly come from monoclonal antibody labeling visualized by electron microscopy. Monoclonal antibodies proved particularly informative because they mainly give a characteristic labeling pattern which consists of a single stripe in each half sarcomere (Figure 3) (Squire et al., 1987; Furst et al., 1988; Itoh et al., 1988; Whiting et al., 1989). Different antibodies produce different stripes at locations throughout the sarcomere, confirming the suggestion that titin is also present in the A-band (Figure 3). One antibody, T20, was particularly important because it was specific for titin but bound only to the Z-line (Furst et al., 1988). This demonstrated that titin spans the entire I-band and thus showed that nebulin was probably not a linkage to the Z-line. The fact that titin was sufficiently accessible to allow antibody binding throughout the sarcomere, while in most cases each epitope occurred only once per half sarcomere, suggested that individual titin molecules span half the sarcomere (Squire et al., 1987; Furst et al., 1988; Whiting et al., 1989).

The behavior of particular titin epitopes as sarcomere length varied was different depending on whether the epitope was in the I- or A-band. Throughout most of the I-band the distances from epitopes to both the Z- or M-lines increased as sarcomere length was increased. Thus titin in this region appeared to move independently of thin or thick filaments and in a manner consistent with an elastic character (Furst et al., 1988; Itoh et al., 1988; Whiting et al., 1989). In contrast, in the A-band the distance of each epitope from the M-line was constant as sarcomere length was varied, which indicates that titin here is in some way directly attached to the thick filament. The accessibility of the A-band epitopes to antibodies further suggests that the location of titin is on the outside of the filament, rather than it being a thick filament core protein. Since the thick filament has three-fold rotational symmetry, this would suggest that there are three (or a multiple of three) titin molecules per half thick filament (Whiting et al., 1989). A further indication of a superficial location involves the behaviour of titin as thick filaments are progressively depolymerized from their ends by increasing salt, since titin retracts to the Z-line before the filaments completely dissolve at the M-line (Higuchi et al., 1992).

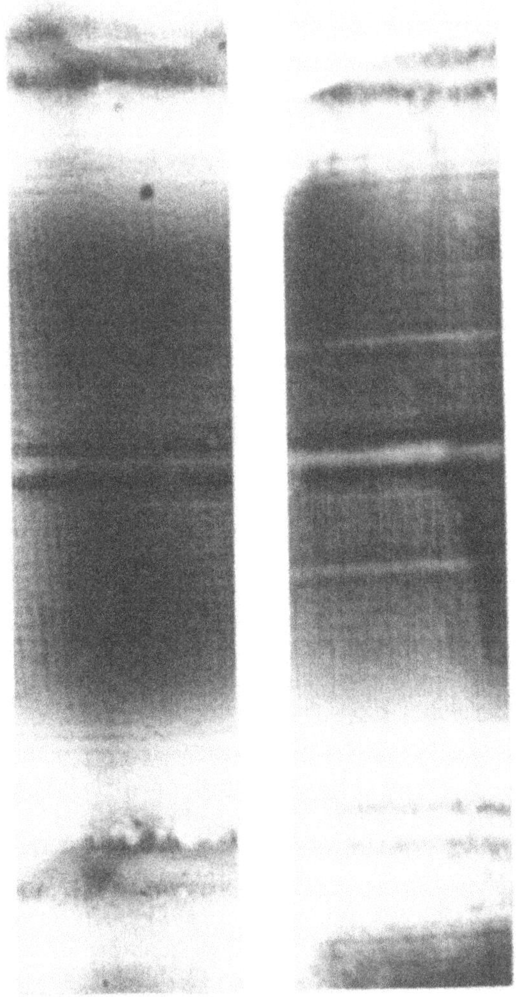

Figure 3. Titin monoclonal antibody labelling of myofibrils after negative staining. The antibody labeled myofibril is on the right and shows a single stripe of labeling (white in the figure) in each half sarcomere (reproduced from Squire et al., 1987). The unlabeled control is on the left. Magnification ×46,000.

F. Functions

Elastic Region

The filaments formed by the I-band sections of titin molecules are elastic connections between the ends of thick filaments and the Z-line (Figure 4) and the available evidence suggests that these have two functions. First, they center thick filaments between adjacent Z-lines. This ensures a similar amount of overlap of the two ends of the thick filaments with their respective thin filaments in opposite halves of the sarcomere. The central location is important, since active force is directly proportional to the number of myosin cross-bridges. A non-central location would result in a force imbalance at opposite ends of the thick filament. This in turn would result in the thick filaments being dragged to one end of the sarcomere.

The two filament model of the sarcomere is thus fundamentally unstable and striking confirmation of this was provided by Horowits and Podolsky and coworkers in two independent experiments. They first showed that if muscle is bombarded with high energy electrons, titin (and nebulin) are selectively cleaved by radiation scission, since radiation damage is related to target size (Horowits et al., 1986). When such muscles were subsequently activated, the thick filaments were not found in the central position but at one end of the sarcomere. In a second set of experiments it was found possible to produee the non-central A-band merely by prolonged and vigorous contraction (Horowits and Podolsky, 1988). The fact that this effect could be produced in the native muscle under extreme conditions therefore suggests that A-band centering by titin operates mainly during the relaxation phase of muscle.

The second function of the elastic filaments is to transmit passive tension, indeed the evidence suggests that they are the main tension-bearing elements in relaxed fibers (Magid and Law, 1985). Similarly, the energy for passive shortening of extended muscle arises from the elastic energy stored in the I-band region of titin molecules.

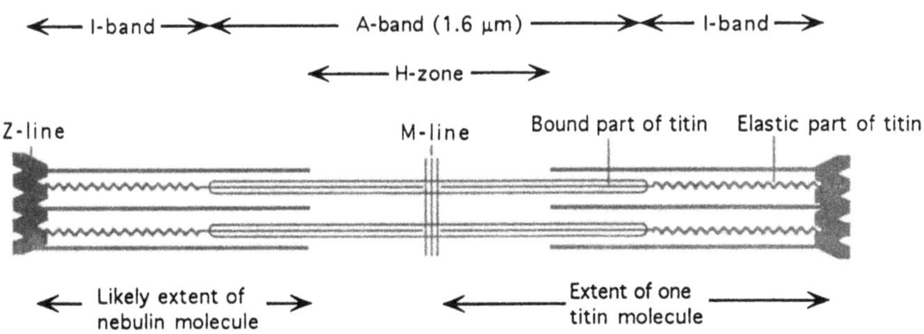

Figure 4. Diagram showing the probable position of the titin molecule and the likely extent of the nebulin molecule within the sarcomere.

A-Band Zone of Titin

The accessibility of titin to antibodies, while not conclusive, argues against the idea that titin is a core protein on the inside of the thick filament. The precise localization of titin on the filament is not known, but the simplest arrangement would be on the outside of the filament shaft, bound to the tails of the myosin molecules that form the shaft. *In vitro* binding experiments with myosin and its proteolytic fragments, both with whole titin and with constructs expressed from titin cDNAs (see below), are compatible with this idea.

What role thick filament titin has is also not yet established, but more than half of the molecule, probably comprising more than 20,000 amino acid residues, appears to be attached in this way. One plausible function that has been proposed is to regulate the assembly of myosin and the other thick filament proteins. The means by which thick filament assembly is controlled has been a long-standing subject of speculation, since the sharpness of the A-band ends in electron micrographs indicates the presence of exact numbers of myosin molecules in the filaments of many types of striated muscles. For instance, in vertebrate striated muscle there are thought to be exactly 294 myosin molecules in each filament (Offer, 1987). Since the length of the myosin molecule (~170 nm) is much smaller than the length of the filament (1.6 μm in this case), it was unclear how the filament length is terminated correct to one myosin molecule.

The discovery that single titin molecules span half the thick filament suggests an obvious mechanism by which filament assembly may be precisely regulated, since a series of unique interactions with myosin and the other filament proteins could be specified throughout the A-band (Whiting et al., 1989). Thus A-band titin may act as a giant template or what has been called a "protein-ruler" to regulate assembly. Other examples of this type of mechanism are the protein-ruler in the tail of lambda bacteriophage (Katsura, 1987) and the RNA strand in tobacco mosaic virus (Fraenkel-Conrat and Williams, 1955). The giant protein nebulin may also be a protein-ruler in muscle thin filaments (see below). One criticism that has been made of the protein-ruler hypothesis for A-band titin function (Epstein and Fischman, 1991) is that antibody labeling of titin and myosin suggests thick filaments assemble independently of titin in embryonic muscle cells (Schultheiss et al., 1990). However, since it is only in the highly ordered structure of the adult myofibril that the A-band ends are sharply defined, it seems possible that exact assembly of the thick filament is achieved only when the titin is subsequently bound.

The fact that titin molecules also extend through the I-band requires that there be a fundamental difference in the titin molecule near the junction of the A- and I-bands for this model to be correct. The nature of this difference is not yet known. It could be that the I-band domains of titin do not bind myosin or it may be that they do but that the stagger between successive myosin molecules is incorrect for filament formation. Since purified myosin forms filaments spontaneously *in vitro*, one of the main roles of titin may be to provide an end point for filament assembly. It is reported that an A-band fragment of titin does bind to myosin (Wang et al., 1992), but is not known whether the converse is also true, i.e., that the I-band region does not bind. A 1200-kDa I-band fragment has recently been described (Matsuura et al., 1991), so it should be possible to test this idea. One circumstantial piece of evidence suggesting that titin is likely to be an important factor regulating myofibril assembly is that it is one of the first muscle-specific proteins to be expressed after cell commitment (Furst et al., 1989).

G. Domain Substructure

Electron micrographs of titin show that the main tail section of the molecule has substructure resembling a string of beads. This detail is seen most clearly by negative staining and is visible in both purified preparations and in titin attached to separated thick filaments (Figure 5) (Knight and Trinick, 1984; Trinick et al., 1984; Whiting et al., 1989). The appearance suggested that the bulk of titin might consist of a linear array of similar domains. To a first approximation, the beads are spherical with a diameter of about 4 nm, which implied that such domains would contain approximately 100 amino acid residues. This suggested that the primary sequence of titin might contain a periodicity of ~100 residues (Whiting et al., 1989). However, the structure could not consist of a series of exactly similar domains, since the binding of monoclonal antibodies only once in each half sarcomere showed there had to be unique features at these positions.

Such antibodies were also used to screen a rabbit muscle cDNA expression library from which the first titin cDNAs were identified (Labeit et al., 1990; Labeit et al., 1992). Southern blots with these suggested that titin is highly conserved between species since similar size restriction fragments of genomic DNA were observed. An estimate of the size of the titin mRNA was made on a Northern blot, since a titin chain weight ~3.10^6 implies an mRNA of ~100 kb. Methods are not available to size such large RNAs, but the titin message migrated with limiting mobility on the blot and at a similar rate to a 21 kb lambda bacteriophage restriction fragment. The locus of the titin gene was shown to be on chromosome 2 in humans (Labeit et al., 1990).

Sequences derived from the cDNAs revealed periodicities of ~100 residues, as predicted, but there were two types of such motif. These motifs were termed class I and II and alignments of them are seen in Figure 6. Both motifs therefore probably fold to form separate domains. It can be seen that within each class the residues found at some positions in the motif are highly conserved while at other positions there appears to be little or no conservation. This lack of exact similarity is presumably linked with the fact that most titin monoclonal antibodies bind only once in each molecule. Database searches with the individual class I and II repeats and with consensus sequences showed that both motifs were similar to previously studied proteins found mainly outside cells. Class I were similar to type III

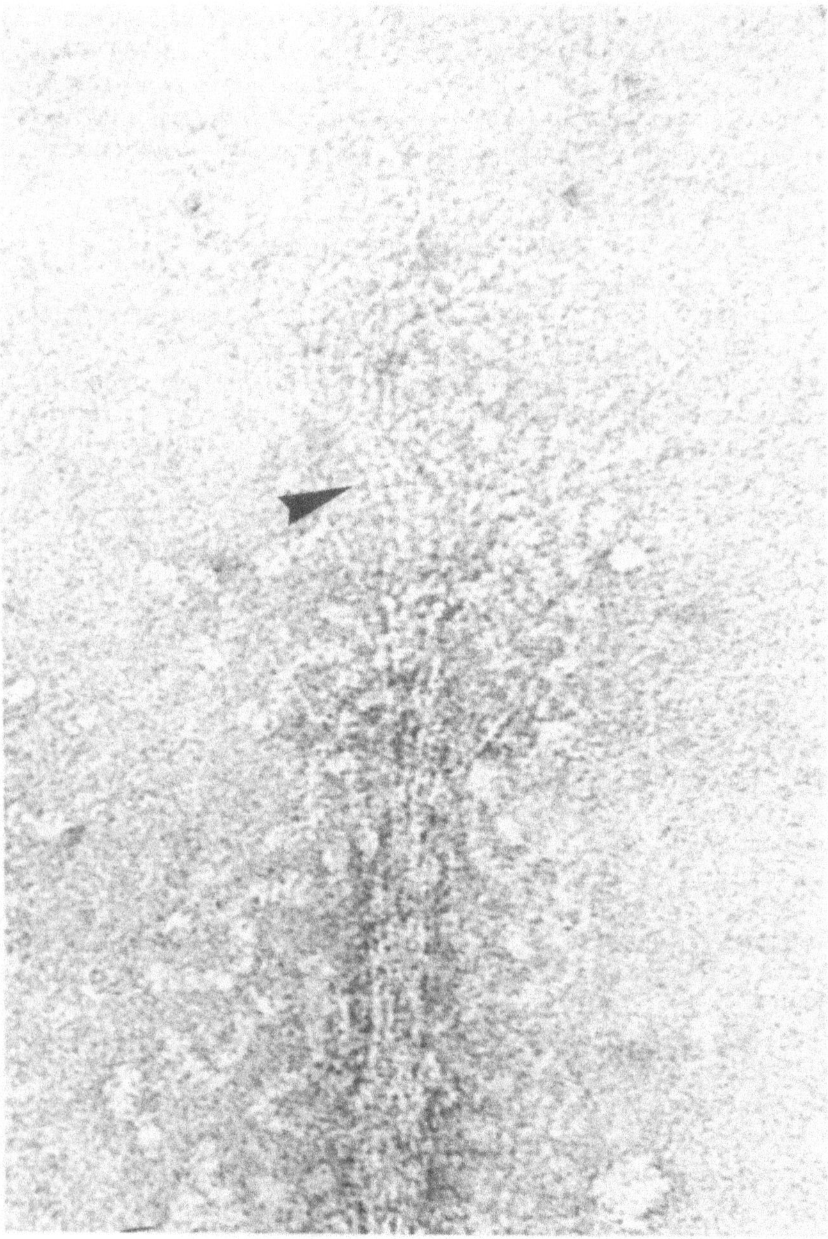

Figure 5. Negatively stained thick filament showing what are thought to be titin molecules exhibiting domains (arrowhead). Magnification ×480,000. (Reproduced from Knight and Trinick, 1984).

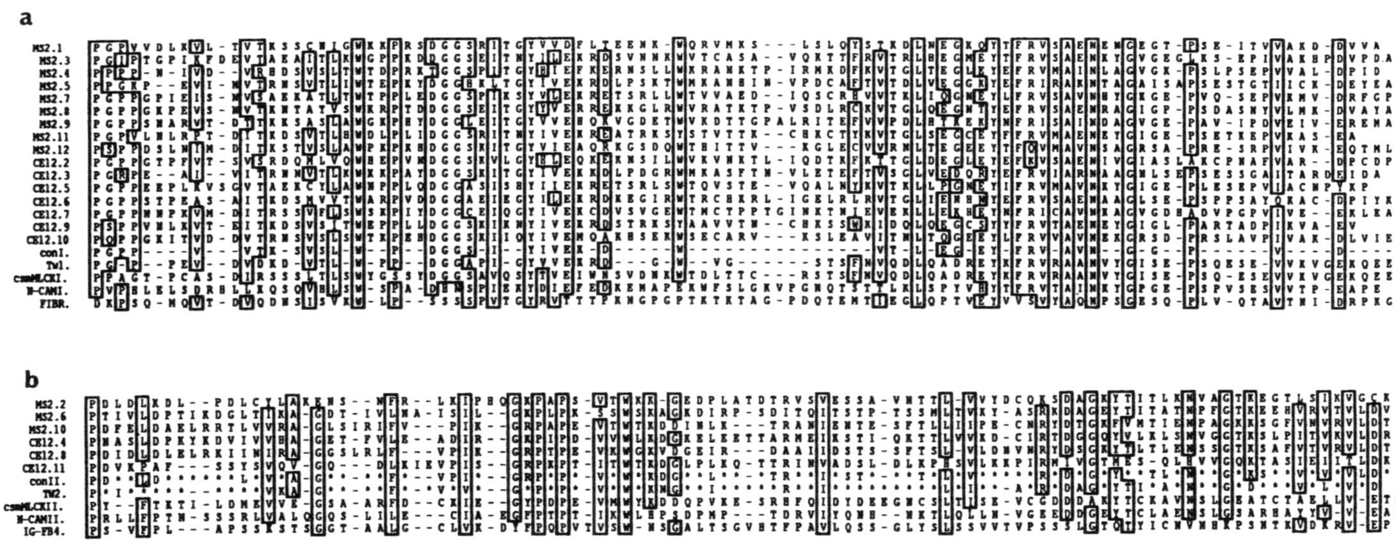

Figure 6. Alignments of class I and II titin sequence motifs. (**a**) Sixteen class I repeats are shown together with similar repeats in twitchin, smMLCK, N-CAM and fibronectin. (**b**) Six class II repeats are shown. Con indicates consensus. (Reproduced from Labeit et al., 1990.)

fibronectin, while class II were similar to the C-2 group of the immunoglobulin superfamily.

H. Domain Super-Repeat

In addition to the similarities between the domains within the two classes formed by these two motifs, there is also a longer range periodicity (Labeit et al., 1992). This long range periodicity has been termed the super-repeat and consists of eleven domains and approximately 1100 amino acid residues (Figure 7). What this means is that local regions of sequence are more similar to regions 1100 residues away

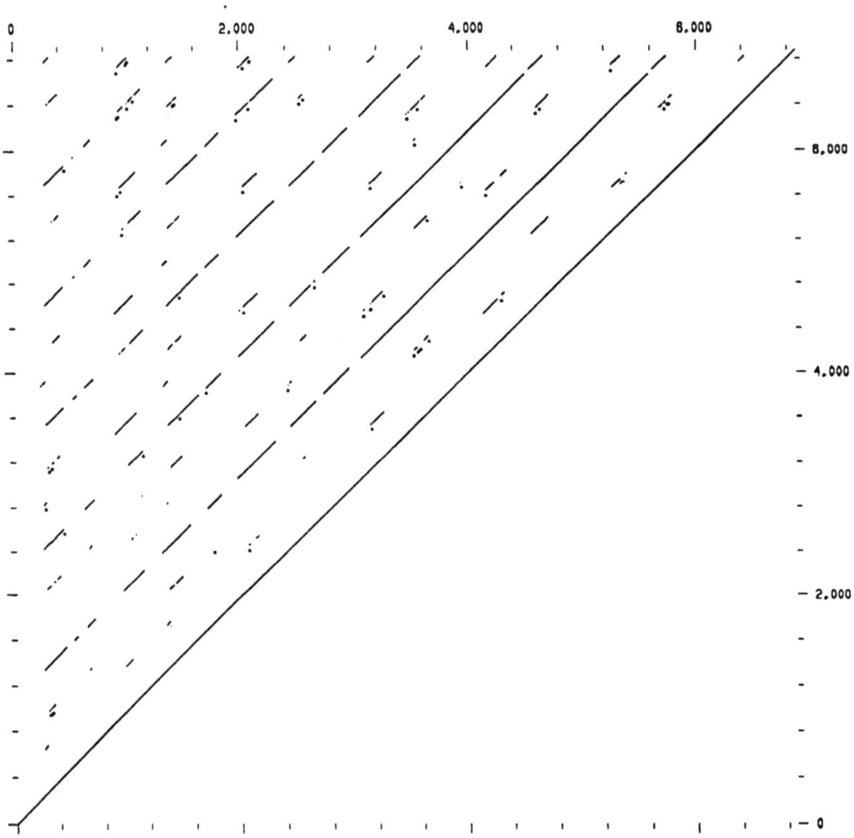

Figure 7. A-band titin sequence self-comparison showing 1100 residue super-repeat. This plot was generated by the UWGCG Compare program (Devereux et al., 1984) using ~7000 residues of A-band titin sequence (Labeit et al., 1992). At the high stringency used (70%) the most prominent features are the strong lines every 1100 residues indicating the presence of the super-repeat.

than they are to sequences at analogous positions in domains of the same class within the super-repeat. The pattern of domains in the super-repeat is I–I–I–II–I–I–I–II–I–I–II. This is present throughout much of the ~10,000 residues of titin so far sequenced which is exclusively from the A-band. However this pattern breaks down before the end of the thick filament is reached and near the bare zone.

Since the interdomain spacing is known to be approximately 4 nm from electron microscopy, the super-repeat should span ~44 nm in the thick filament, assuming that titin is not extensively wound around the filament. Forty-four nm is very close to the 43-nm periodicity known from X-ray fiber diffraction to be present in the thick filament, this being the repeat distance of the helix that describes the layout of the myosin molecules (Figure 8). Another and more direct indication of a 43-nm period in this region of titin comes from monoclonal antibodies that label epitopes that occur more than once. As already described, most titin monoclonal antibodies label at only one axial position in each half sarcomere. However, a few give rise to multiple labeling stripes and of these three, T30, T31 and T32, are interesting because the stripes they produce are ~43 nm apart (Furst et al., 1989). These stripes coincide with known A-band striations that arise due to the presence of non-myosin thick filaments components such as C-protein (see below).

The discovery of a 43-nm super-repeat in A-band titin therefore reinforces the idea of an intimate association between this region of the molecule and the thick filament. One aspect of the filament structure that may be explained by the super-repeat is the binding of C-protein at the 43-nm positions. Myosin heads emerge from the filament at approximately one third of this interval and it is unclear what is different about every third position in order to specify C-protein binding (Craig and Offer, 1976). The presence of the 43-nm super-repeat suggests, in principle, how the assembly of C-protein onto the filament may be specified (Figure 8).

I. *In Vitro* Studies of Titin Interactions

The idea that A-band titin is bound to the thick filament is also supported by *in vitro* binding studies with purified proteins. Use of classical analytical ultracentrifuge methodology for such work is complicated by the fact that titin and several of the other thick filament proteins with which it may interact self-aggregate at low ionic strength. Most work thus far has therefore used other methods. Titin caused aggregation of synthetic thick filaments made from purified myosin, although the ratio of titin to myosin in these aggregates was up to 10-fold higher than *in vivo* (~1:50 on a molar basis (Whiting et al., 1989)), possibly due to the rather low ionic strength used (60 mM) (Maruyama et al., 1989a). Other studies have been performed using solid phase assays which have the advantage that the proteins can first be immobilized (e.g., on microtitre plates) at high ionic strength and then bathed at physiological ionic strength when studying its interactions (Soteriou et al., 1993b; Wang et al., 1992). Binding of titin to myosin has been reported using

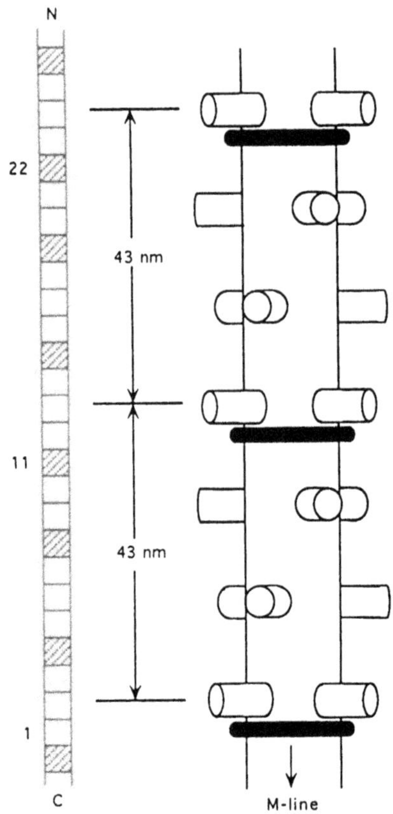

Figure 8. Diagram showing likely correspondence between titin super-repeat and thick filament 43 nm cross-bridge repeat. Clear blocks represent class II domains and hatched blocks class I. The projections on the filament each represent the heads of one myosin molecule. The arrangement of C-protein is not known, but is shown here as a collar. The location of titin on the filament and the stagger of the super-repeat vs. the myosin are also not known.

such assays, and evidence as to which particular part of the myosin molecule is involved was gained by using proteolytic fragments of myosin. Both these studies indicated binding to the LMM portion of the tail of myosin, which is consistent with the idea that titin is attached to the thick filament shaft. Binding of the A-band region of titin to myosin has also been reported (Wang et al., 1992), although it is not clear whether the converse is true, that the I-band section does not bind. An interaction with the S1 fragment of myosin which forms the head of the intact molecule was also reported, but the significance of this observation is not yet clear.

Binding of titin to C-protein, X-protein, and AMP-deaminase was also reported using such an assay, using antibodies to detect bound titin (Soteriou et al., 1993b).

As will be described below, C-protein and X-protein are both thick filament components composed of class I and II sequence repeats. Estimates of the apparent dissociation constant in these assays were all ~10^{-7} M. Evidence of an interaction with AMP-deaminase was sought because antibody labeling studies have shown that this enzyme may be attached to thick filaments *in vivo*, but bound in the I-band beyond where the myosin-containing part of the filament is thought to end (Cooper and Trinick, 1984). The functional significance of the immobilization of this enzyme is unknown, but it may improve its metabolic efficiency. A possible anchorage site for titin in the Z-line may be indicated by the binding to α-actinin (Wang and Jeng, 1992). These studies with whole titin or its fragments are broadly consistent with similar assays using small fragments expressed from cDNAs (Labeit et al., 1992) which showed binding to LMM and C-protein.

J. Titin Isoforms

Two recent studies have suggested that there may be isoforms of titin and that these may have a profound effect on the resting stiffness of the muscle type. (Wang et al., 1991; Horowits, 1992). Skinned fibers from several rabbit muscles were shown to produce widely varying resting tension when stretched, and differences in the mobilities of the titin bands on SDS polyacrylamide gels of the muscles suggest different titin isoforms. For instance, the resting tension of soleus fibers is several-fold less than psoas fibers at the same sarcomere length and the soleus titin migrates on gels slightly more slowly than titin from psoas muscle. The tendency of the A-band to move from its central position in the sarcomere during prolonged active contraction is greatest at short sarcomere lengths where the tension in the titin molecules will be lower and at longer sarcomere lengths the central location is more stable (Horowits, 1992). Plots of resting tension vs. sarcomere length show a smooth rise to a peak and then a small decline to a plateau. Using antibody labeling it was shown that the peak correlated with maximum extension of the I-band titin. At sarcomere lengths beyond this point (~4 μm) A-band epitopes that were fixed with respect to the M-line become detached and start to behave in an elastic manner (Wang et al., 1991). These data suggest that the wide variations observed in passive tension result primarily from differences in titin isoforms. They also show that the whole titin molecule is elastic, but that elasticity in the A-band portion of the molecule is seen only when it is not attached to myosin and the other thick filament components. Appearance of elasticity in A-band epitopes at long sarcomere lengths was also reported by Pollack and coworkers (Trombitas et al., 1991).

K. Mechanism of Titin Elasticity

The molecular mechanism of elasticity in titin is almost completely unexplored and raises a number of interesting questions. Elasticity in cases such as connective tissue occurs through reorientation of the long axes of the collagen molecules onto the fiber axis as the fiber is extended. Thus the molecules themselves do not change their lengths and can be inextensible. In the case of titin, the molecule is parallel (or almost so) to the long axis of the myofibril, so the molecular length must increase as the sarcomere length increases. It is therefore necessary to consider intramolecular mechanisms of elasticity. There is no sequence information from the I-band region of the titin molecule yet, but this appears similar to the A-band region in the electron microscope, and twitchin (see below) is also composed of class I and II domains and can be found in the A- or I-band depending on muscle type. It therefore seems reasonable to suppose that titin will be composed of similar types of domains throughout.

Experiments on the extensibility of different titin isoforms such as those above have been performed under a wide range of sarcomere lengths and it is probably necessary to distinguish between at least two types of elastic mechanism to encompass this range. The normal range of sarcomere lengths *in vivo* is between approximately 2.1 μm and 2.6 μm and, since the A-band is 1.6 μm long, this therefore involves an approximate doubling of the length of the I-band. It seems possible that a 100% increase in the length of I-band titin could be accomplished by some sort of domain reorientation, without unfolding of the domains themselves.

At the other extreme, the maximum extension of the I-band region of titin occurs at a sarcomere length of roughly 4 μm in vertebrate skeletal muscle, beyond which point the A-band region of the molecule is recruited. This implies a roughly 500% increase in the length of the I-band and it is difficult to envisage how such a large extension in this region of titin could be accomplished without some unfolding of the domains. This argument is made with the proviso that there are not substantial linker regions between the domains that could become extended, however, such linkers are not found in the A-band section of the molecule. Compatible with the idea of unfolding, calculations suggested that the work necessary to extend I-band titin is comparable with the free energy of unfolding of titin, as determined by monitoring denaturation in guanidine-HCl by tryptophan fluorescence or circular dichroism (Soteriou et al., 1993a).

There is some evidence that domain unfolding can take place when titin is extended *in vitro*, since molecules deposited on a substrate by spraying from an atomizer appear different from those put down by layering (Trinick et al., 1984). Spraying is accompanied by large shearing forces whereas there will be much less shearing in layered molecules. In the electron microscope the sprayed molecules are sometimes seen as sections of the normal ~4 nm wide configuration joined by very thin connections. These connecting regions are at the limit of detectability of the rotary shadowing method and it seems possible that they are single polypetide chains from unfolded domains. To what extent such domain unfolding is relevant *in vivo* is open to question, since such large sarcomere lengths are not normally found *in vivo*.

In the context of elasticity, it may also be noted that even at slack length of sarcomeres, there may be residual tension in the I-band region of titin. This is because there is electrostatic repulsion between thick and thin filaments and work must therefore be done to interpose the two lattices. Such repulsion is indicated by the fact that at any particular sarcomere length, the filament lattices are in a stable equilibrium at a constant distance from one another and do not collapse. Since work must be done to interpose the lattices there may be a longtitudinal repulsive force between thick and thin filaments, as well as the transverse force that keeps them apart. This longitudinal force would be balanced by the residual tension in the titin when the sarcomere is exerting no net tension. The magnitude of such a longtitudinal repulsive force between thick and thin filaments is not known, but it may be significant since the A-band is found in the center of the sarcomere even at slack length.

Finally, one can speculate that the elasticity seen in titin may be relevant to other proteins composed of similar domains, e.g., in extracellular matrix proteins. C-2 immunoglobulin and type III fibronectin domains are common building blocks of proteins in the extracellular environment, often occurring together in cell-surface proteins such as the neural cell adhesion protein, N-CAM (see Williams and Barclay, 1988 for review). It is not known if molecules such as N-CAM are elastic and such a phenomenon would be technically difficult to observe. It may also be noted that except in a few cases such as where it has an RGD sequence (see e.g., Ruoslahti and Pierschbacher, 1987), the type III fibronectin domains have no known function. In respect of understanding protein elasticity, work on titin is likely to have unique advantages, since the tension in individual molecules can be calculated and the huge size and ordered disposition in the sarcomere allows distances inside molecules to be monitored, even by light microscopy.

One difference between the intra- and extracellular proteins with class I and II domains that may be relevant to the subject of elasticity concerns S–S bonds. In extracellular space C-2 immunoglobulins have conserved cysteine residues that form disulphide bridges, but S–S bonds do not generally occur in intracellular proteins. In titin and the other muscle proteins with these domains (see below), cysteines are conserved only in some of the domains. In telokin, which is the only intracellular Ig domain to have been solved to atomic resolution (see below), the conserved cysteines that would be required to form the S–S bond are present, but the bond is not formed (Holden et al., 1992). Such bonds would prevent the full extension of molecules by domain unfolding, as may be possible *in vitro* with titin.

L. Intracellular Muscle Subgroup of the Ig Superfamily

Motifs similar to class I and II repeats found in titin had previously been found in another intracellular muscle protein, twitchin, discovered in the nematode *Ceanorhabditis elegans* (Benian et al., 1989). Until the sequencing of twitchin, both class I and II were thought to exist only outside cells. In many proteins they are found in the same molecule, for instance in the extracellular region of N-CAM. However, there is now a growing group of muscle proteins known to be composed of these motifs. Twitchin was first identified through a *C. elegans* mutant, *unc-22* (Moerman et al., 1988); this has a twitching phenotype, which led to the name twitchin. The published twitchin sequence has a molecular weight of 668,520 Da, but this is now thought to be incomplete (Benian et al., 1991). The published sequence consists almost exclusively of the class I and II motifs, of which there are 31 and 26 copies, respectively. These are mainly organized in the regular pattern I-I-II.

Subsequently, there has been work reported on several invertebrate proteins that have different names, but which may all be the same as twitchin: a protein termed mini-titin with an estimated chain weight of 700,000 Da was isolated from *Locusta migratoria* (Nave and Weber, 1990) and from *Ascaris lumbricoides* and *C. elegans* (Nave et al., 1991). Labeling with antibodies to mini-titin showed it to be present in several invertebrate phyla, including insects (*Drosophila melanogaster*) and crustacea (*Astacus*) (Nave and Weber, 1990). Electron microscopy after rotary metal shadowing showed the Locusta mini-titin to be a string-like molecule with a diameter of ~4 nm, similar to titin, but with a length of 260 nm; thus its length was reduced compared to titin roughly in proportion to their difference in molecular weights. Other proteins that are probably the same as twitchin/mini-titin are p800 (from *Lethocerus* (Lakey et al., 1990)), and projectin (from *Apis mellifera* (Saide et al., 1990), *Drosophila* (Ayme-Southgate et al., 1991; Vigoreaux et al., 1991), and crayfish (Hu et al., 1990). Partial sequences of P800 (Lakey et al., 1990) and projectin (Ayme-Southgate et al., 1991) have confirmed the presence of class I and II motifs and antibodies to synthetic twitchin peptides cross-reacted with mini-titin (Nave et al., 1991). Most estimates of the molecular weight of this protein are in the range 600,000 to 800,000 Da, but a value of 1,200,000 Da was obtained for crayfish projectin (Hu et al., 1990).

The location of this protein in muscle appears to vary with the muscle type. In insect flight muscle, antibodies label mainly in the narrow I-band (*Lethocerus, Apis*, and *Drosophila*) while in insect leg and *C. elegans* bodywall muscles the location is in the A-band (Moerman et al., 1988; Lakey et al., 1990; Saide et al., 1990; Vigoreaux et al., 1991). This significance of this dual location is unclear but it is paralleled by titin spanning both A- and I-bands in vertebrate muscle.

M. Kinase-Like Domain in Twitchin and Titin

Although the published sequence of twitchin consists almost exclusively of repeats of the class I and II motifs, an ~200 residue section towards the C-terminus does not conform to this pattern. Database searches with this 200 residue section showed it to be very similar to the catalytic domain of smooth muscle myosin kinase (smMLCK, Figure 9) (Benian et al., 1989). This led these authors to examine further the published partial sequence of smMLCK, where previously unrecognized class I and II motifs were discovered flanking the catalytic domain (Benian et al., 1989). Subsequently, the further sequencing of smMLCK revealed three class II and one

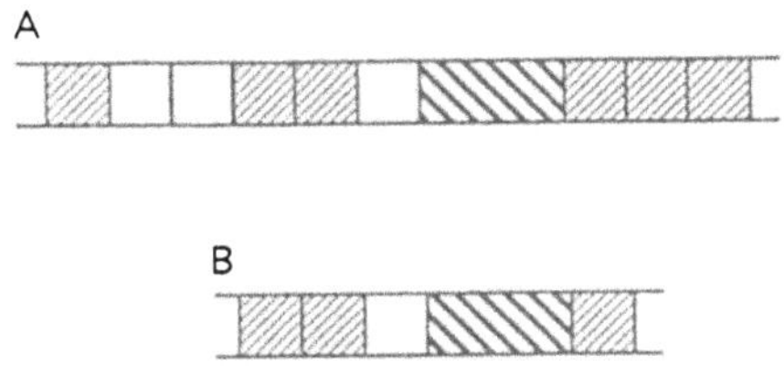

Figure 9. Diagram showing the similarity between the domain arrangements of smMLCK and near the C-termini of twitchin and titin. Open boxes represent class I domains and lightly hatched boxes class II. The heavily hatched boxes are the catalytic domain. (**A**) Titin, (**B**) smMLCK.

class I domains (Olson et al., 1990). Recent comparison with the non-muscle homologue (nmMLCK (Shoemaker et al., 1990)) suggests that this sequence is incomplete and that there may be several further class II domains towards the N-terminus (Gibson and Higgins, 1992).

An smMLCK catalytic-like sequence has also been found in titin (Labeit et al., 1992). The cDNA which led to this discovery was identified by a monoclonal antibody, AB5; that labels muscle near the bare zone, which is towards the location of the C-terminus of the titin molecule *in situ*. The pattern of class I and II motifs surrounding the titin catalytic domain was also the same as in smMLCK and twitchin (Figure 9). The relatively high degree of conservation (~40%) between the catalytic regions of smMLCK, twitchin and titin and the conserved flanking patterns of class I and II repeats suggests important conserved functions, especially considering the large evolutionary distance between nematode, chicken, rabbit, and smooth and striated muscles. Phosphorylation of myosin by smMLCK is the primary activation event for contraction of smooth muscle, which led to the suggestion that twitchin might control contraction in *C. elegans* (Benian et al., 1989). However, neither titin nor myosin is freely diffusible in myofibrils and the *in vivo* substrate for the catalytic domain is unclear. There are reports of autophosphorylation in skeletal MLCK (Gao et al., 1992), twitchin (Maroto et al., 1992), and titin (Takanoohmuro et al., 1992), so it is possible that this is the main reaction that all three proteins catalyse *in vivo*.

N. Other Titin-Related Muscle Proteins

As well as titin, twitchin and MLCK, there are several other intracellular muscle proteins that have also now been shown to be composed of class I and II motifs. Some of these give rise to a pattern of eleven equally spaced stripes that are found in each half of the A-band (Figure 10). None of them have well understood functions, but the spacing between the stripes is 43 nm and this is the repeat distance of the helix that describes the arrangement of myosin molecules in the filament.

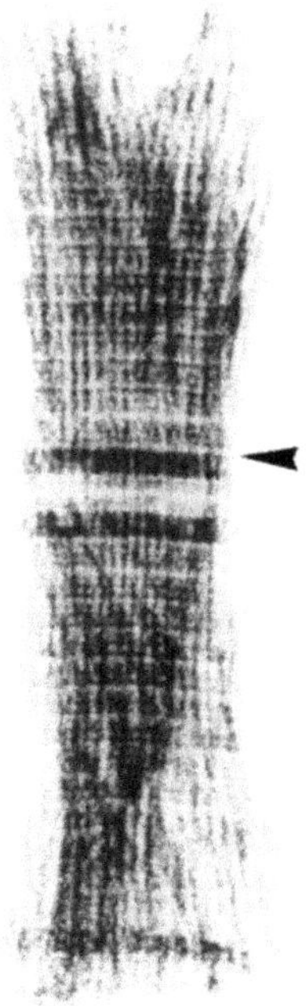

Figure 10. Negatively stained A-segment (assembly of thick filaments in register). Arrowhead indicates stripe number one of the eleven stripe accessory protein pattern extending at intervals of 43 nm from the bare zone. Magnification ×110,000. (J. Trinick, unpublished data.)

The pattern of proteins giving rise to these stripes is complex and not fully characterized, but it varies with species and muscle fiber type (Figure 11) (Bennett et al., 1986). For instance, in rabbit psoas muscle, which contains mainly fast white fibers, stripes number four to eleven (counting from the M-line) result from the presence of C-protein which is a single chain molecule of ~140 kDa (Starr and Offer, 1971; Offer et al., 1973). However, even in some psoas muscle fibers C-protein is sometimes present only at stripe 5-11. In rabbit soleus muscle, which contains slow red fibers, stripes 3-11 are due to the presence of X-protein (152 kDa; Starr and Offer, 1983) which is a C-protein isoform and is also known as red C-protein. Another component, H-protein (69 kDa) (Starr and Offer, 1983), gives rise to stripe 3 in psoas muscle but is absent in soleus. In chicken pectoralis muscle,

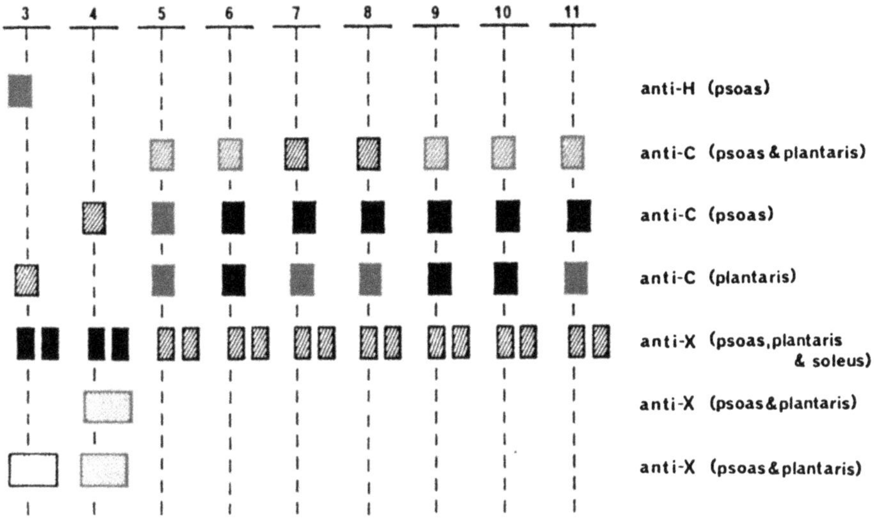

Figure 11. Diagram showing the distribution of C-, X-, and H-proteins. (Reproduced from Bennett et al., 1986, with permission.)

which contains fast white fibers, there is a component termed 86-kDa protein and antibodies to this label stripes 3-11 (Bahler et al., 1985).

Sequences of chicken and beef C-protein consist of eleven class I and II domains in the pattern I–II–II–II–II–II–I–I–II–I–II (Einheber and Fischman, 1990; Furst et al., 1992; Vaughan, 1992). Slow C-protein (X-protein) is similar to C-protein, but has a different pattern of domains, and H-protein and 86 kDa are composed of class I and II domains and have very similar sequences suggesting they are the same protein in rabbits and birds respectively (despite the fact that their molecular weights were estimated to be 69 and 86 kDa) (Vaughan, 1993). In addition to non-myosin thick filament proteins in the cross-bridge regions, there are other proteins located at the M-line (Figure 10, see Squire et al., 1987 for review). These are assumed to be responsible for maintaining thick filaments in register, although this has not been proven. One of these, M-protein (165 kDa), has been sequenced and consists of class I and II arranged as II-II-I-I-I-I-I-II-II-II-II (Noguchi et al., 1992).

The feature uniting this intracellular group of the immunoglobulin superfamily, in addition to being muscle proteins, is that they are all constituents of thick filaments or are likely to interact with myosin. Thus, by inference, some of the class I and or class II domains bind myosin (Labeit et al., 1990). The picture of the intracellular group that is emerging is one of considerable complexity, in terms of its number of members, their locations and the variation of location with differences in fiber type and species. However, most of the data are consistent with the underlying role of the group being involved with myosin assembly.

O. Structure of Class I and Class II Domains and Their Evolution

Class II

Circular dichroism spectroscopy suggested that titin contained little or no α-helix (Trinick et al., 1984) and was likely to be composed predominantly of β-structure (Maruyama et al., 1986). X-ray crystallography has recently solved the structure of both domain types and confirmed these predictions. From sequence similarities, it appeared likely that the structure of class II domains would be an immunoglobulin-like fold, which is largely composed of β-structure. The first intracellular Ig superfamily domain to be solved is from smooth muscle and is called telokin (Holden et al., 1992). Telokin is identical to the C-terminal domain of smMLCK (avian gizzard) but also exists separately. Indeed the separately expressed version is present in gizzard at ~15 μM, considerably higher than the concentration of MLCK (Ito et al., 1989). Its function is not known, but it is present in several smooth muscles (gizzard, trachea, ileum, and uterus) although it was not found in aorta or non-muscle tissues (Gallagher and Herring, 1991). Telokin mRNA may arise beginning at a promoter within an intron in the MLCK gene (Gallagher and Herring, 1991).

The ribbon diagram of telokin is seen in Figure 12a and consists of seven strands of anti-parallel β-pleated sheet that wrap around to form a barrel (Holden et al., 1992). The molecule consists of 154 amino acid residues of which 103 were visible in the electron density map. The unseen regions were from the N- and C-termini, 32 and 19 residues respectively, but the central portion was well ordered to 2.8 Å resolution. The overall dimensions are approximately 47 Å × 24 Å × 31 Å. The barrel can be thought of as two layers of β-sheet that are nearly parallel to one another, with one layer containing four β-strands and the other three. This topology is very similar to the constant domains of immunoglobulin molecules. One interesting point concerns the conserved cysteine residues: extracellular immunoglobulins normally contain a disulphide linkage between β-stands B and F (notation from Williams and Barclay, 1988). Telokin has cysteine residues at the identical positions to those found in immunoglobulin constant domains. These are buried in the core of the protein but there is no evidence for a disulphide linkage.

Class I

The structure of a type III fibronectin domain has been solved by X-ray diffraction and refined to 1.8 Å (Leahy et al., 1992). The crystal that was solved was a section of polypeptide expressed from the cDNA of the extracellular matrix protein, tenascin. Figure 12b shows the ribbon diagram of this domain, which consists of seven β-strands arranged into two β-sheets, one containing four strands and the other three. This topology is very similar to that of immunoglobulin constant domains described above, with the exception that the C′ strand occurs in different

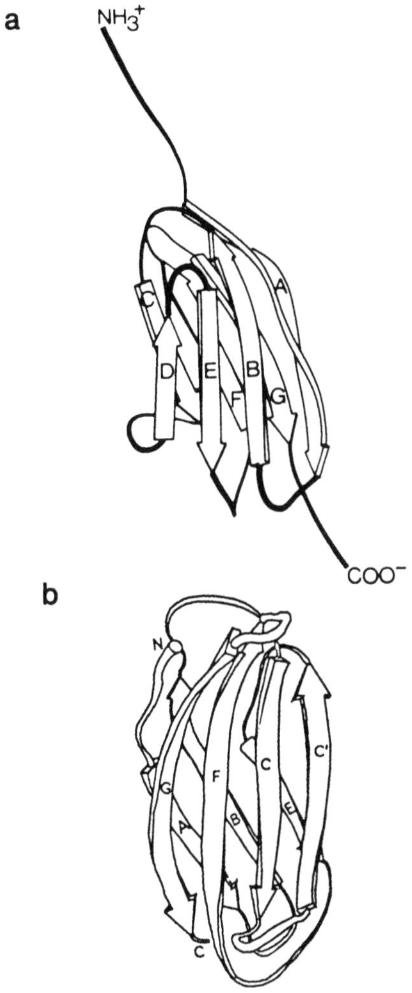

Figure 12a. Telokin ribbon diagram. (Reproduced from Holden et al., 1992, with permission.) (**b**) Ribbon diagram of type III fibronectin-like domain in tenascin. Note that strand D is labeled C′ here and that the structure is seen from the opposite side from the Figure 12a diagram. (Reproduced from Leahy et al., 1992, with permission.)

β-sheets in the two molecules. The connectivity between the strands is seen in Figure 13. The possibility that type III fibronectin might have a structure similar to the immunoglobulin fold had previously been made on the basis of sequence similarities, which led to the suggestion there might be a common ancestral gene (Bazan, 1991). Whether this is indeed the case or whether the molecules evolved

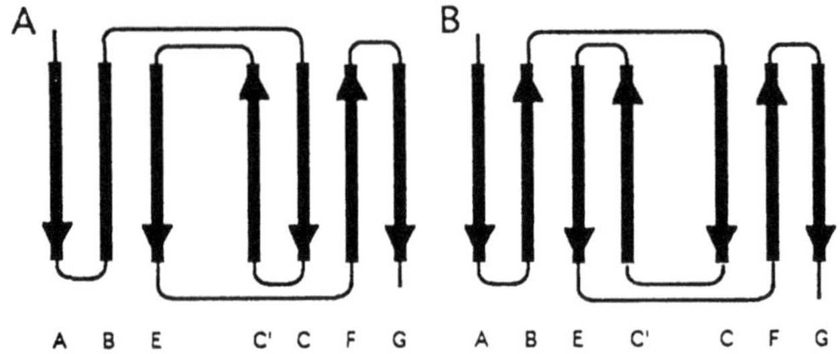

Figure 13. Strand organization in type III fibronectin and C-2 Igs. The three-dimensional relationship between the β-sheets can be reconstructed for each structure by closing the two sheets as if they on were an open book. (Reproduced from Leahy et al., 1992, with permission.)

separately to form similar structures in what has been termed convergent evolution is not yet resolved.

A model has also been proposed for the way in which the type III fibronectin domains are arranged relative to one another in molecules such as tenascin consisting of multiple domains of this sort (Leahy et al., 1992). Since the N- and C-termini are at opposite ends of the long axis of the domain (as in the Ig fold), it is not difficult to see how the domains may be concatenated to form fibrous molecules. However, since the interdomain spacing in fibronectin is less than the distance between the N- and C-termini determined from the X-ray structure, it seems likely that the long axes of the domains are not parallel to the long axis of the molecule. It was also suggested that alternate domains are rotated by 180°.

The question arises whether the intracellular muscle proteins having immunoglobulin- and fibronectin-like domains evolved before the extracellular proteins with these domains, or whether they arose afterwards. In evolutionary terms, the immune system is generally considered to be a relatively recent innovation, probably predated by extracellular matrix proteins of the immunoglobulin superfamily such as N-CAM. The most primitive versions of such matrix proteins so far discovered are from *C. elegans*, but since this organism also expresses twitchin it is not yet possible to decide whether the muscle proteins came first. The bacterial chaperonin PapD also has domains with topology that is very similar to Ig constant domains, which suggests that it too is a member of the Ig superfamily (Holmgren and Branden, 1989). If this proves to be true the origins of the Ig superfamily are much older, but it seems possible that convergent evolution also explains the similarity between PapD and Igs.

It is also of interest to ask in what order the intracellular muscle proteins appeared. From sequence comparisons of the class I, II, and catalytic domains of titin, twitchin and smMLCK phylogenetic trees were constructed (Higgins et al., 1994). These suggested that titin and twitchin diverged from one another at least as early as the divergence of vertebrates and nematodes. It was concluded that titin appeared first, twitchin second, and MLCKs third. Thus smMLCK appears to be a truncated version of titin, rather than titin (or twitchin) being an elongated version of an MLCK. This result is surprising considering that in some respects smooth muscle cells seem intermediate between non-muscle cells and striated muscle.

III. NEBULIN

A. Identification, Isolation, and Molecular Weight

Nebulin is another massive protein for which evidence is accumulating of an important role controlling myofibril assembly. Nebulin was identified on large-pore polyacrylamide gels at the same time as titin and was initially designated as a third component of titin (Wang et al., 1979). Subsequently, it became clear that it was immunologically distinct from titin and was given the name nebulin because antibodies to it bound at, or near a poorly understood structure called the N-line found in the I-band (Wang and Williamson, 1980).

Work on nebulin is considerably less advanced than on titin, due mainly to the fact that it has not so far been possible to isolate it without the use of denaturing solvents. Renaturation from such solvents has also not been possible. A protein of molecular weight similar to nebulin has recently been isolated from intestinal brush border under non-denaturing conditions (Eilertson and Keller, 1992), however, since antibodies did not cross-react with the muscle nebulin it is not yet clear whether this is the same protein. Thus rather little is known about the shape of the nebulin molecule or its secondary structure. It can, however, be isolated in denatured form by gel filtration chromatography in SDS. Its size separates it from titin, which is substantially larger, and all other muscle proteins which are much smaller. As with titin, calibration of SDS polyacrylamide gels in the megadalton range is problematical since there are no naturally occurring molecular weight standards. However, use of cross-linked myosin heavy chain oligomers suggests the chain molecular weight of nebulin is approximately 800 kDa (Hu et al., 1989).

B. Antibody Labeling

Although it has not been possible to isolate native nebulin, antibodies have been raised to the SDS-purified protein and these have been used in a variety of ways. Labeling of muscle with such antibodies gives rise to transverse striations that are found throughout most of the thin filament region (Wang and Williamson, 1980; Furst et al., 1988; Maruyama et al., 1989b; Pierobon-Bormioli et al., 1989). However, there were conflicting claims in these reports as to whether the distances of the stripes were fixed with respect to the Z-line as sarcomere length was varied. Initially, on the basis on fluorescence light microscopy, it was reported that nebulin appeared to be located at a structure in the I-band called the N-2 line (Wang and

Williamson, 1980). Since the N-line behaved in an elastic manner as sarcomere length was varied, it was suggested that nebulin was elastic and formed a connection between the Z-line and titin emerging from the ends of thick filaments (Wang, 1984). Subsequently, this question was re-examined using electron microscopy of labelled single fibers and it was concluded that the distances of antibody stripes from the Z-line are fixed (Wang and Wright, 1988). This led to the proposal that nebulin is in some way directly attached to thin filaments.

Is was also suggested that individual nebulin molecules might span the entire thin filament and thereby regulate exact assembly of the filament (Wang and Wright, 1988). Electron microscopy of many types of striated muscle reveals a very sharply defined H-zone (Figure 4), which indicates that the thin filaments are composed of exact numbers of actin, troponin and tropomyosin subunits. These proteins are capable of spontaneously assembling to form thin filaments *in vitro*, but the length of these "synthetic" filaments is heterogeneous. An obvious role of a single molecule spanning the entire thin filament would be to act as giant template or protein-ruler. This proposed role of nebulin is therefore analogous to the role put forward for A-band titin as a protein-ruler regulating myosin assembly.

C. Nebulin Sequence

Antibodies have also been used to identify nebulin cDNAs in expression libraries (Stedman et al., 1988; Zeviani et al., 1988). One of the cDNAs found in this way was 2 kb long and was subsequently completely sequenced (Labeit et al., 1991). The derived 560 residue sequence has provided most of the information that is so far available about the substructure of the nebulin molecule. It consists entirely of repetition of a motif containing approximately 35 amino acid residues. Alignments of these repeats are seen in Figure 14 which also shows that roughly one quarter of the residues in the motif are conserved. The most strongly conserved features are

```
S1.R1   T I D A M S L E H T R N V N Q I Q S D N V Y K D E Y N S F L K G I G W I
S2.R1   - P D A L N V E H S R N A M Q I Q S D N L Y K S D F T N W M K G I G W V
S3.R1   - P D A M S F E L A K N M M Q I Q S D N Q Y K A D Y A D F M K G I G W L
S1.R2   P I G S L E V E K V K K A G D A L N E R K Y R Q H P D T V K F T S V
S2.R2   P I E S L E V E K A K K A G E I L S E K K Y R Q H P E K L K F T Y A
S1.R3   - P D S M G M M L A Q H N T K Q L S D L N Y K V E G E K L K H K Y T I D
S2.R3   - M D T M E Q A L N K S N K L N M D K R L Y T E K W N K D K T T I H V M
S1.R4   - P E L P Q F I Q A K V N A L N M S D A H Y K A D W K K T I R K A Y D L R
S2.R4   - P D T P D I L L S R V N Q I T M S D K L Y K A G W E E E K K K G Y D L R
S1.R5   - P D A I P I V A A K S S R N I A S D C K Y K E A Y E K A K G K Q V G F L S
S2.R5   - P D A I A I K A A R A S R D I A S D Y K Y K K A Y E Q A K G K H I G F R S
S1.R6   L Q D D P K L V H Y M N V A K I Q S D R E Y K K G Y E A S K T K Y H T
S2.R6   L E D D P K L V H F M Q V A K M Q S D R E Y K K G Y E K S K T S F H T
S1.R7   P L D M V S V T A A K K S Q E V A T N A N Y R Q S Y H H Y T L L
S2.R7   P V D M L S V V A A K K S Q E V A T N A N Y R N V I H T Y N M L
```

Figure 14. Nebulin sequence alignments showing the similarities between 35-residue repeats. S1.R1, etc. refer to the positions of the repeats in three successive super-repeats. (Reproduced from Labeit et al., 1991.)

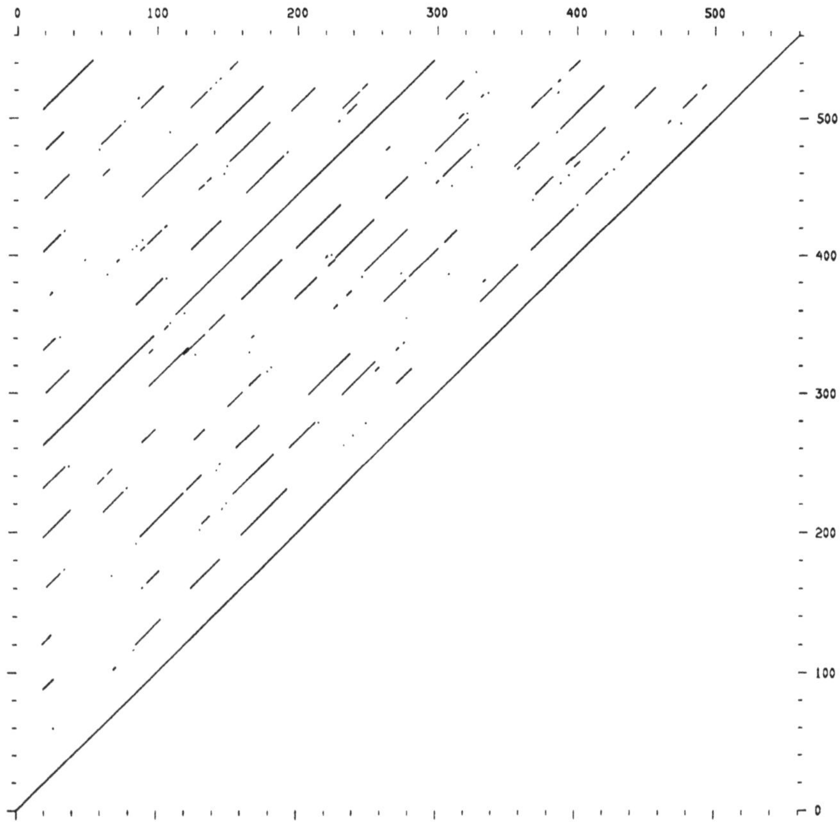

Figure 15. Graphical self comparison of 560 residue nebulin sequence showing 35-residue repeats (weak lines) and 245-residue super-repeat (strong lines). (Reproduced from Labeit et al., 1992.)

pairs of residues, PD at the beginning of the motif and SD and YK spaced two residues apart in the middle.

As well as the 35 residue periodicity, there is also a longer-range repeat containing seven of the 35 residue repeats (i.e., 245 residues). This long-range repeat is analogous to the super-repeat seen in titin and is therefore termed the nebulin super-repeat (Figure 15). Local sequences at comparable positions in different super-repeats are thus more similar than at comparable positions in different 35-residue repeats within the same super-repeat. Sequence database searches with the individual 35 residue repeats, or with a consensus motif, did not reveal significant similarity to known proteins. However, seven-fold periodicities are rare in nature and the presence of one in the super-repeat reinforces the suggestion that nebulin is a thin filament protein, since the molar ratio of actin:troponin:tro-

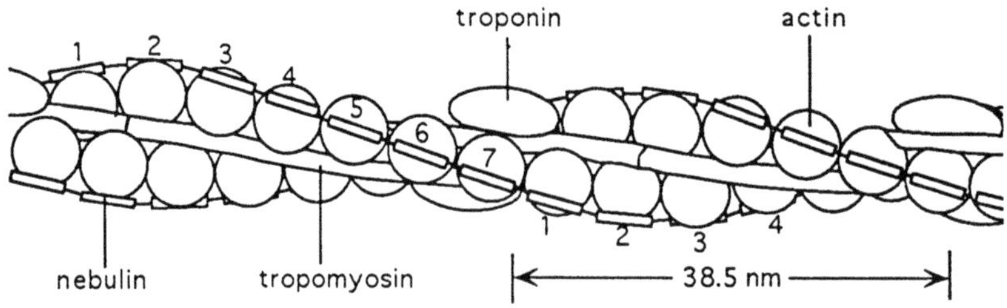

Figure 16. Model of nebulin in thin filament showing the way in which successive 35-residue domains are proposed to interact with successive actin domains in the thin filament. (Reproduced from Trinick, 1992.)

pomyosin in the filament is 7:1:1. There is a preliminary report that 80% of the complete nebulin molecule, which contains ~8000 residues, consists entirely of these 7 × 35 residue patterns, however none of these other data have yet been published (Wang et al., 1990).

D. Secondary Structure and Model of Nebulin *In Situ*

How the 35 residue motif is folded in the native state is not known with certainty, but the pattern conserved residues is suggestive of an α-helical conformation (Labeit et al., 1991). Also the α-helix breakers P and G are rare in the middle of the motif. However, the nebulin molecule cannot be a continuous helix, since there is a conserved P at the beginning of each 35-residue repeat. It also seems unlikely that the molecule is dimerized in the helical regions into the coiled-coil conformation common in fibrous proteins. The coiled-coil is stabilized by the so-called heptad pattern, HxxxHxx, of hydrophobic residues positioned at the interface between the coils (Crick, 1953). There is no heptad pattern in the published 560 residue partial sequence, which led to the proposal that the molecule exists as a single α-helix, which has rarely been found separately in nature (Labeit et al., 1991). One other possibility is that the molecule may be dimerized, but through salt bridges between the coils, rather than hydrophobic residues.

The idea of a largely helical nebulin molecule is also attractive because it is compatible with a number of other considerations. The sevenfold character of the super-repeat suggests that successive 35-residue repeats bind to successive actin subunits in the thin filaments (Figure 16) (Labeit et al., 1991). In this way nebulin would follow the long-pitch helical structure of the filament similar to tropomyosin, with equivalent interactions at each actin binding site. The 245 residues of the super-repeat would therefore need to span the 38.5 nm physical repeat along the filament. This is slightly greater than the 36.8 nm that would be spanned by a continuous α-helix, which suggests that the first few residues at the beginning of each 35-residue repeat may have a more extended conformation. The length the super-repeat required to go around the filament in this arrangement also increases as its radial distance from the long axis of the filament increases. (By about 5% in going from 0 to 4 nm radially; but note that there is little or no information as to the radial location of nebulin in the thin filament.)

The sequence data therefore suggest a nebulin molecule consisting of a series of discrete largely α-helical domains that bind to successive actin subunits in the thin filament. This configuration is also consistent with the idea that individual nebulin molecules spans the entire thin filament. The estimated chain molecular weight of 700–800 kDa would result in a molecule 1 μm long if entirely helical, which is comparable to the length of the vertebrate skeletal muscle thin filament. The screw symmetry of the filament necessitates two, or a multiple of two, nebulin molecules per filament and this is compatible with the estimate that approximately 4% of myofibrillar protein is made up of nebulin (Wang and Wright, 1988). The proposed model also accounts for why, if nebulin is a thin filament protein, it has not been seen in 3-dimensional reconstructions of the filament from electron micrographs. The diameter of a single α-helix would be about 1 nm, whereas resolution of these reconstructions is at best ~3 nm (Moore et al., 1970; Seymour and O'Brien, 1980).

E. Nebulin as a Protein Ruler

Accurate control of thin filament length *in vivo* may therefore be achieved by assembling the actin onto the nebulin until the end of the nebulin molecule is reached. However, since purified actin forms spontaneously into filaments, there is likely to be some mechanism by which the end of the nebulin molecule (or some associated protein here) terminates assembly. There is no proof yet that nebulin controls assembly, but there is some independent evidence consistent with this idea. A prediction of the model is that in muscles where thin filament length is different but accurately specified, the size of the nebulin molecule should vary proportionately. That this is indeed the case was demonstrated by electrophoresis of chicken, rabbit, and beef muscle on SDS-gels. These muscles have thin filaments that are 1.05, 1.1, and 1.3 μm long, respectively, and the mobility of the nebulin band on the gels was seen to vary roughly in inverse proportion to these values (Labeit et al., 1991). A similar correspondence of nebulin size with filament length was also reported in chicken, rabbit, human, and snake muscles (Kruger et al., 1991). The data therefore suggest that nebulin is a family of molecules whose size varies between perhaps 700 and 900 kDa in muscles whose thin filaments range in length between 1.05 μm and 1.3 μm (Kruger et al., 1991).

The discovery of nebulin size variants therefore supports the idea that nebulin acts to regulate thin filament assembly and it is interesting to note that the converse of this argument is also true. In some muscles thin filament length is not precisely specified and in at least in one such case nebulin is known not to be present. Nebulin is known to be absent in cardiac muscle (Locker and Wild, 1986) and thin filaments from this muscle have been shown to vary in length by ~30% (Robinson and Winegrad, 1979).

Nebulin monoclonal antibody labeling patterns have also been reported that give further evidence of a close association with thin filaments (Kruger et al., 1991). Three different antibodies were used to label three different rabbit skeletal muscles. The labeling patterns observed showed considerable variation, but in all cases the striation patterns could be indexed on a lattice of approximately 40 nm and the positions of the stripes were fixed with respect to the Z-line. Since the thin filament has a physical repeat of 38.5 nm (determined accurately from X-ray diffraction), the data suggest the nebulin molecule has repeating structure with this period. This in turn is consistent with the proposal that the 245 residue super-repeat spans 38.5 nm. One of these antibodies bound close to the end of the thin filament (0.9 μm from the Z-line). Since an *in vitro* interaction with α-actinin, which is located in the Z-line *in situ*, has also been reported (Nave et al., 1990), the presence of nebulin throughout the thin filament region has been demonstrated.

F. *In Vitro* Binding Studies

Although it has not been possible to isolate native nebulin, some information about the interactions it makes has been gained from binding studies with constructs expressed from cDNAs. A series of constructs having from 2 to 15 of the 35-residue repeats were expressed in *E. coli* and purified (Jin and Wang, 1991a,b). Binding of these to actin, troponin and tropomyosin was monitored by co-sedimentation and solid-phase assays. All but the smallest two-module fragment bound to actin with high affinity (apparent dissociation constants in the range 10^{-7} to 10^{-8} M). One of the fragments, NB5, also bound to myosin and C-protein. This led to the suggestion that nebulin may have a regulatory as opposed to a purely structural function, but such a role remains to be substantiated. NB5 was known to be from the N-terminal end of the nebulin molecule and since antibodies to this fragment labeled near the tips of the thin filaments, the polarity of the molecule *in situ* was established.

G. Sequence Conservation and Interactions With Other Thin Filament Proteins

Although the data are consistent with a nebulin molecule composed of a series of actin binding domains, it seems likely that the interaction with the thin filament is more complex. The seven actin subunit repeat in the filament is defined by the regulatory proteins troponin and tropomyosin and there is no seven-fold character to filaments made only of actin. This suggests that nebulin may also interact with one or both of troponin or tropomyosin. The nebulin sequence is also very highly conserved and between 200 residues common to both the human and rabbit molecules there were only four non-conservative substitutions (Labeit et al., 1991). This shows that the molecule is likely to make interactions throughout the 38.5 nm spanned by the super-repeat. Since the actin binding site is likely to consist of only a few residues and to be conserved, and since troponin molecules do not span 38.5 nm in the filament, an interaction with tropomyosin was predicted (Labeit et al., 1991). Binding of nebulin constructs to tropomyosin (or troponin) was not observed; however, interaction of the constructs with actin was inhibited by the regulatory proteins (Jin and Wang, 1991a). Clearly, further work is needed to explore this area.

IV. INTERMEDIATE FILAMENTS—DESMIN

Most of the recent developments concerning the assembly and maintenance of sarcomere structure have concerned titin (and related proteins) and nebulin, but in addition to these there are other important elements involved. The M- and Z-lines that anchor and align the thick and thin filaments are dealt with only briefly here, but have been reviewed by Squire et al., 1987. Connections between myofibrils are formed from the intermediate filament protein, desmin, and it is these linkages that are thought to be responsible for the A- and I-bands in adjacent myofibrils being in register in many muscle types. This allows myofibrils to act in concert and to by-pass defective sarcomeres. It also gives rise to the characteristic appearance of cross-striated muscle. The desmin linkages between myofibrils are at the Z-line (for review, see Lazarides, 1982).

In addition to forming these transverse linkages between myofibrils, desmin filaments are thought to run parallel to the fiber axis and to connect adjacent Z-lines in the same myofibril (Tokuyasu, 1983; Wang and Ramirez-Mitchell, 1983). The role of these filaments is not understood, but it seems possible that they are responsible for the continuity of myofibrils after complete extraction of the A-band (Huxley and Hanson, 1954). If this is the case, then the desmin filaments may not be parallel to the fiber axis at short sarcomere lengths, since such "ghost myofibrils" are elastic (Huxley and Hanson, 1954) and intermediate filaments are not thought to be elastic. The continuity of ghost myofibrils is probably not provided by titin molecules, since these retract to the Z-line before complete extraction of the A-band (Higuchi et al., 1992).

NOTE ADDED IN PROOF

Since this article was written there have been a number of advances which have consolidated and extended understanding in the areas discussed. The following briefly outlines some of these developments and lists some of the relevant papers.

Suzuki et al. (Biochem. J., 116, 406–410) reported electron microscopy of titin molecules and fragments showing that the full length of the unstretched molecule is likely to be ~1250 nm. These micrographs were also interesting because they suggest that a region in the I-band part of the molecule is particularly extensible or fragile. Further evidence on the extensibility of titin *in situ*, both in the A- and I-bands, was reported by Wang et al. (Biophys. J., 64, 1161–1177, 1994). The complete titin sequence is likely to be available soon and consists mainly of class II (Ig-like) domains in the I-band (Politou et al. Biochemistry 33, 4730–4737, 1994). The sequence of the C-terminal part of titin has been published and also mainly consists of class II domains. These are separated by linkers, one of which contains phosphorylation sites (Gautel et al. EMBO J., 12, 3827–3834). Ay-mesouthgate et al., (J. Cell Biol., 128, 393–403, 1995) reported that projectin (twitchin) from both synchronous and asynchronous *Drosophila* muscles has functional kinase domains. The complete sequence and new electron micrographs of C-protein were published (Furst et al., J. Cell Sci., 102, 769–778, 1992), and the major myosin binding domain was shown to be near the C-terminus (Okagaki et al., J Cell Biol., 123, 619–626 1993).

Further evidence has been obtained supporting the model of nebulin *in situ* in which a series of discrete single α-helical domains span the entire thin filament. Pfuhl et al., (EMBO J., 13, 1782–1789, 1994) solved the atomic structures of synthetic nebulin peptides by NMR. These were helical in SDS, which was suggested to mimic the charged environment of actin filaments at low radius close to the filament long axis. There have been preliminary reports of the isolation of nebulin without use of denaturing solvents (King, Biophys. J., 68, A286, 1995) and of the ability of phalloidin to "unzip" nebulin from thin filaments (Ao and Lehrer, Biophys. J., 68, A286, 1995).

ACKNOWLEDGMENTS

I thank N. Barclay, C. Chotia, C. Cohen, T. Gibson, A. Houmeida, P. Knight, G. Offer, A. Soteriou, and L. Tskhovrebova for discussion and/or comments on the manuscript; and H. Erickson, H. Holden, P. Knight, G. Offer, and I. Rayment for figures.

REFERENCES

Auber, J. & Couteaux, R. (1963). Ultrastructure de la strie Z dans des muscles de dipteres. J. Microscopie 2, 309–324.

Ayme-Southgate, A., Vigoreaux, J., Benian, G., & Pardue, M.L. (1991). Drosophila has a twitchin/titin-related gene that appears to encode projectin. Proc. Natl. Acad. Sci. USA 88, 7973–7977.

Bahler, M., Eppenberger, H.M., & Wallimann, T. (1985). Novel thick filament protein of chicken pectoralis muscle: the 86 kd protein. II Distribution and localization. J. Mol. Biol. 186, 393–401.

Bazan, J.F. (1991). WKS motifs and the cytokine receptor framework of tissue factor. Trends Biochem. Sci. 16, 329.

Benian, G.M., Kif, J.E., Neckelmann, N., Moermann, D.G., & Waterston, R.H. (1989). Sequence of an unusually large protein implicated in regulation of myosin activity in C. elegans. Nature 342, 45–50.

Benian, G.M., L'Hernault, S.W., Fox, L.A., Tucker, C.Y., & Sale, W.S. (1991). Structure of twitchin, an unusually large sarcomeric protein of C. elegans. J. Cell Biol. 115, 29a.

Bennett, P., Craig, R., Starr, R., & Offer, G. (1986). The ultrastructural location of C-protein, X-protein and H-protein in rabbit muscle. J. Musc. Res. Cell Motil. 7, 550–567.
Cooper, J. & Trinick, J. (1984). Binding and location of AMP deaminase in rabbit psoas muscle myofibrils. J. Mol. Biol. 177, 137–152.
Craig, R. & Offer, G. (1976). The location of C-protein in rabbit skeletal muscle. Proc. R. Soc. Lond. B. 192, 451–461.
Crick, F.H.C. (1953). Acta Cryst. 6, 689–697.
Devereux, J., Haeberli, P., & Smithies, O. (1984). A comprehensive set of sequence-analysis programs for the vax. Nuc. Acids Res. 12, 387–395.
Dos Remedios, C.G. & Gilmour, D. (1978). Is there a third type of filament in striated muscle? J. Biochem. (Tokyo) 84, 235–238.
Eilertson, K.J. & Keller, T.C.S. (1992). Identification and characterization of two huge protein components of brush border cytoskeleton. Evidence for a cellular isoform of titin. J. Cell Biol. 119, 549–557.
Einheber, S. & Fischman, D.A. (1990). Isolation and characterization of a cDNA clone encoding avian skeletal-muscle C-protein. An intracellular member of the immunoglobulin superfamily. Proc. Nat. Acad. Sci. USA 87, 2157–2161.
Epstein, H.F. & Fischman, D.A. (1991). Molecular analysis of protein assembly in muscle development. Science 251, 1039–1044.
Fraenkel-Conrat, H. & Williams, R.C. (1955). Reconstitution of active tobacco mosaic virus from its inactive protein and nucleic acid components. Proc. Natl. Acad. Sci. USA 41, 690–698.
Funatsu, T., Higuchi, H., & Ishiwata, S. (1990). Elastic filaments in skeletal muscle revealed by selective removal of thin filaments with plasma gelsolin. J. Cell Biol. 110, 53–62.
Furst, D.O., Nave, R., Osborn, M., & Weber, K. (1989). Repetitive titin epitopes with a 42 nm spacing coincide in relative position with A-band striations also identified with major myosin-associated proteins. J. Cell Sci. 94, 119–125.
Furst, D.O., Osborn, M., Nave, R., & Weber, K. (1988). The organization of titin filaments in the half-sarcomere revealed by monoclonal antibodies in immunoelectron microscopy: a map of ten nonrepetitive epitopes starting at the Z line extends close to the M line. J. Cell. Biol. 106, 1563–1572.
Furst, D.O., Osborn, M., & Weber, K. (1989). Myogenesis in mouse embryo: differential onset of myogenic proteins and the involvement of titin in myofibril assembly. J. Cell Biol. 109, 517–527.
Furst, D.O., Vinkemeier, U., & Weber, K. (1992). Mammalian skeletal-muscle C-protein—purification from bovine muscle, binding to titin and the characterization of a full-length human cDNA. J. Cell Sci. 102, 769–778.
Gallagher, P.J. & Herring, B.P. (1991). The carboxyl terminus of the smooth muscle kinase is expressed as an independent protein. J. Biol. Chem. 266, 23945–23952.
Gao, Z.H., Moomaw, C.R., Hsu, J., Slaughter, C.A., & Stull, J.T. (1992). Autophosphorylation of skeletal-muscle myosin light chain kinase. Biochemistry 31, 6126–6133.
Gibson, T. & Higgins, D.G. (1993). Myosin light chain kinase: no end in sight. DNA Sequence 3, 333–335.
Grove, B.K., Kurer, V., Lehner, C., Doetschman, T.C., Perriard, J.C., & Eppenberger, H.M. (1984). A new 185,000 Dalton skeletal muscle protein detected by monoclonal antibodies. J. Cell Biol. 98, 518–524.
Hainfeld, J.F., Wall, J.S., & Wang, K. (1988). Titin: quantitative mass measurements by scanning transmission electron microscopy and structural implications for the sarcomere matrix of skeletal muscle. FEBS. Lett. 234, 145–148.
Higgins, D.G., Labeit, S., Gautel, M., & Gibson, T. (1994). The evolution of titin and related giant muscle proteins. J. Molec. Evol. 38, 395–404.

Higuchi, H., Suzuki, T., Kimura, S., Yoshioka, T., Maruyama, K., & Umazume, Y. (1992). Localization and elasticity of connectin (titin) filaments in skinned frog-muscle fibers subjected to partial depolymerization of thick filaments. J. Musc. Res. Cell Motil. 13, 285–294.

Holden, H.M., Masaaki, I., Hartshorne, D.J., & Rayment, I. (1992). X-ray structure of telokin, the C-terminal domain of myosin light chain kinase at 2.8 Å resolution. J. Mol. Biol. 227, 840–851.

Holmgren, A. & Branden, C.-I. (1989). Crystal structure of chaperone protein PapD reveals an immunoglobulin fold. Nature 342, 248–251.

Horowits, R. (1992). Passive force generation and titin isoforms in mammalian skeletal muscle. Biophys. J. 61, 392–398.

Horowits, R., Kempner, E.S., Bisher, M.E., & Podolsky, R.J. (1986). A physiological role for titin and nebulin. Nature 323, 160–164.

Horowits, R. & Podolsky, R.J. (1988). Thick filament movement and isometric tension in activated skeletal muscle. Biophys. J. 54, 165–171.

Hu, D.H., Kimura, S., & Maruyama. K. (1986). SDS electrophoresis studies of connectin-like high molecular weight proteins of various types of vertebrate and invertebrate muscles. J. Biochem. 99, 1485–1492.

Hu, D.H., Kimura, S., & Maruyama, K. (1989). Myosin oligomers as the molecular mass standard in the estimation of molecular mass of nebulin (approximately 800 Kda) by sodium dodecyl sulfate-polyacrylamide gel-electrophoresis. Biomed. Res. 10, 165–168.

Hu, D.H., Matsuno, A., Terakado, K., Matsuura, T., Kimura, S., & Maruyama, K. (1990). Projectin is an invertebrate connectin (titin): isolation from crayfish claw muscle and localization in crayfish claw muscle and insect flight muscle. J. Muscle. Res. Cell. Motil. 11, 497–511.

Huxley, H.E. & Hanson, J. (1954). Changes in the cross-striations of muscle during contraction and stretch and their interpretation. Nature 173, 973–976.

Ito, M., Dabrowska, R., Guerriero, V., & Hartshorne, D.J. (1989). Identification in turkey gizzard of an acidic protein related to the C-terminal portion of smooth muscle myosin light chain kinase. J. Biol. Chem. 264, 13971–13974.

Itoh, Y., Suzuki, T., Kimura, S., Ohashi, K., Higuchi, H., Sawada, H., Shimizu, T., Shibata, M., & Maruyama, K. (1988). Extensible and less-extensible domains of connectin filaments in stretched vertebrate skeletal muscle sarcomeres as detected by immunofluorescence and immunoelectron microscopy using monoclonal antibodies. J. Biochem. (Tokyo) 104 (4), 504–508.

Jin, J.P. & Wang, K. (1991a). Cloning, expression, and protein interaction of human nebulin fragments composed of varying numbers of sequence modules. J. Biol. Chem. 266, 21215–21223.

Jin, J.P. & Wang, K. (1991b). Nebulin as a giant actin-binding template protein in skeletal muscle sarcomere—Interaction of actin and cloned human nebulin fragments. FEBS Lett. 281, 93–96.

Katsura, I. (1987). Determination of bacteriophage-lambda tail length by a protein ruler. Nature 327, 73–75.

Kimura, S., Matsuura, T., Ohtsuka, S., Nakauchi, Y., Matsuno, A., & Maruyama, K. (1992). Characterization and localization of alpha-connectin (titin-1)—An elastic protein isolated from rabbit skeletal muscle. J. Muscle Res. Cell Motil. 13 (1), 39–47.

Knight, P. & Trinick, J. (1984). Structure of the myosin projections on native thick filaments from vertebrate skeletal-muscle. J. Mol. Biol. 177, 461–482.

Kruger, M., Wright, J., & Wang, K. (1991). Nebulin as a length regulator of thin filaments of vertebrate skeletal muscles—Correlation of thin filament length, nebulin size, and epitope profile. J. Cell Biol. 115, 97–107.

Kurzban, G.P. & Wang, K. (1988). Giant polypeptides of skeletal muscle titin: sedimentation equilibrium in guanidine hydrochloride. Biochem. Biophys. Res. Commun. 150, 1155–1161.

Labeit, S., Barlow, D.P., Gautel, M., Gibson, T., Holt, J., Hsieh, C.L., Francke, U., Leonard, K., Wardale, J., Whiting, A., & Trinick, J. (1990). A regular pattern of 2 types of 100-residue motif in the sequence of titin. Nature 345, 273–276.

Labeit, S., Gautel, M., Lakey, A., & Trinick, J. (1992). Towards a molecular understanding of titin. EMBO J. 11, 1711–1716.

Labeit, S., Gibson, T., Lakey, A., Leonard, K., Zeviani, M., Knight, P., Wardale, J., & Trinick, J. (1991). Evidence that nebulin is a protein-ruler in muscle thin filaments. FEBS Lett. 282, 313–316.

Lakey, A., Fergusson, C., Labeit, S., Reedy, M., Larkins, A., Butcher, G., Leonard, K., & Bullard, B. (1990). Identification and localization of high molecular weight proteins in insect flight and leg muscle. EMBO J. 9, 3459–3467.

Lazarides, E. (1982). Intermediate filaments: a chemically heterogeneous, developmentally regulated class of proteins. Ann. Rev. Biochem. 51, 219–250.

Leahy, D.J., Hendrickson, W.A., Aukhil, I., & Erickson, H.P. (1992). Structure of a fibronectin type III domain from tenascin phased by MAD analysis of the selenomethionyl protein. Science 258, 987–991.

Locker, R.H. & Leet, N.G. (1976). Histology of highly stretched beef muscle. J. Ultrastruct. Res. 56, 31–38.

Locker, R.H. & Wild, D.J.C. (1986). A comparative study of high molecular weight proteins in various types of muscles across the animal kingdom. J. Biochem. 99, 1473–1484.

Magid, A. & Law, D.J. (1985). Myofibrils bear most of the resting tension in frog skeletal muscle. Science 230, 1280–1282.

Maroto, M., Vinos, J., Marco, R., & Cervera, M. (1992). Autophosphorylating protein kinase activity in titin-like arthropod projectin. J. Mol. Biol. 224, 287–291.

Maruyama, K., Itoh, Y., & Arisakam, F. (1986). Circular dichroism spectra show abundance of β-structure in connectin, a muscle elastic protein. FEBS Lett. 202, 353–355.

Maruyama, K., Kimura, S., Yoshidomi, H., Sawada, H., & Kikuchi, M. (1984). Molecular size and shape of β-connectin, an elastic protein of striated muscle. J. Biochem. (Tokyo) 95, 1423–1493.

Maruyama, K., Matsuno, A., Higuchi, H., Shimaoka, S., Kimura, S., & Shimizu, T. (1989b). Behavior of connectin (titin) and nebulin in skinned muscle-fibers released after extreme stretch as revealed by immunoelectron microscopy. J. Mus. Res. Cell Motil. 10, 350–359.

Maruyama, K., Natori, K., & Nonomura, Y. (1976). New elastic protein from muscle. Nature 262, 58–60.

Maruyama, T., Nakauchi, Y., Kimura, S., & Maruyama, K. (1989a). Binding of connectin to myosin filaments. J. Biochem. 105, 323–326.

Matsuura, T., Kimura, S., Ohtsuka, S., & Maruyama, K. (1991). Isolation and characterization of 1,200-kDa peptide of alpha-connectin. J. Biochem. (Tokyo) 110, (4), 474–478.

McNeill, P.A. & Hoyle, G. (1967). Evidence for super-thin filaments. Am. Zool. 7, 483–498.

Moerman, D.G., Benain, G.M., Barstead, R.J., Schriefer, L.A., & Waterson, R.H. (1988). Identification and intracellular localization of the *unc-22* gene product in C. elegans. Genes Dev. 2, 93–105.

Moore, P.B., Huxley, H.E., & DeRosier, D.J. (1970). Three-dimensional reconstruction of F-actin, thin filaments and decorated thin filaments. J. Mol. Biol. 50, 279–295.

Nave, R., Furst, D., Vinkemeier, U., & Weber, K. (1991). Purification and physical properties of nematode mini-titins and their relation to twitchin. J. Cell Sci. 98, 491–496.

Nave, R., Furst, D.O., & Weber, K. (1989). Visualization of the polarity of isolated titin molecules: a single globular head on a long thin rod as the M-band anchoring domain. J. Cell Biol. 109, 2177–2187.

Nave, R., Furst, D.O., & Weber, K. (1990). Interaction of alpha-actinin and nebulin in vitro. Support for the existence of a fourth filament system in skeletal muscle. FEBS. Lett. 269, 163–166.

Nave, R. & Weber, K. (1990). A myofibrillar protein of insect muscle related to vertebrate titin connects Z-band and A-band: purification and molecular characterization of invertebrate mini-titin. J. Cell. Sci. 95, 535–544.

Noguchi, J., Yanagisawa, M., Imamura, M., Kasuya, Y., Sakurai, T., Tanaka, T., & Masaki, T. (1992). Complete primary structure and tissue expression of chicken pectoralis M-protein. J. Biol. Chem. 267, 20302–20310.

Offer, G. (1987). Myosin filaments. In: Fibrous Protein Structure (Squire, J.M., & Vibert, P.J., eds.) pp. 307–356. Academic Press, London.

Offer, G., Moos, C., & Starr, R. (1973). A new protein from the thick filaments of vertebrate skeletal myofibrils. J. Mol. Biol. 74, 653–676.

Olson, N.J., Pearson, R.B., Needleman, D.S., Hurwitz, M.Y., Kemp, B.E., & Means, A.R. (1990). Regulatory and structural motifs of chicken gizzard myosin light chain kinase. Proc. Nat. Acad. Sci. USA 87 (6), 2284–2288.

Pardo, J.V., Siliciano, J.D., & Craig, S.W. (1983). A vinculin-containing cortical lattice in skeletal-muscle—transverse lattice elements (costameres) mark sites of attachment between myofibrils and sarcolemma. Proc. Nat. Acad. Sci. USA 80, 1008–1012.

Pierobon-Bormioli, S., Betto, R., & Salviati, G. (1989). The organization of titin (connectin) and nebulin in the sarcomeres—an immunocytolocalization study. J. Musc. Res. Cell Motil. 10, 446–456.

Pringle, J.W.S. (1977). Mechanical properties of insect flight muscle. In: Insect Flight Muscle (Tregear, R., ed.), pp. 177–196, North Holland, Amsterdam, New York, Oxford.

Robinson, T. & Winegrad, S. (1979). Measurement and dynamic implications of thin filament lengths in heart muscle. J. Physiol. 286, 607–619.

Ruoslahti, E. & Pierschbacher, M.D. (1987). New perspectives in cell-adhesion—RGD and integrins. Science 238, 491–497.

Saide, J.D., Chin-Bow, S., Hogan-Sheldon, J., & Busquets-Turner, L. (1990). Z-band proteins in the flight muscle and leg muscle of the honeybee. J. Musc. Res. Cell Motil. 11, 125–136.

Schultheiss, T., Lin, Z.X., Lu, M.H., Murray, J., Fischman, D.A., Weber, K., Masaki, T., Imamura, M., & Holtzer, H. (1990). Differential distribution of subsets of myofibrillar proteins in cardiac nonstriated and striated myofibrils. J. Cell. Biol. 110, 1159–1172.

Seymour, J. & O'Brien, E. (1980). Position of tropomyosin in muscle thin filaments. Nature 283, 680–682.

Shoemaker, M.O., Lau, W., Shattuck, R.L., Kwiatowski, A.P., Matrasian, P.E., Guarra-Santos, L., Wilson, E., Lukas, T.J., van Eldick, L.J., & Watterson, D.M. (1990). Use of DNA sequence and mutant analyses and antisense deoxynucleotides to examine the molecular basis of nonmuscle myosin light chain kinase autoinhibition, calmodulin recognition, and activity. J. Cell Biol. 111, 1107–1125.

Sjostrand, F. (1962). Connections between A- and I-band filaments in striated frog muscle. J. Ultrastructure Res. 7, 225–246.

Small, J.V. (1987). The cytoskeleton: selected topics. In: Fibrous Protein Structure (Squire, J.M., & Vibert, P.J., eds.), pp. 357–388, Academic Press, London.

Soteriou, A., Clarke, A., Trinick, J., & Martin, S. (1993a). Titin folding energy and elasticity. Proc. R. Soc. Lond. B 254, 83–86.

Soteriou, A., Gamage, M., & Trinick, J. (1993b). A survey of the interactions made by titin. J. Cell Sci. 104, 119–123.

Squire, J.M., Luther, P.K., & Trinick, J. (1987). Muscle myofibril architecture. In: Fibrous Protein Structure (Squire, J.M., & Vibert, P.J., eds.), pp. 423–450, Academic Press, London.

Starr, R. & Offer, G. (1971). Polypeptides of intermediate molecular weight in myosin preparations. FEBS Lett. 15, 40–44.

Starr, R. & Offer, G. (1983). H-protein and X-protein. Two new components of thick filaments of vertebrate skeletal muscle. J. Mol. Biol. 170, 675–698.

Stedman, H., Browning, K., Oliver, N., Oronzi-Scott, M., Fischbeck, K., Sarkar, S., Sylvester, J., Schmickel, R., & Wang, K. (1988). Nebulin cDNAs detect a 25-kilobase transcript in skeletal muscle and localize to human chromosome 2. Genomics 2, 1–7.

Takanoohmuro, H., Nakauchi, Y., Kimura, S., & Maruyama, K. (1992). Autophosphorylation of beta-connectin (titin-2) in vitro. Bioch. Biophys. Res. Comm. 183, 31–35.

Tokuyasu, K.T. (1983). Visualization of longitudinally-oriented intermediate filaments in frozen-sections of chicken cardiac-muscle by a new staining method. J. Cell Biol. 97, 562–565.

Trinick, J. (1981). End-filaments: a new structural element of vertebrate skeletal muscle thick filaments. J. Mol. Biol. 151, 309–314.

Trinick, J. (1992). Understanding the functions of titin and nebulin. FEBS Lett. 307 (1), 44–48.

Trinick, J., Knight, P., & Whiting, A. (1984). Purification and properties of native titin. J. Mol. Biol. 180, 331–356.

Trombitas, K., Baatsen, P.H.W.W., Kellermayer, M.S.Z., & Pollack, G.H. (1991). Nature and origin of gap filaments in striated muscle. J. Cell Sci. 100, 809–814.

Vaughan, K.T., Weber, F.E., Einheber, S., & Fischman, D.A. (1993). Molecular cloning of chicken MyBP-H (86 kD protein) reveals conserved protein structure in the myosin-binding protein (MyBP) family containing Ig C2 and FN III motifs. J. Biol. Chem. 268, 3670–3676.

Vigoreaux, J.O., Saide, J.D., & Pardue, M.L. (1991). Structurally different drosophila striated muscles utilize distinct variants of Z-band-associated proteins. J. Mus. Res. Cell Motil. 12 (4), 340–354.

Walcott, B. & Ridgeway, E.B. (1967). Structure of myosin extracted striated muscle fibres. Am. Zool. 7, 499–504.

Wang, K. (1984). Cytoskeletal matrix in striated muscle. The role of titin, nebulin and intermediate filaments. In: Contractile Mechanisms in Muscle (Pollack, G.H., & Sugi, H., eds.), pp. 285–303, Plenum, New York.

Wang, K., Knipfer, M., Huang, Q.Q., Hsu, L., van Heerden, A., Browning, K., Quian, X., & Stedman, H.J. (1990). Structural motifs of nebulin as predicted by sequences of human nebulin cDNA. J. Cell Biol. 111, 428a.

Wang, K., McCarter, R., Wright, J., Beverly, J., & Ramirez-Mitchell, R. (1991). Regulation of skeletal muscle stiffness and elasticity by titin isoforms—A test of the segmental extension model of resting tension. Proc. Natl. Acad. Sci. USA 88 (16), 7101–7105.

Wang, K., McClure, J., & Tu, A. (1979). Titin: Major myofibrillar components of striated muscle. Proc. Natl. Acad. Sci. USA 76, 3698–3702.

Wang, K. & Ramirez-Mitchell, R. (1983). A network of transverse and longtitudinal intermediate filaments is associated with sarcomeres of adult vertebrate skeletal muscle. J. Cell Biol. 96, 562–570.

Wang, K., Ramirez-Mitchell, R., & Palter, D. (1984). Titin is an extraordinarily long, flexible and slender myofibrillar protein. Proc. Natl. Acad. Sci. USA 81, 3685–3689.

Wang, K. & Williamson, C.L. (1980). Identification of an N2 line protein in striated muscle. Proc. Nat. Acad. Sci. USA 77, 3254–3258.

Wang, K. & Wright, J. (1988). Architecture of the sarcomere matrix of skeletal muscle: immunoelectron microscopic evidence that suggests a set of parallel inextensible nebulin filaments anchored at the Z line. J. Cell. Biol. 107, 2199–2212.

Wang, S.M. & Jeng, C.J. (1992). Interaction between titin and alpha-actinin. Biomed. Res. 13, 197–202.

Wang, S.M., Jeng, C.J., & Sun, M.C. (1992). Studies on the interaction between titin and myosin. Histol. Histopathol. 7, 333–337.

Whiting, A., Wardale, J., & Trinick, J. (1989). Does titin regulate the length of muscle thick filaments? J. Mol. Biol. 205, 263–268.

Williams, A.F. & Barclay, A.N. (1988). The immunoglobulin superfamily—domains for cell surface recognition. Ann. Rev. Immunol. 6, 381–405.

Zeviani, M., Darras, B.T., Rizzuto, R., Salviati, G., Betto, R., Bonilla, E., Miranda, A.F., Du, J., Samitt, C., & Dickson, G. (1988). Cloning and expression of human nebulin cDNAs and assignment of the gene to chromosome 2q31-q32. Genomics 2, 249–256.

THE SEQUESTRATION OF mRNA IN THE CYTOSKELETON AND OTHER SUBCELLULAR STRUCTURES

Ian F. Pryme, Arild Johannessen, and Anni Vedeler

Advances in Molecular and Cell Biology
Volume 12, pages 41–74.

ISBN: 1-55938-845-5

I. INTRODUCTION

It has been recognized for many years that in eukaryotic cells mRNA molecules, following their synthesis and processing in the nucleus and subsequent transport through the nuclear membrane, are sequestered into polysomes where the information contained in the RNA molecule is translated into an amino acid sequence. There is evidence suggesting that an association of mRNAs with the cytoskeleton might be important in the transport of mRNA from nucleus to cytoplasm. Until the late 1970s, it appeared that polysomes existed either free in the cytosol (free polysomes) or were associated with the membranes of the rough endoplasmic reticulum (ER) (membrane-bound polysomes). In the late 1970s, however, the development of high voltage electron microscopy enabled the cytoplasmic ground substance to be studied in detail and it became evident that ribosomes, in addition to being associated with ER membranes, were also observed to be associated with fine filamentous structures now known to be components of the cytoskeleton. Since these initial observations a variety of morphological and biochemical studies have shown that mRNA, polysomes, and translational factors are closely associated with the cytoskeleton. The significance of the interaction between components of the translational apparatus and the cytoskeleton is not yet understood but it would seem highly likely that such interactions are important both in the organization and control of protein synthesis. Recent information, for example, indicates that polysomes associated with the cytoskeleton (cytoskeletal-bound polysomes) synthesize particular protein species, thus an association of a population of polysomes with the cytoskeleton may provide a further mechanism for the sorting and topographical segregation of mRNAs. This review discusses current knowledge of the subcellular compartmentalization of mRNAs.

II. mRNA IN THE NUCLEUS

Before release from the nucleus mRNA molecules are subject to posttranscriptional processing: capping, polyadenylation and splicing. Capping occurs very rapidly on the nascent primary transcript. The cap structure consists of a terminal nucleotide, 7-methylguanylate (m^7G) in a 5′–5′ linkage with the initial nucleotide of the mRNA chain, in addition methylation of the ribose ring(s) in the first nucleotide(s) occurs. The biological significance of this modification can be summed up as follows: (a) the structure seems important in the initiation of translation (Bannerjee, 1980), (b) the cap structure prevents the mRNA molecule from degradation by 5′-nucleases (Furuichi et al., 1977; Shimotahno et al., 1977), and (c) it seems necessary for splicing of the HnRNA molecule (Konarska et al., 1985).

A 24-kDa protein is found in association with the cap structure (Trachsel et al., 1980; Shatkin, 1985), and this protein has been shown to associate with intermediate filaments (Zumbe et al., 1982). Since it has been suggested that translation is both dependent on a cap structure and cytoskeletal binding, the biological relationship between these two findings could be the 24-kDa protein (Agutter, 1988). In the nucleus the cap associated proteins could link the HnRNA molecule to the nuclear matrix (Agutter, 1988). Evidence for a more direct role for cap in transport is found in experiments with cap analogues where mRNA transport is dramatically affected ($m^7GpppG >> m^{2,2,7}GpppG, m^{2,7}GpppG$) indicating that a transacting recognition factor may be involved in export (Hamm and Mattaj, 1990).

A highly conserved nucleotide sequence (AAUAAA), which constitutes the polyadenylation signal is found at the 3′-end of the primary transcript (Proudfoot and Brownlee, 1976). This signal is necessary for both cleavage of the HnRNA by an endonuclease, and then for the addition of 100–250 adenosines by poly(A) polymerase to the 3′-end of HnRNA (Hofer et al., 1982; McDevitt et al., 1984). Adenylation occurs quite rapidly after transcription and normally precedes splicing. The best documented role for polyadenylation is maintenance of mRNA stability, and it has been shown that the half life for mRNA correlates positively with poly(A) length (Huez et al., 1981). The poly(A) tail could also participate in splicing (Bina et al., 1980; Bachmann et al., 1984a).

The actual role for the poly(A) tail in transport of mRNA from the nucleus to the cytoplasm is still under debate (Agutter, 1988), because poly(A)-mRNA, like histone mRNA is readily transported (Ballantine and Woodland, 1985), and when polyadenylation is inhibited adenovirus mRNA still enters the cytoplasm and polysomes (Zeevi et al., 1982). In contrary there is also evidence suggesting that poly(A) is important in the transport of mRNA. The nuclear envelope binding site has a general affinity for oligopurinosines, including poly(A) (Agutter et al., 1977), and inhibition of polyadenylation has effects on efflux *in vitro* that cannot be ascribed to inhibition of adenylation (Agutter et al., 1979; Kletzein, 1980). In spite of the arguments against involvement of poly(A) in mRNA transport, it has been postulated by Agutter (1988) that poly(A) has an effect on transport of mRNA from the nucleus to the cytoplasm.

Current knowledge concerning the mechanisms by which introns are removed from HnRNA to form mRNA has been excellently reviewed by (Green, 1986; Padgett et al., 1986). Splicing has a clear role in the transport of mRNA from the nucleus to the cytoplasm. This has been demonstrated in experiments where genes are microinjected into the nucleus. These genes are efficiently transcribed, capped, and polyadenylated irrespective of the numbers of introns, but unless all introns are fully removed no mRNA enters the cytoplasm (Gruss et al., 1979). There are exceptions however, some late virus transcripts, for instance the late transcripts from SV40 enter the cytoplasm without being completely spliced (Wickens and Gurdon, 1983), and studies with the HIV *rev* protein suggest that this protein promotes transport of unspliced viral envelope mRNA (Chang and Sharp, 1989).

In spite of these exceptions the general rule is that mRNA is always completely spliced before transport to the cytoplasm occurs, thereby preventing transport of immature mRNA transcripts from the nucleus.

III. TRANSPORT OF mRNA FROM THE NUCLEUS TO THE CYTOPLASM

Transport of mRNA from the nucleus to the cytoplasm is a heterogenous mechanism which involves several proteins and factors which influence these proteins. The first events relevant for the transport of mRNA after transcription are modifications of the HnRNA molecule, which can be summarized as "capping", "splicing", and "polyadenylation". The mRNA molecule is then transferred to the nuclear pore complex where translocation takes place with hydrolysis of NTPs as the energy source. This is followed by transport of the mRNA molecule to the cytoplasm, most probably in association with intermediate filaments.

During these processes several control steps regulate the transport of the mRNA molecule. After processing the molecule must be released from the nuclear matrix, and transferred to the nuclear pore complex. At the pore complex the mRNA molecule binds to a receptor before it is extruded through the pore. This translocation process is controlled by phosphorylation, and involves different proteins.

The nuclear envelope consists of a double lipid bilayer which separates nucleus from the cytoplasm. Pores with a diameter of 90–100 nm penetrate both membranes in the envelope, and form a cylinder with its axis orthogonal to the membrane plane. One widely accepted model of the pore complex involves three different rings inserted into the envelope. The outer rings face either the cytoplasm or nucleoplasm (Miller et al., 1991), while a central large plug is often observed between the inner and outer ring associated with a middle ring of spokes (Akey, 1989).

Determination of the composition of the nuclear pore complex has proven difficult due to the inability to obtain pure preparations of the pore complex. In spite of these problems some proteins have been shown to be associated with the nuclear pore complex. A 210 KDa glycoprotein (gp210) is found bound to the nuclear membrane (Gerace et al., 1982), and it has been proposed that this protein anchors the pore proper to the channel through the nuclear envelope (Miller et al., 1991). Other proteins associated with the nuclear pore complex compromise a group of O-linked N-acetylglucosamine nucleoporins (Davis and Blobel, 1987; Snow et al., 1987), where p62 is the major protein (Miller et al., 1991). These GlcNac-bearing proteins are peripheral rather than integral membrane proteins (Miller et al., 1991), and are present both on the nuclear and the cytoplasmic face of the nuclear pore, but not in the lumen of the envelope (Hanover et al., 1987).

The nuclear lamina in rat liver consists of the three lamins A, B, and C (Aaronson and Blobel, 1975), and is associated with the inner surface of the nuclear membrane. Lamin B is much more tightly bound to the inner nuclear membrane than lamins A and C, and a lamin B receptor has been characterized in the inner nuclear membrane (Worman et al., 1988). Lamin A and C can form large and stable homo-oligomeres, while lamin B can form hetero-oligomeres with lamin A and C (Aebi et al., 1986). The nuclear lamina therefore presumably consist of A/C oligomeres associated to lamin B monomers, which again are bound to the inner nuclear membrane through a lamin B receptor (Agutter, 1991). During mitotic prophase the nuclear lamina is disrupted due to phosphorylation of the lamin A/C binding domain in lamin B (Georgatos et al., 1988).

The actual existence of a nucleoskeleton is still under debate, but if it exists the nucleoskeleton and the metaphase chromosome scaffold at least share some common protein factors (Agutter, 1991). Unlike the cytoskeleton the nucleoskeleton is not well characterized, but there is an isolated preparation (the nuclear reticulum (Agutter, 1991)) which is a strong candidate. The major protein component of the preparation is actin (Bladon et al., 1988), and perhaps DNA topoisomerase II (Berrios et al., 1985; Schröder et al., 1987), but there might be other proteins associated with this structure. One important feature of the nuclear reticulum preparation, however, is that HnRNA is integrated into it (Long et al., 1979; Van Eekelen and Van Venrooij, 1981).

The cytoplasmic face of the nuclear envelope and nuclear pore complex appear to be anchored to cytoplasmic intermediate filaments with its associated proteins (Akey, 1989; Carmo-Foneca et al., 1987; Fujitani et al., 1989; Georgatos and Blobel, 1987; Georgatos and Blobel, 1987). Intermediate filament proteins like vimentin and cytokeratin have been seen to emanate from both the cyto- and nucleoplasmic sides of the pore complex (Akey, 1989; Carmo-Foneca et al., 1987; Fujitani et al., 1989; Georgatos and Blobel, 1987; Georgatos and Blobel, 1987; Miller et al., 1991), and fibrils have actually been seen passing along the walls of the envelope lumen, and continuing with fibrils that extend across the cytoplasmic face of the envelope (Fujitani et al., 1989). In detergent extracts intermediate filaments have been shown to associate with the nuclear lamina suggesting that intermediate filaments from the cytoplasm extend through the nuclear envelope to the lamina (Georgatos and Blobel, 1987a,b).

Different preparations of the nucleoskeleton have been shown to have precursors of mRNA integrated into its structure (Long et al., 1979; Van Eekelen and Van Venrooij, 1981), and release from this structure seems to be an important step in the regulation of mRNA transport (Agutter, 1988). By using microfilament disrupting agents, and inhibitors of DNA topoisomerase it has been shown that immature precursor messengers are attached to the actin filaments of the nucleoskeleton, while mature mRNA in the nucleus is attached to areas containing DNA topoisomerase II (Schröder et al., 1987). The transfer from actin binding sites to DNA topoisomerase II binding sites might therefore be an essential step during release (Agutter, 1991). The selectivity by which mRNA is to be transported presumably hinges on the dynamics of DNA topoisomerase II (Agutter, 1991).

Translocation of mRNA over the nuclear envelope occurs through the nuclear pore complex (Dworetzky and Feldherr, 1989), and is an energy dependent process

(Agutter et al., 1976; Bernd et al., 1982). The enzyme responsible for this is a broad specificity NTPase which hydrolyses ATP, dATP, GTP, and to some extent CTP and UTP (Agutter, 1988). The NTPase is located on the inner face of the nuclear envelope, probably in the lamina or at the nucleocytoplasmic face of the inner nuclear membrane (Kondor-Koch et al., 1982; Vorbrodt and Maul, 1980). Evidence for the importance of the NTPase has been demonstrated in several experiments involving stimulation of the NTPase by RNA and poly(A) (Agutter et al., 1977; Bernd et al., 1982), NTPase dependent export of mRNA (Riedel et al., 1987), and increased RNA transport which is followed up by an increase in NTPase activity (Clawson et al., 1980; Purello et al., 1982).

Other components concerned in mRNA transport are intramolecular signals in the messenger and the signal receptor in the pore complex. Such a signal is found in the poly(A) tail, which most mRNA contains, and a single receptor binding site with a fair degree of specificity is found in the nuclear envelope (Agutter, 1991). Binding of poly(A) mRNA to this binding site activates the NTPase (McDonald and Agutter, 1980; Bernd et al., 1982).

The binding of mRNA to the poly(A) binding site is affected by phosphorylation (Agutter, 1991). Phosphorylation studies have shown that poly(A) addition inhibits kinase activity, and stimulates phosphorylase activities (McDonald and Agutter, 1980), thereby demonstrating an importance in translocation. The main protein to be phosphorylated is a 110 kDa molecule located in the nuclear envelope (Bachmann et al., 1984b; Agutter, 1985). By affinity labelling poly(A) a protein complex can be detected with a molecular weight of 270 kDa (Pronchow et al., 1990). This complex can be split by RNases and proteinases producing several polypeptide families similar to the breakdown product of the 110 kDa protein (Agutter, 1991). Interestingly one of the products of proteolysis of the 270 kDa protein complex has the same molecular weight as the NTPase (Agutter, 1991; Schröder et al., 1986).

With these findings in mind the scenario for mRNA translocation is as follows: the mRNA molecule binds to its receptor in the nuclear envelope, where the receptors affinity for its ligand can be modified by phosphorylation. The translocation process through the pore complex is an energy requiring process, and is accomplished by an NTPase.

IV. mRNA IN POLYSOMES

A. Free and Membrane-Bound Polysomes

Up to the late 1970s it was apparent that after their transport into the cytoplasm mRNA molecules were sequestered into two compartments of polysomes: cytosolic or free polysomes, and those associated with the rough ER, i.e., membrane-bound polysomes. From early studies the free polysomes were assigned a function concerned with the synthesis of intracellular proteins while membrane-bound polysomes were considered to represent the site of synthesis of proteins destined to be secreted from the cell (Svardal and Pryme, 1980). It is now evident, however, that the situation is much more complex and that a wide variety of proteins which are not ultimately secreted from the cell are also synthesized on membrane-bound polysomes. Observations suggesting this include (a) an active protein synthesis occurs on membrane-bound polysomes in cells that are not engaged to any significant extent in the process of secretion, and (b) a massive proliferation of the ER with associated membrane-bound polysomes is observed during the active development of non-secretory as well as secretory tissues. It is now apparent that a major function of interaction between ribosomes and ER is to effect a topographical segregation of populations of ribosomes which are programmed to synthesize different classes of protein. In 1975 Blobel and Dobberstein postulated the now well-documented signal hypothesis; mRNAs, coding, for example, secretory proteins, contain a stretch of codons at the 5′ end of mRNA immediately following the initiation site, which upon translation give rise to a sequence of amino acids termed the signal sequence. This, in conjunction with the signal recognition particle and docking protein in the rough ER, directs the polypeptide chain to the ER and ribosomes become associated to the membrane through receptor proteins (ribophorins I and II and p34). The nascent polypeptide chain then passes through the membrane and into the lumen of the ER. This sequence rich in hydrophobic amino acids has subsequently been demonstrated in a series of mRNA molecules coding for proteins which either become associated with, are inserted into, or cross ER membranes. This mechanism provides a method for the topographical segregation of proteins.

Since the initial evidence was provided which showed an interaction between polysomes and ER membranes a series of studies have shown that the rough ER is compartmentalized into subfractions, each of which appears to have individual properties (for reviews see Pryme, 1986; 1989a,b). Three populations of ER membranes containing associated polysomes have been described: light rough (LR), heavy rough (HR) and nuclear-associated endoplasmic reticulum (NER). These ER subfractions possess a variety of separate characteristics: size differences in poly A mRNA content (Pryme, 1989b); the LR and HR subfractions differ in the manner by which polysomes are associated with the ER membrane (Pryme, 1988), since it appears that HR and LR membranes contain loosely-bound and tightly bound polysomes respectively; the HR subfraction has a fourfold higher content of light chain immunoglobulin mRNA than LR membranes (Pryme, 1988); the turnover of 60S ribosomes is much faster in HR than LR membranes (Pryme, 1988); nascent polypeptide chains on polysomes associated with NER were labeled with [^{3}H]-glucosamine to a much greater extent than those on LR/HR membranes, indicating that NER is enriched in mRNA species which code for proteins that are glycosylated at an early stage (Pryme and Svardal, 1978). Using spin probes the LR and HR membranes isolated from Krebs II ascites and L-929 cells have recently been shown to differ with respect to lipid microviscosity (Maltseva et al., 1991a,b).

Shore and Tata (1977) demonstrated that some ER cosedimented with the low speed nuclear pellet prepared following the homogenization of rat liver and this was termed rapidly sedimenting endoplasmic reticulum (RSER). The RSER membranes were separated from rough ER membrane species by a low speed centrifugation; translation of mRNA isolated from these membrane fractions in a heterologous in vitro system followed by immunoprecipitation, showed that albumin mRNA was equally distributed between the RSER and rough ER while mRNA coding for mitochondrial proteins was enriched in the rough ER.

Taking into account the observations described above it would appear that the ER is compartmentalized into subfractions and that different proteins are synthesized in individual membrane compartments. This is presumably part of the mechanism which assures the corrected targeting of newly synthesized proteins. How individual mRNA species are sorted into correct membrane compartments is not yet understood.

B. A Class of Polysomes Is Associated With Cytoskeletal Filaments

The first evidence for the presence of a third population of polysomes in addition to free and membrane-bound populations appeared in 1976 when Wolosewick and Porter using high voltage techniques of electron microscopy (HVEM) obtained a detailed structure of the cytoplasmic ground substance. Using these techniques they observed the presence of ribosomes and polysomes in close proximity to finely dispersed cytoplasmic filaments and also in pockets at junctions where filaments cross or join one another (Wolosewick and Porter, 1976; 1979). These observations indicated the existence of polysomes occurring in association with cytoskeletal filaments, and these have been termed cytoskeletal-bound polysomes.

Following the initial observations using HVEM (which indicated polysome-cytoskeleton association) more recent evidence has been obtained in several laboratories using a variety of electron microscopical, biochemical, and histochemical studies (for reviews see Nielsen et al., 1983; Hesketh and Pryme, 1991). One of the important initial observations was made by Osborn and Weber (1977) who showed that treatment of cultured cells with low concentrations of non-ionic detergents caused solubilization of the plasma membrane releasing cytoplasmic constituents and integral membrane proteins but leaving the cytoskeleton as an insoluble matrix (referred to in this review as the cytomatrix). Ribosomes were shown to be present in this non-ionic detergent insoluble cytomatrix by electron microscopy and the results were taken as evidence that these ribosomes were associated with the cytoskeleton (Lenk et al., 1977). Further studies using electron microscopy showed that in both lens cells (Ramaekers et al., 1983) and ascidian eggs (Moon et al., 1983) ribosomes were present either attached to, or closely associated with, cytoskeletal filaments. Using quick-freeze deep-etch techniques in combination with electron microscopy Heuser and Kirschner (1980) showed that clusters of ribosomes appeared to be attached to filaments in the cell matrix. This was confirmed histochemically using acridine orange (Fulton et al., 1980). Autoimmune serum obtained from lupus erythromatosus patients contains anti-ribosomal antibodies, and the use of this antiserum on the non-ionic detergent insoluble cytomatrix showed an aligned punctate labeling indicating the association of ribosomal material with filamentous structures (Toh et al., 1980). Recent results obtained by Hesketh et al. (1991a) are also consistent with an association between ribosomes and filamentous structures; the staining of 3T3 fibroblasts with anti-ribosomal subunit antibodies gave rise to linear arrays of beaded or punctate patterns of ribosome distribution in non-ionic detergent extracted cells. Further evidence has been obtained from experiments where specific antibodies which recognize either the 5′mRNA cap binding protein (Zumbe et al., 1982) or initiation factor eIF-2 (Heuijerjans et al., 1989) were shown to label a filamentous component of the cytomatrix. Immunostaining experiments using an antibody against a nuclear envelope antigen which binds to mRNA also showed a staining pattern suggestive of an association with microfilaments (Schröder et al., 1988). Recent studies by Singer et al. (1989), where a combination of high resolution electron microscopy and *in situ* hybridization was used, showed that mRNA species coding for actin, tubulin, and vimentin occur in clusters around filamentous structures present in the cytomatrix. Experiments on skeletal muscle using anti-ribosomal subunit antibodies (Horne and Hesketh, 1990a,b) have provided data showing ribosomes to be present in an organized and repetitive manner along myofibrils. The above-mentioned electron microscopical and immunohistochemical studies have thus provided a wealth of evidence indicating that components of the protein synthetic apparatus in cells are distributed in a manner suggesting an association with filaments.

One of the problems in making a correct interpretation of the results obtained from the above studies has been the possibility that polysomes or mRNA may be "artifactually" retained within the cytoskeletal network in non-ionic detergent extracted cells. Thus these experiments did not provide conclusive evidence that actively translated mRNA is in direct association with the cytoskeleton. Evidence against an artifactual trapping of mRNA in the cytomatrix was obtained by Pondel and King (1988), who, using *Xenopus* eggs, were unable to demonstrate a trapping of either endogenous mitochondrial mRNAs or exogenous globin mRNA during the preparation of cytomatrix. Similarly, *in situ* hybridization studies where ascidian eggs were employed have shown that non-ionic detergent treatment does not affect the overall distribution of either total or specific mRNAs in the cell (Jeffrey, 1984). Adams et al. (1983) using a gel filtration approach, found no evidence for non-specific polysome-protein interactions occurring in the presence of Triton X100. This indicates that non-physiological ribosome-cytoskeleton interactions do not occur under conditions where non-ionic detergent is employed for the purpose of cytomatrix preparation. Other evidence suggesting that it is unlikely that ribosomes are retained in the cytoskeleton by some form of nonspecific trapping is the fact that more than 70% of cell protein and virtually 99% of lactic dehydrogenase activity is released by non-ionic detergent treatment (Vedeler et al., 1991a). The fact that identical amounts of (free) polysomes were released from identical portions of cells either by treatment with non-ionic detergent or by disrupting cells using nitrogen cavitation (Vedeler et al., 1991a) also suggests that the observed presence of polysomes and associated factors in the cell matrix reflects the situation in the cell.

C. Attachment of Cytoskeletal-Bound Polysomes to Microfilaments

The identification of the filament type to which cytoskeletal-bound polysomes are associated has come from a variety of studies. Morphological observations using cultured cells clearly indicated the presence of polysomes in areas of the cytoplasm free of intermediate filaments (Lenk et al., 1977; Fulton et al., 1980), suggesting that polysomes do not interact with this class of filaments. It has also been observed that although polysomes can be extracted by salt from the cytomatrix of ascites cells vimentin is not affected in the same manner (Traub and Nelson, 1982; Vedeler et al., 1991a). Furthermore, eIF2 has been shown to be associated with filaments in extracts prepared from three different cell lines each lacking vimentin (Heuijerjans et al., 1989). In addition in these three cell lines both polysomes and specific mRNAs for actin and histone H2 are distributed between soluble and non-ionic detergent insoluble fractions as in normal vimentin-containing HeLa cells. Colchicine has no effect on polysome distribution in cultured cells (Lenk et al., 1977) suggesting that cytoskeletal-bound polysomes are not associated with microtubules. It has been shown that colchicine, which promotes microtubule disruption, causes collapse of the intermediate filament network (Hynes and Destree, 1978) suggesting that the microtubule and intermediate filament systems are linked. The fact that colchicine does not release polysomes from the cell matrix (Vedeler et al., 1991a) supports the notion that cytoskeletal-bound polysomes are not associated with microtubules or intermediate filaments. There is, therefore, substantial evidence which suggests that cytoskeletal-bound polysomes are not associated to any great extent with either microtubules or intermediate filaments.

On the other hand, however, much evidence has accumulated during recent years suggesting an association of polysomes with microfilaments. Based on size measurements obtained from electron microscopy, Ramaekers et al. (1983) using lens cells, tentatively identified the filaments as microfilaments. Both cytochalasin B and D, which cause a perturbation of microfilaments (Cooper, 1987), have been shown to cause a release of polysomes or mRNAs from the insoluble cytomatrix in a variety of cell lines (Lenk et al., 1977; Ramaekers et al., 1983; Bird and Sells, 1986; Ornelles et al., 1986; Bagchi et al., 1987; Vedeler et al., 1991a). Seely and Aggeler (1991), studying the modulation of milk protein synthesis in mammary epithelial cells, showed that treatment of cells with cytochalasin D resulted in the reduced synthesis of most proteins, β-casein being first affected following disruption of the microfilament system. Using HeLa cells Ornelles et al. (1986) showed that cytochalasin D, which caused a simultaneous release of both mRNA and polysomes, resulted in a reversible inhibition of protein synthesis. In myoblasts cytochalasin D causes the release of nascent myosin heavy chains (Isaacs and Fulton, 1987). DNAse I, which also causes actin depolymerization, has been shown to induce the release of polysomes (Adams et al., 1983). Vedeler et al. (1991a) demonstrated that phalloidin treatment, which stabilizes microfilaments (Dancker et al., 1975), resulted in a 130% increase in the amount of actin present in the cytoskeletal fraction of Krebs II ascites cells but only a small increase in cytoskeletal-bound polysomes was observed (12%). These results could be explained by the repolymerization of actin at the ends of filaments devoid of polysomes. In experiments where 3T3 fibroblasts were extracted with non-ionic detergent in buffer containing either 25 or 130 mM KCl it was shown that the presence of the higher salt concentration led to an increased release of both polysomes (+50%) and actin (+76%) from the cytomatrix to the non-ionic detergent soluble fraction (Hesketh and Pryme, 1988). In cytochalasin B treated cells the amount of actin released by extraction with detergent at a low salt concentration was equivalent to that measured following extraction with 130 mM KCl, indicating that extraction at the latter salt concentration caused a breakdown of actin filaments similar in degree to that produced by the microfilament perturbing agent (Hesketh and Pryme, 1988). These results again favored the suggestion that polysomes are associated with microfilaments. Earlier results had clearly demonstrated that salt concentrations of 100–150 mM destabilize actin filaments (Kasai, 1969) but not intermediate filaments (Steinart et al., 1982).

Further supporting evidence for an association between polysomes and microfilaments has come from studies following cytomegalovirus infection (Jones and Kilpatrick, 1988) where a rapid depolymerization of microfilaments was observed early after virus infection and followed later by a repolymerization. These changes were accompanied by first a decrease and then an increase in the proportion of polysomes associated with the cytomatrix. A direct interaction between components of the translatory apparatus and microfilaments is supported by the observation that EF-1a possesses actin-binding properties (Yang et al., 1990). Considerable evidence has therefore accumulated suggesting that a population of polysomes is associated with microfilaments. Estimates of the percentage of total cellular polysomes in the cytoskeletal-bound population varies between 25 and 40% in different cell lines (Ramaekers et al., 1983; Hesketh and Pryme, 1988; Vedeler et al., 1991a).

V. FREE, CYTOSKELETAL-BOUND, AND MEMBRANE-BOUND POLYSOMES ISOLATED AS SEPARATE POLYSOME POPULATIONS

Early studies where non-ionic detergent extraction was used in order to study "cytoskeletal-bound" polysomes were performed on HeLa cells (Lenk et al., 1977;

Ornelles et al., 1986). These workers demonstrated that monosomes and ribosomal subunits rather than polysomes were released using non-ionic detergent alone. Using the same cell line but with slightly modified extraction conditions, Bird and Sells (1986) and Katze et al. (1989) showed that 20–30% of total cellular polysomes were released upon the lysis of cells with non-ionic detergent alone, indicating the existence of free polysomes. These observations were in agreement with a series of other results where either Potter-Elvehjem (Blobel and Potter, 1966, 1967), Thomas homogenization (Attardi et al., 1969), nitrogen cavitation (Pryme, 1974; Pryme and Svardal, 1978; Svardal et al., 1981; Pryme 1989a,b), or non-ionic detergent (Birckbichler and Pryme, 1973; Pryme et al., 1973; Abraham et al., 1973; Ramaekers et al., 1983; Bagchi et al., 1987; Lequang and Gauthier, 1989; Bouhtiauy et al., 1989) was used for the isolation of free polysomes. There are thus many observations consistent with the existence of free polysomes. However, the actual definition of free polysomes is still somewhat unclear since it is quite possible that such polysomes may well be loosely associated with the cytoskeleton such that they are easily dislodged upon lysis of the cell, independent of the method used for cell disruption.

It is likely that under some extraction conditions free polysomes may be contaminated to some degree with polysomes dislodged from the cytoskeleton during the manipulation of cells in the isolation procedure. A direct comparison of non-ionic detergent extraction and subcellular fractionation has shown that 54% of histone mRNA was recovered in the free (cytosolic) fraction obtained following cell disruption by Dounce homogenization, while only 16% of the same mRNA species was present in the soluble (free) fraction released by non-ionic detergent (Zambetti et al., 1985). It is thus apparent that the actual methods used for fractionation into subcellular compartment has effects on the ultimate distribution of specific mRNAs. Results reported by Vedeler et al. (1991a), however, showed that almost identical amounts of free polysomes were released when two quite different techniques were employed, namely solubilization of the plasma membrane with non-ionic detergent and nitrogen cavitation, where the use of detergent was avoided. Recently Biegel and Pachter (1992) have demonstrated that when NIH 3T3 cells grown in monolayer were treated with non-ionic detergent *in situ* such that an initial perturbation of the cytoskeleton was avoided then 50% of the polyribosome population was not extracted by 1.0 M KCl, suggesting a strong form of interaction between polysomes and cytomatrix.

In their initial studies Lenk et al. (1977) showed that polysomes which were retained in the cytomatrix could be released by treatment with a combination of non-ionic (1% Tween 40) and ionic detergents (1% deoxycholate). These and many subsequent authors took this as conclusive evidence for an association of polysomes with the cytoskeleton. This methodology, however, is virtually identical to that described some years ago for the separation of free polysomes and membrane-bound polysomes where it was shown that non-ionic detergents alone are insufficient to release membrane-bound polysomes, the ionic detergent deoxycholate being required to effectively solubilize this class of polysomes from the rough ER. That the polysomes solubilized by deoxycholate contained mRNA species expected to be translated in membrane-bound polysomes was shown by *in vitro* protein synthesis: in contrast to free polysomes, membrane-bound polysomes were shown to be able to direct the synthesis of immunoglobulin light chain polypeptides (Pryme et al., 1973; Pryme, 1974; 1989a). Taking these results into account it would appear likely that membrane-bound polysomes are present in the cytomatrix prepared according to the technique of Lenk et al. (1977) and are then subsequently released upon treatment with deoxycholate. That the use of a combination of detergents alone in a two-step process does not allow one to separate cytoskeletal-bound from membrane-bound polysomes is clearly exemplified by the results of Zambetti et al. (1985). These workers prepared polysomes using two different methods, namely Dounce homogenization and detergent extraction. They found that 96% of the mRNA coding for the membrane antigen HLA B7 was present in membrane-bound polysomes isolated from ER membranes following Dounce homogenization, however, they showed that up to 97% of the same mRNA species was located in the "cytoskeletal-bound" fraction prepared using the two-step detergent technique (Lenk et al., 1977; Cervera et al., 1981). It is thus evident that the so-called "cytoskeletal-bound" polysome fraction released by deoxycholate is enriched in mRNA species expected to be found in membrane-bound polysomes.

The presence of ER residues in the detergent-insoluble cytomatrix has been confirmed by electron microscopy (Dang et al., 1983; Ramaekers et al., 1983). Other results from immunohistochemical studies where antibodies to ER components were used have clearly shown that membrane material is indeed retained in the non-ionic detergent insoluble cytomatrix (Dang et al., 1983; Mirande et al., 1985). Furthermore, the detection of large amounts of labeled phospholipids in the material obtained following deoxycholate treatment of the cytomatrix (Lenk et al., 1977; Vedeler et al., 1991a) clearly demonstrates the presence of solubilized membranous elements in the so-called cytoskeletal fraction as defined by Lenk et al. (1977). Polysomes released by double-detergent treatment following an initial extraction of cells with non-ionic detergent in the presence of low salt are, therefore, clearly a mixture of those initially derived both from the cytoskeleton and the rough ER.

Based on three pieces of information—first, that cells can be lysed by treatment with non-ionic detergent releasing significant amounts of polysomes (Birckbichler and Pryme, 1973); second, that by increasing the salt concentration from 25 to 130 mM during non-ionic detergent treatment of cells a breakdown of microfilaments occurred with a simultaneous liberation of polysomes (Hesketh and Pryme, 1988); and third, that polysomes are released from rough ER membranes upon solubilization with deoxycholate (Birckbichler and Pryme, 1973; Pryme et al., 1973; Pryme, 1974; Svardal et al., 1981)—Vedeler et al. (1990, 1991a) developed a sequential detergent/salt extraction technique which enabled total cellular polysomes to be separated into free, cytoskeletal-bound and membrane-bound polysome populations. Briefly, cells were first treated with non-ionic detergent in the presence of 25 mM KCl in order to cause lysis of the plasma membrane. The resulting extract was shown to contain almost 99% of the cytosolic marker enzyme lactate dehydrogenase, large amounts of tubulin and non-filamentous actin, and was rich in labeled phospholipid. The insoluble cytomatrix was then treated with 130 mM KCl and the extract was shown to have a low phospholipid content but was enriched in the cytoskeletal components actin and cytokeratin. Treatment of the 130 mM KCl extracted non-ionic detergent insoluble residue with a combination of Triton X-100 and deoxycholate solubilized almost 75% of cellular phospholipid and the extract had a high vimentin (intermediate filament) content. Based on SDS-PAGE the three fractions exhibited different protein compositions (Vedeler et al., 1991b), the 130 mM KCl extract being particularly enriched in a 35-kDa protein. The distinct nature of these fractions is further exemplified by the different appearance of polysome profiles obtained following sucrose density gradient centrifugation (Kirkeeide et al., 1992, 1993; Vedeler at al., 1991a). The fact that treatment of cells with cytochalasin B reduced the amount of polysomes recovered in the 130 mM salt extract but had no effect on those in the Triton/deoxycholate fraction indicated that polysomes were released from microfilaments rather than rough ER membranes (Vedeler et al., 1991a). Interestingly, using Krebs II ascites cells intact polysomes were released into the free polysome fraction such that perturbation of microfilaments did not result in a degradation of polysomes. Other workers have shown that pre-treatment of cells with either cytochalasin B or D caused an increase in the proportion of polysomes recovered in the non-ionic detergent soluble fraction and a simultaneous reduction in amount of polysomes associated with the cytomatrix (Ramaekers et al., 1983; Ornelles et al., 1986). Finally, results from vaccinia virus infected cells have shown that up to 40% of viral mRNAs are associated with polysomes in the free fraction released by gentle treatment with non-ionic detergent (Lemieux and Beaud, 1982).

VI. EVIDENCE FOR AN ASSOCIATION OF mRNAs AND TRANSLATIONAL FACTORS WITH THE CYTOSKELETON

Lenk and Penman (1979) showed that in virus-infected guanidine-treated cells there was a release of uridine labeled mRNA from the cell matrix with the result that neither viral nor host mRNAs were translated. These observations led to the suggestion that attachment of mRNA to the cytoskeleton must occur before the onset of translation can take place. One cannot rule out the possibility, however, that the above results can be explained by a release of mRNAs caused by a partial breakdown of the cytoskeleton rather than being due to a regulatory mechanism involved in the control of protein synthesis. In this context it is important to note that infection with certain viruses results in widespread changes in cytoskeletal organization (Cervera et al., 1981; Jones and Kilpatrick, 1988; Katze et al., 1989).

Host mRNAs, however, are retained on the cytoskeleton following infection by either adenovirus or influenza virus infection, where no change in the integrity of the cytoskeleton is observed (Katze et al., 1989). In addition, as mentioned above, the cytomatrix contains residual ER material, thus it cannot be ruled out that the observed release of mRNAs coding for viral glycoproteins may in fact have occurred from rough ER membranes rather than the cytoskeleton. Furthermore, it seems that there is a retention of some host mRNAs (including actin and tubulin species) in the cytomatrix following viral infection despite the fact that they are not actively translated (Lemieux and Beaud, 1982; Bonneau et al., 1985; Katze et al., 1989). Meadus et al. (1990) showed that the mRNA coding for p40 was present in the cytomatrix isolated from both myocytes and myotubes, but active translation of the mRNA was only observed in myoblasts.

Experiments with EDTA, which cause ribosomal subunits to dissociate and separate from the mRNA, and inhibitors of protein synthesis such as fluoride and pactamycin have also provided evidence for an attachment of mRNA to the cytomatrix (van Venrooij et al., 1981; Bag and Pramanik, 1987). Since these workers demonstrated an almost total lack of release of mRNA it seems likely that mRNAs are retained on both the cytoskeleton and rough ER membranes when not being actively translated. These results suggest that certain mRNA species may be either present in cytoskeletal-bound polysomes or remain associated with the cytoskeleton in an untranslated state, for example as prosomes, a class of ribonucleoprotein particles which have been shown to be associated with intermediate filaments (Grossi de Sa et al., 1988a,b; Scherrer, 1990; Coux et al., 1992).

A series of authors using various cell types have shown that up to 70% of total cellular mRNA can be recovered in a so-called cytoskeletal-bound form either retained in the cytomatrix or released by Triton/deoxycholate treatment (Lenk and Penman, 1979; Cervera et al., 1981; van Venrooij et al., 1981; Lemieux and Beaud, 1982; Moon et al., 1983; Jeffrey, 1984; Bonneau et al., 1985; Bag and Praminik, 1987; Yisraeli and Melton, 1988; Katze et al., 1989; Meadus et al., 1990). A major drawback of these experiments, however, similar to the analysis of polysome distribution using the two-step detergent technique, is the definition of the insoluble cytomatrix as a cytoskeletal fraction and any mRNAs associated with it or with material released by Triton/deoxycholate treatment as being cytoskeletal-bound. As discussed above such fractions are heterogeneous making it very difficult to make definite conclusions with respect to the possibility that mRNA species are associated with the cytoskeleton.

A more specific approach in an effort to distinguish between cytoskeletal-associated material from that bound to the rough ER has been the use of cytochalasin B or D in order to promote a release of microfilament-associated material from the cytomatrix. Using such methodology and hybridization to poly-T as an assay for total cellular mRNA, Ornelles et al. (1986) demonstrated that about 80% of mRNA was released from the cell matrix following pretreatment of cells with cytochalasin D. In similar experiments Yisraeli and Melton (1988) observed that cytochalasin B treatment of *Xenopus* oocytes resulted in release of specific Vg1 mRNA, and in HeLa cells Bird and Sells (1986) showed that pretreatment with cytochalasin B caused a redistribution of actin mRNA such that the bulk was recovered from the soluble rather than the "cytoskeletal" fraction. In contrast, however, the situation is not as clear cut in myoblasts where cytochalasin apparently had no effect on the localization of several mRNAs, although it was shown to result in a total breakdown of microfilaments (Bagchi et al., 1987). It should be pointed out, however, that care must be taken when interpreting results where relatively high concentrations of cytochalasin B have been used since in experiments with Krebs II ascites cells it has been shown that concentrations above 20 μg/ml result in a progressive conversion of HR to LR ER membranes due to a loss of polysomes (Fjose and Pryme, 1984). It is therefore apparent that under certain conditions the use of cytochalasin B may produce spurious results.

Direct evidence favoring an association of specific mRNA species with the cytoskeleton has come from work where fractions enriched in cytoskeletal components were prepared from the cell matrix and shown to be highly enriched in mRNA coding for Vg1 (Pondel and King, 1988). Klymkowsky et al. (1991) showed that fragmentation of cytokeratin filaments only leads to the solubilization of a minor fraction of Vg1 mRNA, suggesting that only small amounts of this mRNA species are anchored to cytokeratin filaments. Bird and Sells (1986), using HeLa cells, showed that considerably more actin mRNA was present in an extract of the cytomatrix than in the non-ionic detergent soluble fraction, the opposite, however, being the case for histone mRNA. Results from experiments with myoblasts indicated that mRNAs for actin, ribosomal protein L32, and histone were essentially recovered in a salt extract of the cytomatrix rather than in the non-ionic detergent soluble fraction (Bagchi et al., 1987). *In situ* hybridization studies at the electron microscopy level (Singer et al., 1989) have recently provided important evidence for the presence of mRNAs on the cytoskeleton. These workers demonstrated that the mRNA for actin is closely associated with filaments. There is thus ample evidence which indicates that mRNAs are associated with the cytoskeleton though the actual nature of such interaction is not yet understood. Polysomes appear to be bound to the cytoskeleton through mRNA and not ribosomes since a removal of ribosomes and initiation factors from the cytomatrix has been demonstrated after heat shock (Welch and Feramisco, 1985) or incubation of cells with fluoride or pactamycin (Howe and Hershey, 1984). Other experiments, however, have clearly shown that not all mRNA is released from the cytomatrix upon inhibition of protein synthesis or dissociation of polysomes (van Venrooij et al., 1981; Bag and Pramanik, 1987) thus the association of ribosomes with both the rough ER and cytoskeletal components is not entirely dependent on the presence of intact mRNA.

It would appear that there are two forms of interaction between components of polysomes and the cytoskeleton, an initial interaction between mRNA and the cytoskeleton followed by subsequent binding through ribosomes. This may occur through a receptor mechanism analogous to the binding of ribosomes to rough ER with the subsequent formation of membrane-bound polysomes (Savitz and Meyer, 1990). Evidence in favor of the possibility that an interaction between polysomes and microfilaments is mediated through protein–protein interaction has come from the work of Yang et al. (1990) who demonstrated that EF-1a has actin-binding properties. Howe and Hershey (1984) found that increased amounts of eIF 2, 3, 4a, and 4b were recovered in the soluble fraction following incubation of cells with cytochalasin B, suggesting that these were originally associated with cytoskeletal-bound polysomes. Using indirect immuno-fluorescent microscopy Shestakova et al. (1991) provided evidence that EF-2 is colocalized with bundles of actin filaments, though they suggest it unlikely that EF-2 is an integral microfilament protein. Interestingly, these authors observed that not all microfilament bundles were stained with EF-2 antibodies, demonstrating that the protein synthetic apparatus is associated with a subset of microfilaments. A subdivision of the microfilament system into compartments is also consistent with results of Vedeler et al. (1991a). These workers showed that stabilization of microfilaments with phalloidin prior to homogenization only had a small effect on the amount of polysomes found associated with microfilaments (12% increase), while the actin content increased by about 130%.

VII. CYTOSKELETAL-BOUND POLYSOMES AND SPECIFICITY OF PROTEIN SYNTHESIS

Since a technique for the isolation of cytoskeletal-bound polysomes as a population separate from free and membrane-bound polysomes has only recently been developed (Vedeler et al., 1990; 1991a), information concerning the details of the involvement of cytoskeletal-bound polysomes in protein synthesis is still somewhat limited. Using this technique free, cytoskeletal-bound and membrane-bound polysomes have been isolated from Krebs II ascites and 3T3 cells (Vedeler et al., 1990; 1991a,b; Hesketh et al., 1991b; Kirkeeide et al., 1991; 1993) and MPC-11 cells (Moss et al., 1994). The technique has been recently adapted such that the three polysome populations have also been isolated from L-929 cells (Figure 1). As earlier observed for Krebs II ascites cells (Vedeler et al., 1991a), pretreatment of L-929 fibroblasts with cytochalasin B also caused a great reduction in the amount of polysomes recovered in the cytoskeletal-bound polysome fraction (Figure 2). Thus a major population of cytoskeletal-bound polysomes in L-929 cells is associated with microfilaments. Interestingly, equivalent amounts of monosomes were recovered in the cytoskeletal fraction isolated from untreated and cytochalasin B treated cells (Figure 2).

That the three polysome populations are not merely artifacts caused by the isolation procedure is supported by the observation that in Krebs II ascites cells the three populations respond differently in response to stimulation of cells with insulin (Vedeler et al., 1990) and culture conditions (Kirkeeide et al., 1992; 1993). Insulin

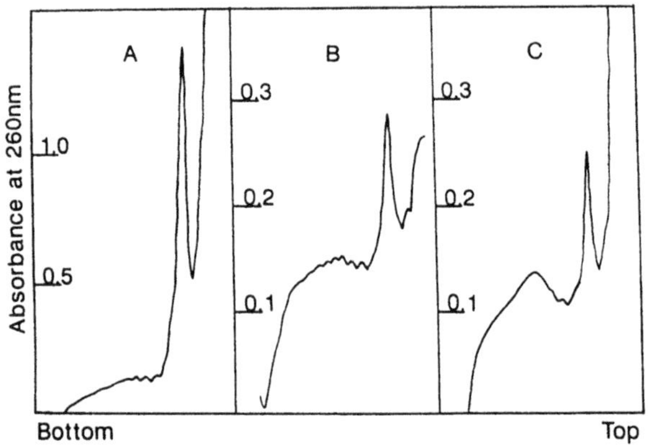

Figure 1. Profiles of polysomes isolated from L-929 cells. Free (**A**), cytoskeletal-bound (**B**), and membrane-bound (**C**) polysomes were isolated from L-929 cells (Pryme and Hesketh, 1987) using the sequential salt/double detergent extraction technique described by Vedeler et al. (1991a). Free polysomes were released using 1% Nonidet P-40 in low salt buffer (25mM KCl), cytoskeletal-bound polysomes were extracted in buffer containing 130mM KCl and membrane-bound polysomes were finally solubilized using a combination of 0.5% Triton X-100 and 0.5% sodium deoxycholate.

has both long term and short term effects on protein synthesis and the cytoskeleton in a variety of cell lines (Almås et al., 1992a,b). A short term effect of insulin was observed on the microfilament system both in Krebs II ascites cells and 3T3 fibroblasts (Vedeler et al., 1991c) in that a rapid polymerization of actin was

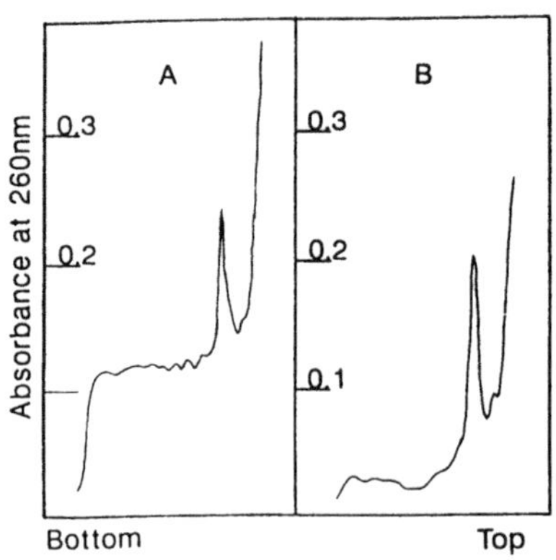

Figure 2. Effect of cytochalasin B treatment on cytoskelatal-bound polysomes. Cytoskeletal-bound polysomes isolated from L-929 cells (see legend to Figure 1 for details) in the absence (**A**) or presence (**B**) of 40μg/ml cytochalasin B added 30 min before harvesting the cells.

observed at the plasma membrane within 5 min of incubation. An increase in actin synthesis occurred after about 20 min. During the course of a 1 h incubation with insulin the relative amount of polysomes in the cytoskeletal-bound fraction increased by 23% while a 27% reduction occurred in membrane-bound polysomes (Vedeler et al., 1990); no significant changes were observed in free polysomes. However, a 2 h incubation with insulin resulted in a 65% increase in polysomes in the cytosol, a 15% increase in cytoskeletal-bound polysomes while a 40% reduction occurred in the amount of membrane-bound polysomes in comparison to control cells.

Morphological changes were seen in Krebs II ascites cells following incubation with insulin for 24 h; up to 80% of the initially rounded cells became attached to the substratum and adopted a flattened morphology associated with normal adhesion and movement (Pryme and Hesketh, 1990). The observed morphological changes were associated with a reorganization of cellular actin. Recent results have shown that the distribution of ribosomes between free, cytoskeletal-bound and membrane-bound polysomes in Krebs II ascites cells is affected by insulin during long term incubation; two populations of cells were isolated from the same stationary culture exposed to the same culture medium (attached and non-attached cells), and these were shown to contain different amounts of cytoskeletal-bound and membrane-bound polysomes (Kirkeeide et al., 1992; 1993). Although the percentage of total polysomes in free polysomes was unaffected, a shift of ribosomes from the cytoskeletal-bound to the membrane-bound polysome population had occurred in attached cells. In the latter cells more than 70% of ribosomes were in polysomes in the membrane-bound fraction (compared to 50% in non-attached cells) indicating a high rate of protein synthesis in attached cells. This may reflect the necessity for adhering and spreading cells to synthesize larger amounts of membrane proteins (Kirkeeide et al., 1992). These results show that ribosomes can be diverted from one population of polysomes into another and this is presumably a reflection of the ability of cells to modify protein synthesis according to specific requirements.

The possibility that ribosomal associated proteins may play a regulatory role in protein synthesis was suggested by Olsen et al. (1988) who showed that the profile of proteins in a high salt wash of total polysomes was different at various stages of the cell cycle. Recent results have shown that there are differences with respect to protein species present in high-salt wash extracts of free, cytoskeletal-bound and membrane-bound polysomes isolated from MPC-11 cells (Moss et al., 1991); proteins of 34, 47, 56, and 120 kDa were unique in the free polysome extract while a 26 kDa protein appeared to be restricted to the membrane-bound fraction. Proteins of 37 and 42 kDa were present in high salt-washes of both free and cytoskeletal-bound polysomes but not in extracts of membrane-bound polysomes while protein bands at 99 and 110 kDa were present in all three fractions. Proteins of 31, 32, 50, and 64 kDa were found in increased amounts in the high salt-wash of free polysomes isolated from MPC-11 cells following short-term insulin stimulation.

Table 1. Compartmentation of Specific mRNAs in Polysome Fractions

		Polysome Fraction	
mRNA probe	*FP*	*CBP*	*MBP*
β-Actin	+	+	
c-myc		+	
β-2-microglobulin			+

Note: Total RNA was isolated from free (FP), cytoskeletal-bound (CBP), and membrane-bound polysome (MBP) fractions. Northern hybridization analysis was performed using probes for specific mRNA species. Data from Hesketh et al. (1991a) and Vedeler et al. (1991b). See also Johannesen et al. (1995).

The 64-kDa protein was not observed in the cytoskeletal-bound polysome extract in unstimulated cells but appeared after incubation of cells with insulin. In the salt wash prepared from membrane-bound polysomes increased amounts of proteins of 26 and 50 kDa were observed after insulin stimulation (Moss et al., 1993). The actual identity and roles of these proteins is not yet known. It is evident, however, that the three polysome populations possess specific complements of associated proteins which can be modified following physiological stimuli.

Using Northern hybridization actin, c-myc and β-2-microglobulin mRNAs were shown to be unequally distributed in the polysome populations (Hesketh et al., 1991b; Vedeler et al., 1991b). β-2-microglobulin mRNA was essentially limited to the membrane-bound fraction, as expected, since the coded protein is exported to the extracellular surface of the plasma membrane, c-myc mRNA was enriched in the cytoskeletal-bound fraction, while the mRNA for actin was found in both free and cytoskeletal-bound polysomes (Table 1). Since the c-myc/actin mRNA ratio in the fractions containing free and cytoskeletal-bound polysomes is quite different (Hesketh et al., 1991b; Vedeler et al., 1991b), it would seem unlikely that cytoskeletal-bound polysomes merely represent a population of free polysomes trapped within the framework of the cytoskeleton. At present it is unclear why actin mRNA is found in both free and cytoskeletal-bound polysomes but several possible explanations can be suggested. First, recovery of actin mRNA in the free polysome fraction may be due to a partial breakdown of the cytoskeleton occurring during early stages of the isolation procedure, releasing a portion of polysomes initially cytoskeletal-bound *in situ* to the free polysome fraction. Such polysomes may represent a population of polysomes loosely associated with microfilaments. Hoock et al. (1991) have demonstrated that β-actin mRNA is localized at the plasma membrane-cytoskeletal interface and this mRNA may therefore be susceptible to release upon solubilization of the plasma membrane with non-ionic detergent. Second, different isoforms of actin may be synthesized in the two fractions.

The existence of cytoskeletal-bound polysomes as a population separate from free and membrane-bound polysomes is now well documented. Whether or not free polysomes *in situ* are in fact in some loose form of association with the cytoskeleton

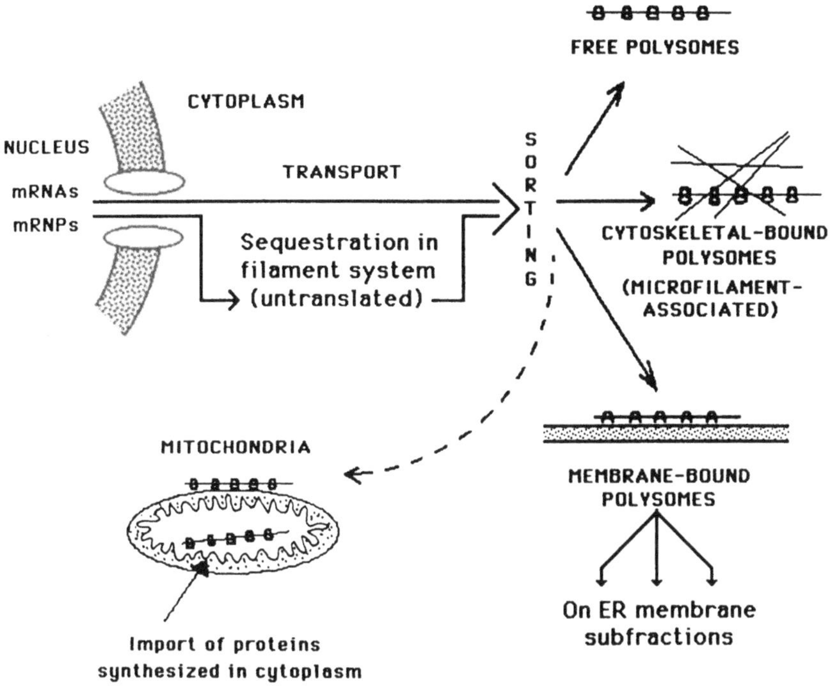

Figure 3. Scheme summarizing the organization of protein synthesis in eukaryotic cells.

is not yet known. Details of the interaction between cytoskeletal-bound polysomes and components of the cytoskeleton are at the present time not well defined. For example, although the microfilament system is without doubt involved, there is at the moment no conclusive evidence indicating that polysomes are also associated with other components of the cytoskeleton such as intermediate filaments. Although recent data has shown that specific mRNA species appear to be compartmentalized into cytoskeletal-bound polysomes, a considerable amount of work still remains before the actual physiological role of this class of polysomes in protein synthesis is finally understood. A simple scheme depicting current ideas on the organization of protein synthesis is shown in Figure 3.

VIII. SPATIAL DISTRIBUTION OF mRNAs AND RIBOSOMES

Several studies from quite different and specialized biological systems have indicated that mRNA and ribosomes are orientated with cellular components in a specific manner (Jeffrey, 1984; Rebagliati et al., 1985). Weeks and Melton (1987), using *Xenopus* eggs, have shown by *in situ* hybridization that the Vg1 mRNA is almost entirely restricted to the cortical cytoplasm of the vegetal half of the egg, and Yisraeli et al. (1990) have suggested that this apparent spatial distribution of mRNA is caused by an association with microfilaments since cytochalasin B but not colchicine affected the intracellular distribution of this mRNA. Since microtubule inhibitors prevented the translocation of the mRNA at an earlier developmental stage it was suggested that transport of mRNA was initially effected by microtubules but at a later time mRNA was transferred to and became associated with microfilaments. Although these and similar experiments (Pondel and King, 1988; Berleth et al., 1988; MacDonald and Struhl, 1988; Wharton and Struhl, 1989) indicate that the spatial distribution of mRNA in fertilized eggs is dependent on the integrity of the cytoskeleton it is evident that future work will be necessary in order that the actual roles of the individual filament systems in the segregation of mRNA species may be defined. The intracellular spatial distribution of mRNAs, however, is not restricted to oocytes; in cultured fibroblasts, for example, *in situ* hybridization experiments have shown that the mRNA for actin is localized in the lamellipodia, mRNA for vimentin in the perinuclear region while that coding for tubulin is segregated in the peripheral cytoplasm (Lawrence and Singer, 1986; Singer et al., 1989). Sundell and Singer (1991) have demonstrated in chicken embryo fibroblasts that the movement of actin mRNA to the cell periphery and its maintenance there is dependent on an intact microfilament system but not microtubules or intermediate filaments. These workers suggest that actin-binding proteins may be of importance in the sorting of mRNAs. Hoock et al. (1991) showed that β-actin and its encoded mRNA are localized at the plasma membrane-cytoskeletal interface within regions of moving cytoplasm. In other studies Garner et al. (1988) have convincingly demonstrated that the mRNA for MAP-2 is located in dendrites in the developing brain but not in neuronal cell bodies. Another example of spatial organization is in the neuron where polysomes were identified in dendrites but not in axons (Steward and Levy, 1982). These results may be related to recent observations by Davis et al. (1987) on hippocampal neurons in dissociated cell culture where transport of RNA from the cell body into dendrites but not axons was demonstrated. Labeled RNA was recovered in a detergent insoluble cytomatrix fraction which may indicate that the cytoskeleton is involved in the transport process.

The presence of specific mRNAs in cytoskeletal-bound polysomes requires a mechanism which directs these RNA species to the cytoskeletal compartment rather than to the ER or the cytoplasmic ("free") compartment. Recent observations indicate that exogenous mRNA microinjected into oocytes becomes correctly localized suggesting that the necessary directing information resides in the mRNA itself; furthermore, it is likely that this information is present in a sequence in the 3′ untranslated region (MacDonald and Struhl, 1988; Yisraeli and Melton, 1988). Interestingly, this region may have a function in controlling mRNA stability

(Bonnieu et al., 1988) and it is possible that the subcellular localization of mRNAs and their stability are functionally related (Zambetti et al., 1990). In specialized cells such as neurons, muscle, and the fertilized egg, it is possible that information in the 3′ untranslated region of certain mRNAs allows specific mRNAs to be translated into their corresponding proteins in certain localized areas of the cytoplasm (Hesketh and Pryme, 1991). This mechanism could therefore be involved in the spatial organization of protein synthesis in the cell.

Immunohistochemical evidence for a spatial distribution of ribosomes in skeletal muscle has been recently provided by Horne and Hesketh (1990a); using anti-60S antibodies they have shown that ribosomes are associated with the myofibrils in addition to being present in the subsarcolemmal cytoplasm. These workers suggest that myofibril-associated ribosomes may be involved in the synthesis of myofibrillar proteins. This possibility is supported by observations of Isaacs and Fulton (1987); using myoblasts they showed that myosin nascent polypeptide chains are synthesized on polysomes essentially located in the cell matrix and that the polysomes were released following pre-treatment with cytochalasin. These latter authors made the interesting suggestion that translation of myosin mRNA on polysomes associated with the cytoskeleton may provide a mechanism by which the site of translation is able to direct the myosin heavy chains into appropriate sites in the cytoskeleton or myofibril. The data discussed above provide strong evidence for the spatial organization of ribosomes within cells and imply some form of association with structural filaments. A mechanism must therefore exist which enables ribosomes to become associated with microfilaments, myofibrils, or dendritic structures.

IX. mRNAs CODING FOR MITOCHONDRIAL PROTEINS

The translational system in eukaryotic mitochondria has many unique characteristics which clearly distinguish it from the protein synthetic apparatus present in the cytoplasm of both prokaryotic and eukaryotic cells. These are briefly discussed below.

The mitochondrial ribosome is a 55-60S unit composed of 28S and 39S subunits (O'Brien and Matthews, 1976). These subunits contain only 25–30% rRNA (Matthews et al., 1982). The mitochondrial mRNAs neither contain a cap structure at their 5′ end like the cytoplasmic eukaryotic messengers (Kozak, 1983), nor a significant leader sequence 5′ to the initiation codon (Ojala et al., 1980; Attardi, 1985) which would make a Shine-Dalgarno base pairing possible between the mRNA and the rRNA, similar to that found in prokaryotes (Kozak, 1983). However, the mitochondrial mRNAs often contain AU-rich untranslated 5′-leader and 3′-trailer sequences (Grivell, 1989) and proteins have been found to bind specifically to the 5′-untranslated region in yeast mitochondria (McMullin, et al., 1990; Papadopoulou et al., 1990). This would indicate that these sequences have some sort of a regulatory function during the translational process in mitochondria. Furthermore it appears that mammalian mRNAs can bind to the 28S ribosomal subunit in the absence of both initiation factors and fMet-tRNA and do not require an AUG codon at their 5′ end (Liao and Spremulli, 1989). As a prerequisite for initiation of translation in mammalian mitochondria Denslow et al. (1989) have suggested that mitochondrial initiation factors are required for the proper recognition and melting of the secondary structure in the 5′-terminal end of mitochondrial mRNAs. Many other aspects of the initiation mechanism remain obscure at the present time. Mitochondrial mRNA species also contain a poly(A) tail (Jacob and Schindler, 1972) although this appears to be shorter than that present in cytoplasmic eukaryotic mRNAs. There is evidence for a non-uniform translation of individual polypeptides in mammalian mitochondria (Rabinovich and Kreinin, 1991) which would imply that the individual mRNAs are subject to transcriptional/translational regulation.

The mitochondrial mRNAs mainly code for proteins, such as ATP synthase and those which are part of the mitochondrial electron-transport complexes, and others which reside in the inner membrane of the organelle. In addition, other mitochondrially coded proteins are required for the expression of mitochondrial genes (van der Klei et al., 1994). The electron-transport complexes contain more than 50 different proteins and of these about 13 are coded for by mitochondrial mRNAs (for a review, see Hurt and van Loon, 1986). The rest are products of nuclear genes and these proteins thus have to be imported into the organelle (Schatz and Butow, 1983). A typical example is mammalian cytochrome oxidase which is comprised of 13 subunits; 10 are coded in nuclear genes and 3 in mitochondrial genes (Kadenbach et al., 1983, 1987). The mitochondrially encoded subunits possess catalytic properties whereas the subunits coded in the nucleus have regulatory functions (Kadenbach and Merle, 1981). Both sets of cytochrome oxidase subunits are regulated by the level of activity in the neurons. The degree of coordination between the genomes is unknown at the present time, although it appears that nuclear gene expression proteins might be transported into mitochondria where they stimulate synthesis of cytochrome oxidase subunits encoded by mitochondrial DNA (Hevner and Wong-Riley, 1991). Others have found that the cytochrome c oxidase subunits encoded by the two genomic systems are expressed coordinately at the mRNA level (Hood, 1990). This is not necessarily in contradiction to the former finding. In general it would be important for the cell to co-regulate transcription and translation in the two coding systems. In mammalian mitochondria most proteins regulating transcription are coded in the nucleus. It has been reported that a mitochondrially synthesized protein of 22 kDa isolated from mouse liver might be released from the mitochondria (Abraham et al., 1981). One possible role for this protein could be to regulate and/or synchronize transcription in the nucleus of one or more genes that code for subunits of proteins destined for mitochondria.

Many questions concerning details of protein synthesis in mitochondria are at present unanswered. One question is the following: are all mRNAs or individual mRNAs present in mitochondria regulated by hormones or other modulators of transcription/translation? It is known that certain hormones, such as thyroid hormone, have an effect on the levels of mitochondrial mRNAs. In addition to increasing the rate of transcription, thyroid hormone could also act by increasing the stability of mRNA (Mutvei et al., 1989). Insulin, on the other hand, seems to stimulate the overall rate of synthesis of proteins in mitochondria (McKee and Grier, 1990). It is thus clear that the total amount of mRNA in the organelle can be influenced by hormones but it is not yet known whether a certain population of mRNAs can be specifically modulated. Another important question in this connection is whether or not there is a compartmentalization of protein synthesis in mitochondria. At the moment it appears most likely that the majority of polysomes in the mitochondria are membrane-bound (Marzuki and Hibbs, 1986) and mainly synthesize hydrophobic proteins which are part of the inner mitochondrial membrane. These proteins are presumably co-translationally inserted into the inner membrane (van der Klei et al., 1994).

It has been reported that cytosolic polysomes are located on the outer mitochondrial membrane and that this binding is confined to protein import sites which are contact sites between the outer and inner mitochondrial membranes (van der Klei et al., 1994). The binding is mediated by bound, nascent polypeptides as would have been expected since those polysomes constitute a subpopulation of cytoplasmic membrane-bound polysomes synthesizing some of the mitochondrial subunits coded for in the nucleus. The signal for targeting to mitochondria resides in the N-terminal end of the protein (van der Klei et al., 1994). The localization of membrane-bound polysomes in mitochondria is more uncertain. In yeast they have been found both bound to the cristae and at contact sites (Vignais et al., 1969; Watson, 1972).

Actin has been reported to be present in mitochondria both in the matrix space and on the internal surface of the inner membrane (Etoh et al., 1990). Vimentin and actin were found to be enriched in the matrix and the intermembrane space in mitochondria of rat liver (Bendixen and Vedeler; unpublished results). It is possible that the mitochondrially coded proteins required for expression of mitochondrial genes are synthesized on cytoskeletal-bound polysomes in the matrix.

ACKNOWLEDGMENTS

The authors' work was partly supported by a grant from the Norwegian Research Council for Science and the Humanities.

REFERENCES

Aaronson, R.P. & Blobel, G. (1975). Isolation of nuclear pore complexes in association with a lamina. Proc. Natl. Acad. Sci. USA 72, 1007–1011.

Abraham, A.K., Pryme, I.F., Åbro, A., & Dowben, R.M. (1973). Polysomes in various phases of the cell cycle in synchronized plasmacytoma cells. Exp. Cell Res. 82, 95–102.

Abraham, A.K., Vedeler, A., Reistad, R., & Pihl, A. (1981). Polypeptide synthesis by mitoplasts isolated from mouse liver. Exp. Cell Res. 132, 265–272.

Adams, A., Fey, E.G., Pike, S.F., Taylorson, C.J., White, H.A., & Rabin, B.R. (1983). Preparation and properties of a complex from rat liver of polyribosomes with components of the cytoskeleton. Biochem. J. 216, 215–226.

Aebi, U., Kohn, J., & Buhle, L. (1986). The nuclear lamina is a meshwork of intermediate-type filaments. Nature 323, 560–564.

Agutter, P.S. (1985). The nuclear envelope NTPase and RNA translocation. In: UCLA Symposium (Clawson, G.A., & Smuckler, E.A., eds.), pp. 561–578. Alan R. Liss, New York.

Agutter, P.S. (1988). Nucleo-cytoplasmic transport of mRNA: its relationship to RNA metabolism, subcellular structures and other nucleocytoplasmic exchanges. Prog. Mol. Subcell. Biol. 10, 15–96.

Agutter, P.S. (1991). Between nucleus and cytoplasm, pp. 1–139. Chapman and Hall, London.

Agutter, P.S., Harris, J.R., & Stevenson, I. (1977). RNA stimulation of mammalian liver nuclear envelope NTPase. Biochem. J. 162, 671–679.

Agutter, P.S., McArdle, H.J., & McCaldin, B. (1976). Evidence of involvement of nuclear envelope NTPase in nucleocytoplasmic translocation of ribonucleoproteins. Nature 263, 165–167.

Agutter, P.S., McCaldin, B., & McArdle, H.J. (1979). Importance of nuclear envelope NTPase in nucleocytoplasmic RNA transport. Biochem. J. 182, 811–819.

Akey, C.W. (1989). Interaction and structure of the nuclear core complex revealed by cryo-electron microscopy. J. Cell. Biol. 109, 955–970.

Almås, B., Pryme, I.F., Vedeler, A., & Hesketh, J.E. (1992a). Insulin: Signal transduction and short-term effects on the cytoskeleton and protein synthesis. Int. J. Biochem. 24, 183–191.

Almås, B., Vedeler, A., & Pryme, I.F. (1992b). The effects of insulin, cycloheximide and phalloidin on the content of actin and p35 in extracts prepared from the nuclear fraction of Krebs II ascites cells. Mol. Cell. Biochem. 115, 187–194.

Attardi, B., Cravioto, B., & Attardi, G. (1969). Membrane-bound ribosomes in HeLa cells. I. Their proportion to total cell ribosomes and their association with mRNA. J. Mol. Biol. 44, 47–70.

Attardi, G. (1985). Animal mitochondrial DNA: An extreme example of genetic economy. Int. Rev. Cytol. 93, 93–145.

Bachmann, M., Schröder, H.C., Messer, M., & Müller, W.E.G. (1984a). Base-specific ribonucleases potentially involved in HnRNA and poly(A) metabolism. FEBS Lett. 171, 25–30.

Bachmann, M., Bernd, A., Schröder, H.C., Zahn, R.K., & Müller, W.E.G. (1984b). The role of protein phosphokinase and protein phosphatase during the nuclear envelope NTPase reaction. Biochim. Biophys. Acta 773, 308–316.

Bag, J. & Pramanik, S. (1987). Attachment of mRNA to the cytoskeletal framework and translational control of gene expression in rat L6 muscle cells. Biochem. Cell Biol. 65, 565–575.

Bagchi, T., Larson, D.E., & Sells, B.H. (1987). Cytoskeletal association of muscle-specific mRNAs in differentiating L6 rat myoblasts. Exp. Cell Res. 168, 160–172.

Ballantine, J.E.M. & Woodland, H.R. (1985). Polyadenylation of histone mRNA in Xenopus oocytes and embryos. FEBS Lett. 180, 224–228.

Bannerjee, A.K. (1980). 5′ terminal cap structure in eucaryotic messenger RNAs. Microbiol. Rev. 44, 175–205.

Berleth, T., Burri, M., Thoma, G., Bopp, D., Richstein, S., Frigerio, G., Noll, M., & Nusslein-Volhard, C. (1988). The role of localization of bicoid RNA in organizing the anterior pattern of the Drosophila embryo. EMBO J. 7, 1749–1756.

Bernd, A., Schröder, H.C., Zahn, R.K., & Müller, W.E.G. (1982). Modulation of the nuclear envelope NTPase by poly(A) rich mRNA and microtubule protein. Eur. J. Biochem. 129, 43–49.

Berrios, M., Osherhoff, N., & Fisher, P.A. (1985). In situ localization of DNA topoisomerase II, a major polypeptide component of the Drosophila nuclear matrix. Proc. Natl. Acad. Sci. USA 82, 4142–4146.

Biegel, D. & Pachter, J.S. (1992). mRNA association with the cytoskeletal framework likely represents a physiological binding event. J. Cell Biochem. 48, 98–106.

Bina, M., Feldmann, R.J., & Deeley, R.G. (1980). Could poly(A) align splicing sites? Proc. Natl. Acad. Sci. USA 77, 1278–1282.

Birckbichler, P.J. & Pryme, I.F. (1993). Fractionation of membrane-bound polysomes, free polysomes and nuclei from tissue cultured cells. Eur. J. Biochem. 33, 368–373.

Bird, R.C. & Sells, B.H. (1986). Cytoskeleton involvement in the distribution of mRNP complexes and small cytoplasmic RNAs. Biochim. Biophys. Acta 868, 215–225.

Bladon, T., Brasch, K.R., & Brown, D.L. (1988). Changes in structure and protein composition of bovine lymphocyte nuclear matrix during concanavalin-A-stimulated mitogenesis. Biochem. Cell Biol. 66, 40–53.

Blobel, G. & Dobberstein, B. (1975). Transfer of proteins across membranes. I. Presence of proteolytically processed and unprocessed nascent immunoglobulin light chain on membrane-bound ribosomes of murine myeloma. J. Cell Biol. 67, 835–851.

Blobel, G. & Potter, V.R. (1966). Relation of ribonuclease and ribonuclease inhibitor to the isolation of polysomes from rat liver. Proc. Natl. Acad. Sci. USA 55, 1283–1288.

Blobel, G. and Potter, V.R. (1967). Studies on free and membrane-bound ribosomes in rat liver. I. Distribution as related to total cellular RNA. J. Mol. Biol. 26, 279–301.

Bonneau, A.M., Darveau, A., & Sonenberg, N. (1985). Effect of viral infection on host protein synthesis and mRNA association with the cytoplasmic cytoskeletal structure. J. Cell Biol. 100, 1209–1218.

Bonnieu, A., Piechaczyk, M., Marty, L., Cuny, M., Blanchard, J.M., Fort, P., & Jeanteur, P. (1988). Sequence determinants of c-myc mRNA turn-over: influence of 3′ and 5′ non-coding regions. Oncogene Res. 3, 155–166.

Bouhtiauy, I., Choukri, Y., Turpin, C., & Gauthier, D. (1989). Immunological and chemical characterization of hamster brain polyribosomes-cytomatrix complexes. Neurochem. Res. 14, 635–640.

Carmo-Foneca, M., Cidadao, A.J., & David-Ferreira, J.F. (1987). Filamentous cross-bridges link intermediate filaments to the nuclear pore complex. Eur. J. Cell Biol. 45, 282–290.

Cervera, M., Dreyfuss, G., & Penman, S. (1981). Messenger RNA is translated when associated with the cytoskeletal framework in normal and VSV-infected HeLa cells. Cell 23, 113–120.

Chang, D.D. & Sharp, P.A. (1989). Regulation by HIV Rev depends upon recognition of splice sites. Cell 59, 789–795.

Clawson, G.A., James, J., Woo, C.H., Friend, D.S., Moody, D., & Smuckler, E.A. (1980). Pertinence of nuclear envelope NTPase to transport of RNA from rat liver nuclei. Biochemistry 19, 2756–2762.

Cooper, J.A. (1987). Effects of cytochalasin and phalloidin on actin. J. Cell Biol. 105, 1473–1478.

Coux, O., Nothwang, H.G., Scherrer, K., Bergsma-Schutter, W., Arnberg, A.C., Timmins, P.A., Langowski, J., & Cohen-Addad, C. (1992). Structure and RNA content of the prosomes. FEBS Lett. 300, 49–55.

Dancker, P., Low, I., Hasselback, W., & Wieland, T. (1975). Interaction of actin with phalloidin: Polymerization and stabilization of F-actin. Biochim. Biophys. Acta 400, 407–414.

Dang, C.V., Yang, D.C., & Pollard, T.D. (1983). Association of methionyl-tRNA synthetase with detergent-insoluble components of the rough endoplasmic reticulum. J. Cell Biol. 96, 1138–1147.

Davis, L., Banker, G.A., & Steward, O. (1987). Selective dendritic transport of RNA in hippocampal neurons in culture. Nature 330, 477–479.

Davis, L.I. & Blobel, G. (1987). Nuclear pore complex contains a family of glycoproteins that includes p62: glycosylation through a previously unidentified cellular pathway. Proc. Natl. Acad. Sci. USA 84, 7552–7556.

Denslow, N.D., Michaels, G.S., Montoya, J., Attardi, G., & O'Brien, T.W. (1989). Mechanism of mRNA binding to bovine mitochondrial ribosomes. J. Biol. Chem. 264, 8328–8338.

Dworetzky, S.I. & Feldherr, C.M. (1989). Translocation of RNA-coated gold particles through the nuclear pores of oocytes. J. Cell. Biol. 106, 575–584.

Etoh, S., Matsui, H., Tokuda, M., Itano, T., Nakamura, M., & Hatase, O. (1990). Purification and immunohistochemical study of actin in mitochondrial matrix. Biochem. Int. 20, 599–606.

Fjose, A. & Pryme, I.F. (1984). Altered distribution profiles of endoplasmic reticulum subfractions after incubation of Krebs II ascites cells with different concentrations of cytochalasin B. Cell Biochem. Function 2, 38–42.

Fujitani, Y., Higaki, S., Sawada, H., & Hirosawa, K. (1989). Quick-freeze, deep-etch, visualization of the nuclear pore complex. J. Electron Microsc. 38, 34–40.

Fulton, A.B., Wan, K.W., & Penman, S. (1980). The spatial distribution of polyribosomes in 3T3 cells and the associated assembly of proteins into the skeletal framework. Cell 20, 849–857.

Furuichi, Y., Fiandra, A., & Shatkin, A.J. (1977). 5′-terminal structure and mRNA stability. Nature 266, 235–239.

Garner, C.C., Tucker, R.P., & Matus, A. (1988). Selective localization of messenger RNA for cytoskeletal protein MAP2 in dendrites. Nature 336, 674–677.

Georgatos, S.D. & Blobel, G. (1987a). Lamin B constitutes an intermediate filament attachment site at the nuclear envelope. J. Cell. Biol. 105, 105–115.

Georgatos, S.D. & Blobel, G. (1987b). Two distinct attachment sites for vimentin along the plasma membrane and the nuclear envelope in avian erythrocytes: a basis for a vectorial assembly of intermediate filaments. J. Cell. Biol. 105, 117–125.

Georgatos, S.D., Stournaras, C., & Blobel, G. (1988). Heterotypic and homotypic associations between the nuclear laminins: site-specificity and control by phosphorylation. Proc. Natl. Acad. Sci. USA 85, 4325–4329.

Gerace, L., Ottaviano, Y., & Kondor-Koch, C. (1982). Identification of a major polypeptide of the nuclear pore complex. J. Cell. Biol. 95, 826–837.

Green, M.R. (1986). Pre-mRNA splicing. Ann. Rev. Genet. 20, 671–708.

Grivell, L.A. (1989). Nucleo-mitochondrial interactions in yeast mitochondrial biogenesis. Eur. J. Biochem. 182, 477–493.

Grossi de Sa, M.-F., Martins de Sa, C., Harper, F., Olink-Coux, M., Akhayat, O., Gounon, P., Pal, J.K., Florentin, Y., & Scherrer, K. (1988a). Cytolocalization of prosomes as a function of differentiation. J. Cell Sci. 89, 151–165.

Grossi de Sa, M.-F., Martins de Sa, C., Harper, F., Olink-Coux, M., Huesca, M., & Scherrer, K. (1988b). The association of prosomes with some of the intermediate filament network of the animal cell. J. Cell. Biol. 107, 1517–1530.

Gruss, P., Lai, C.-J., Dhar, R., & Khoury, G. (1979). Splicing as requirement for biogenesis of functional 16S mRNA of SV40. Proc. Natl. Acad. Sci. USA 76, 4317–4321.

Hamm, J. & Mattaj, I.W. (1990). Monomethylated cap structures facilitate RNA transport from the nucleus. Cell 63, 109–118.

Hanover, J.A., Cohen, C.K., Willingham, M.C., & Park, M.K. (1987). O-linked N-acetylglucosamine is attached to the proteins of the nuclear pore. Evidence for cytoplasmic and nucleoplasmic glycoproteins. J. Biol. Chem. 262, 9887–9894.

Hesketh, J.E., Campbell, G.P., & Whitelaw, P.F. (1991a). c-myc Messenger RNA in cytoskeletal-bound polysomes in fibroblasts. Biochem. J. 274, 607–609.

Hesketh, J.E., Horne, Z., & Campbell, G.P. (1991b). Immunohistochemical evidence for an association of ribosomes with microfilaments in 3T3 fibroblasts. Cell Biol. Int. Rep. 15, 141–150.

Hesketh, J.E. & Pryme, I.F. (1988). Evidence that insulin increases the proportion of polysomes that are bound to the cytoskeleton in 3T3 fibroblasts. FEBS Lett. 231, 62–66.

Hesketh, J.E. & Pryme, I.F. (1991). Interaction between mRNA, ribosomes and the cytoskeleton. Biochem. J. 277, 1–10.

Heuijerjans, J.H., Pieper, F.R., Ramaekers, F.C.S., Timmermans, L.J.M., Kuijpers, H., Bloemendal, H., & Van Venrooij, W.J. (1989). Association of mRNA and eIF-2 alpha with the cytoskeleton in cells lacking vimentin. Exp. Cell Res. 181, 317–330.

Heuser, J. & Kirschner, M.W. (1980). Filament organization revealed in platinum replicas on freeze-dried cytoskeletons. J. Cell Biol. 86, 212–234.

Hevner, R.F. & Wong-Riley, M.T. (1991). Neuronal expression of nuclear and mitochondrial genes for cytochrome oxidase (CO) subunits analyzed by in situ hybridization: comparison with CO activity and protein. J. Neurosci. 11, 1942–1958.

Hinshaw, J.E., Carragher, B.O., & Milligan, R.A. (1992). Architecture and design of the nuclear pore complex. Cell 69, 1133–1141.

Hofer, E., Hofer-Warbinek, R., & Darnell, J.E. (1982). Globin RNA transcription: a possible termination site and demonstration of transcriptional control correlated with altered chromatin structure. Cell 29, 887–893.

Hoock, T.C., Newcomb, P.M., & Herman, I.M. (1991). Beta actin and its mRNA are localized at the plasma membrane and the regions of moving cytoplasm during the cellular response to injury. J. Cell Biol. 112, 653–664.

Hood, D.A. (1990). Co-ordinate expression of cytochrome c oxidase subunit III and VIc mRNAs in rat tissues. Biochem. J. 269, 503–506.

Horne, Z. & Hesketh, J. (1990a). Immunological localization of ribosomes in striated rat muscle. Evidence for myofibrillar association and ontological changes in the subsarcolemmal:myofibrillar distribution. Biochem. J. 268, 231–236.

Horne, Z. & Hesketh, J. (1990b). Increased association of ribosomes with myofibrils during the skeletal-muscle hypertrophy induced either by the beta-adrenoceptor agonist clenbuterol or by tenotomy. Biochem. J. 272, 831–833.

Howe, J.G. & Hershey, J.W. (1984). Translational initiation factor and ribosome association with the cytoskeletal framework fraction from HeLa cells. Cell 37, 85–93.

Huez, G., Bruck, C. & Cleuter, Y. (1981). Translational stability of native and deadenylated rabbit globin mRNA injected into HeLa cells. Proc. Natl. Acad. Sci. USA 78, 908–911.

Hurt, E.C. & van Loon, A.P.G.M. (1986). How proteins find mitochondria and intramitochondrial compartments. Trends Biochem. Sci. 11, 204–207.

Hynes, R.O. & Destree, A.S. (1978). 10nm filaments in normal and transformed cells. Cell 13, 151–163.

Isaacs, W. & Fulton, A.B. (1987). Cotranslational assembly of myosin heavy chain in developing cultured skeletal muscle. Proc. Natl. Acad. Sci. USA 84, 6174–6178.

Jackson, D.A. & Cook, P.R. (1985). Transcription occurs at a nucleoskeleton. EMBO J. 4, 919–926.

Jackson, D.A., McCready, S.J., & Cook, P.R. (1981). RNA is synthesized at the nuclear cage. Nature 292, 552–555.

Jacob, S.T. & Schindler, D.G. (1972). Polyriboadenylate polymerase solubilized from rat liver mitochondria. Biochem. Biophys. Res. Commun. 48, 126–134.

Jeffery, W.R. (1984). Spatial distribution of messenger RNA in the cytoskeletal framework of ascidian eggs. Dev. Biol. 103, 482–492.

Johannessen, A., Pryme, I.F., & Vedeler, A. (1995). Changes in distribution of actin mRNA in different polysome fractions following stimulation of MPC-11 cells. Mol. Cell. Biochem. 142, 107–115.

Jones, N.L. & Kilpatrick, B.A. (1988). The effects of human cytomegalovirus infection on cytoskeletal-associated polysomes. Eur. J. Cell Biol. 46, 31–38.

Kadenbach, B., Jarausch, J., Hartmann, R., & Merle, P. (1983). Separation of mammalian cytochrome c oxidase into 13 polypeptides by a sodium dodecyl sulfate-gel electrophoretic procedure. Anal. Biochem. 129, 517–521.

Kadenbach, B., Kuhn-Nentwig, L., & Buge, U. (1987). Evolution of a regulatory enzyme: Cytochrome-c oxidase (complex IV). Curr. Top. Bioenerg. 15, 113–161.

Kadenbach, B. & Merle, P. (1981). On the function of multiple subunits of cytochrome c oxidase from higher eukaryotes. FEBS Lett. 135, 1–11.

Kasai, M. (1969). Thermodynamical aspect of G-F transformations of actin. Biochim. Biophys. Acta. 180, 399–409.

Katze, M.G., Lara, J., & Wambach, M. (1989). Nontranslated cellular mRNAs are associated with the cytoskeletal framework in influenza virus or adenovirus infected cells. Virology 169, 312–322.

Kirkeeide, E-K., Pryme, I.F., & Vedeler, A. (1991). Changes in amount of polysomes in free, cytoskeleton-bound and membrane-bound populations in Krebs II ascites cells subjected to different growth conditions. Biochem. Soc. Trans. 19, 1136–1137.

Kirkeeide, E-K., Pryme, I.F., & Vedeler, A. (1992). Morphological changes in Krebs II ascites tumour cells induced by insulin are associated with differences in protein composition and altered amounts of free, cytoskeletal-bound and membrane-bound polysomes. Mol. Cell. Biochem. 118, 131–140.

Kirkeeide, E-K., Pryme, I.F., & Vedeler, A. (1993). Microfilaments and protein synthesis; Effects of insulin. Int. J. Biochem. 25, 853–864.

Kletzein, R.F. (1980). Nucleocytoplasmic transport of RNA: effect of 3′-deoxy ATP on RNA release from isolated nuclei. Biochem. J. 192, 753–760.

Klymkowsky, M.W., Maynell, L.A., & Nislow, C. (1991). Cytokeratin phosphorylation, cytokeratin filament severing and the solubilization of the maternal mRNA Vg1. J. Cell Biol. 114, 787–797.

Konarska, M.M., Grabowski, P.J., Padgett, R.A., & Sharp, P.A. (1985). Characterization of the branch site in the lariat RNAs produced by splicing of mRNA precursors. Nature 313, 552–557.

Kondor-Koch, C., Riedel, N., Valentin, R., Fasold, H., & Fischer, H. (1982). Characterization of an ATPase on the inside of rat liver nuclear envelopes by affinity labeling. Eur. J. Biochem. 127, 285–290.

Kozak, M. (1983). Comparison of initiation of protein synthesis in procaryotes, eucaryotes, and organelles. Microbiol. Rev. 47, 1–45.

Lawrence, J.B. & Singer, R.H. (1986). Intracellular localization of messenger RNAs for cytoskeletal proteins. Cell 45, 407–415.

Lawrence, J.B., Singer, R.H., & Marselle, L.M. (1989). Highly localized tracks of specific transcripts within interphase nuclei visualized by in situ hybridization. Cell 57, 493–502.

Lemieux, R. & Beaud, G. (1982). Expression of vaccinia virus early mRNA in Ehrlich ascites tumor cells. 2. Part of the polysomes at an early stage of virus infection are not bound to the cytoskeleton. Eur. J. Biochem. 129, 273–279.

Lenk, R. & Penman, S. (1979). The cytoskeletal framework and poliovirus metabolism. Cell 16, 289–301.

Lenk, R., Ransom, L., Kaufmann, Y., & Penman, S. (1977). A cytoskeletal structure with associated polyribosomes obtained from HeLa cells. Cell 10, 67–78.

Lequang, H. & Gauthier, D. (1989). Isolation of hamster brain polyribosomes-cytoskeleton complexes. Neurochem. Res. 14, 239–243.

Liao, H.-X. & Spremulli, L.L. (1989). Interaction of bovine mitochondrial ribosomes with messenger RNA. J. Biol. Chem. 264, 7518–7522.

Long, B.H., Huang, C.-Y., & Pogo, A.O. (1979). Isolation and characterization of the nuclear matrix in Friend erythroleukemia cells: chromatin and HnRNA interactions with the nuclear matrix. Cell 18, 1079–1090.

MacDonald, P.M. & Struhl, G. (1988). cis-acting sequences responsible for anterior localization of bicoid mRNA in Drosophila embryos. Nature 336, 595–598.

Maltseva, E.L., Palmina, N.P., & Pryme, I.F. (1991a). The effect of a phorbol ester on the lipid microviscosity of two fractions of endoplasmic reticulum membranes isolated from Krebs II ascites cells. J. Cell. Biochem. 46, 260–265.

Maltseva, E.L., Palmina, N.P., & Pryme, I.F. (1991b). The mapping of three subfractions of endoplasmic reticulum membranes isolated from L-929 cells by the use of spin probes. Mol. Cell. Biochem. 106, 49–54.

Marzuki, S. & Hibbs, A.R. (1986). Are all mitochondrial translation products synthesized on membrane-bound ribosomes? Biochim. Biophys. Acta 866, 120–124.

Matthews, D.E., Hessler, R.A., Denslow, N.D., Edwards, J.S., & O'Brien, T.W. (1982). Protein composition of the bovine mitochondrial ribosome. J. Biol. Chem. 257, 8788–8794.

McDevitt, M.A., Imperiale, M.J., Ali, H., & Nevins, J.R. (1984). Requirement of a downstream sequence for generation of poly(A) addition site. Cell 37, 933–999.

McDonald, J.R. & Agutter, P.S. (1980). The relationship between polyribonucleotide binding and the phosphorylation and dephosphorylation of nuclear envelope protein. FEBS Lett. 116, 145–148.

McKee, E.E. & Grier, B.L. (1990). Insulin stimulates mitochondrial protein synthesis and respiration in isolated perfused rat heart. Am. J. Physiol. 259, E413–E421.

McMullin, T.W., Haffter, P., & Fox, T.D. (1990). A novel small-subunit ribosomal protein of yeast mitochondria that interacts functionally with an mRNA-specific translational activator. Mol. Cell Biol. 10, 4590–4595.

Meadus, W.J., Pramanik, S., & Bag, J. (1990). Cytoskeleton-bound mRNA for a 40-kDa polypeptide in rat L6 cells is not always translated. Exp. Cell Res. 187, 25–32.

Miller, M., Park, M.K., & Hanover, J.A. (1991). Nuclear pore complex: structure, function, and regulation. Phys. Rev. 71, 909–949.

Mirande, M., Le Corre, D., Louvard, D., Reggio, H., Pailliez, J.-P., & Waller, J.-P. (1985). Association of an aminoacyl-tRNA synthetase complex and of phenylalanyl-tRNA synthetase with the cytoskeletal framework fraction from mammalian cells. Exp. Cell Res. 156, 91–102.

Moon, R.T., Nicosia, R.F., Olsen, C., Hille, M.B., & Jeffery, W.R. (1983). The cytoskeletal framework of sea urchin eggs and embryos: developmental changes in the association of messenger RNA. Dev. Biol. 95, 447–458.

Moss, R., Pryme, I.F., & Vedeler, A. (1991). Differences in patterns of protein isolated from polysomes in free, cytoskeleton-bound and membrane-bound fractions in MPC-11 cells incubated with insulin. Biochem. Soc. Trans. 19, 1138–1139.

Moss, R., Vedeler, A, & Pryme, I.F. (1993). The effect of insulin on proteins associated with free, cytoskeletal-bound and membrane-bound polysome populations. Cell Biol. Int. 12, 1065–1073.

Moss, R., Pryme, I.F., & Vedeler, A. (1994). Free, cytoskeletal-bound and membrane-bound polysome isolated from MPC-11 and Krebs II ascites cells differ in their complement of poly(A) binding proteins. Mol. Cell. Biochem. 131, 131–139.

Mutvei, A., Kuzela, S., & Nelson, B.D. (1989). Control of mitochondrial transcription by thyroid hormone. Eur. J. Biochem. 180, 235–240.

Nielsen, P., Goelz, S., & Trachsel, H. (1983). The role of the cytoskeleton in eukaryotic protein synthesis. (A minireview). Cell Biol. Int. Rep. 7, 245–254.

O'Brien, T.W. & Matthews, D.E. (1976). In: Handbook of Genetics, Vol. 5 (King, R.C., ed.), pp. 535–580, Plenum Publishing Corp., New York.

Ojala, D.K., Montoya, J., & Attardi, G. (1980). The putative mRNA for subunit II of human cytochrome c oxidase starts directly at the translation initiator codon. Nature 287, 79–82.

Olsen, L.C., Pryme, I.F., & Bassøe, C.-F. (1988). Protein factors in extracts of ribosomes isolated from L-929 cells during various phases of the cell cycle. Biofactors 1, 161–165.

Ornelles, D.A., Fey, E.G., & Penman, S. (1986). Cytochalasin releases mRNA from the cytoskeletal framework and inhibits protein synthesis. Mol. Cell Biol. 6, 1650–1662.

Osborn, M. & Weber, K. (1977). The detergent-resistant cytoskeleton of tissue culture cells includes the nucleus and the microfilament bundles. Exp. Cell Res. 106, 339–349.

Padgett, R.A., Grabowski, P.J., Konarska, M.M., Seiler, S., & Sharp, P.A. (1986). Splicing of pre messenger RNA precursors. Ann. Rev. Biochem. 55, 1119–1150.

Papadopoulou, B., Dekker, P., Blom, J., & Grivell, L.A. (1990). A 40 kd protein binds specifically to the 5′-untranslated regions of yeast mitochondrial mRNAs. EMBO J. 9, 4135–4143.

Pondel, M.D. & King, M.L. (1988). Localized maternal mRNA related to transforming growth factor beta mRNA is concentrated in a cytokeratin-enriched fraction from Xenopus oocytes. Proc. Natl. Acad. Sci. USA 85, 7612–7616.

Pronchow, D., Riedel, N., Agutter, P.S., & Fasold, H. (1990). Poly(A) binding proteins located at the inner surface of resealed nuclear envelopes. J. Biol. Chem. 265, 6536–6539.

Proudfoot, N.J. & Brownlee, G.G. (1976). 3′-noncoding sequences in eucaryotic mRNA. Nature 211, 214.

Pryme, I.F. (1974). The synthesis of light chain immunoglobulin in MPC-11 on only one of two fractions of membrane-bound polysomes prepared by a combination of nitrogen cavitation/detergent treatment. FEBS Lett. 48, 200–203.

Pryme, I.F. (1986). Compartmentation of the rough endoplasmic reticulum. Mol. Cell Biochem. 71, 3–18.

Pryme, I.F. (1988). Differences in the association of ribosomes with heavy rough and light rough endoplasmic reticulum membranes of MPC-11 cells. Biochim. Biophys. Acta 967, 218–223.

Pryme, I.F. (1989a). The nuclear-associated endoplasmic reticulum. Int. J. Biochem. 21, 119–125.

Pryme, I.F. (1989b). Domains of rough endoplasmic reticulum. (a review). Mol. Cell. Biochem. 87, 93–103.

Pryme, I.F., Garatun-Tjeldstø, O., Birckbichler, P.J., Weltman, J.K., & Dowben, R.M. (1973). Synthesis of immunoglobulins by membrane-bound polysomes and free polysomes from plasmacytoma cells. Eur. J. Biochem. 33, 374–378.

Pryme, I.F. & Hesketh, J.E. (1987). Cytochalasin B binding and actin content of endoplasmic reticulum subfractions isolated from L-929 cells. Cell Biol. Int. Repts. 11, 615–623.

Pryme, I.F. & Hesketh, J.E. (1990). Insulin induces cell adhesion and normal flattened morphology in Krebs II ascites tumour cells. Cell Biol. Int. Rep. 14, 447–455.

Pryme, I.F. & Svardal, A. (1978). Differences in the incorporation of ^{3}H-glucosamine into nascent polypeptide chains on free polysomes and two fractions of membrane-bound polysomes in mouse myeloma cells. Mol. Biol. Repts. 4, 223–228.

Purello, F., Vigneri, R., Clawson, G.A., & Goldfine, I.D. (1982). Insulin stimulation of nucleoside triphosphatase activity in isolated nuclear envelopes. Science 216, 1005–1007.

Rabinovich, Y.M. & Kreinin, M.O. (1991). Evidence for non-uniform translation of individual polypeptides in rat liver mitochondria. Biochim. Biophys. Acta 1089, 193–196.

Ramaekers, F., Benedetti, E., Dunia, I., Vorstenbosch, P., & Bloemendal, H. (1983). Polyribosomes associated with microfilaments in cultured lens cells. Biochim. Biophys. Acta 740, 441–448.

Rebagliati, M.R., Weeks, D.L., Harvey, R.P., & Melton, D.A. (1985). Identification and cloning of localized maternal RNAs from Xenopus eggs. Cell 42, 769–777.

Riedel, N., Fasold, H., Bachmann, M., & Pronchow, D. (1987). Permeability measurement in closed vesicles from rat liver nuclear envelopes. Proc. Natl. Acad. Sci. USA 84, 3540–3544.

Savitz, A.J. & Meyer, D.I. (1990). Identification of a ribosome receptor in the rough endoplasmic reticulum. Nature 346, 540–544.

Schatz, G. & Butow, R.A. (1983). How are proteins imported into mitochondria? Cell 32, 316–318.

Scherrer, K. (1990). Prosomes, subcomplexes of untranslated mRNP. Mol. Biol. Rep. 14, 1–9.

Schröder, H.C., Bachmann, M., Diehl-Seifert, B., & Müller, W.E.G. (1987). Transport of mRNA form nucleus to cytoplasm. Prog. Nucl. Acid Res. Mol. Biol. 34, 89–142.

Schröder, H.C., Diehl-Seifert, B., Rottmann, M., Messer, R., Bryson, B.A., Agutter, P.S., & Muller, W.E. (1988). Functional dissection of nuclear envelope mRNA translocation system: effects of phorbol ester and a monoclonal antibody recognizing cytoskeletal structures. Arch. Biochem. Biophys. 261, 394–404.

Schröder, H.C., Rottmann, M., Bachmann, M., & Müller, W.E.G. (1986). Purification and characterization of the major NTPase from the rat liver nuclear envelopes. J. Biol. Chem. 261, 663–668.

Schröder, H.C., Trölltsch, D., Friese, K., Bachmann, M., & Müller, W.E.G. (1987). Mature mRNA is selectively released from the nuclear matrix by an ATP/dATP-dependent mechanism sensitive to topoisomerase inhibitors. J. Biol. Chem. 262, 8917–8925.

Seely, K.A. & Aggeler, J. (1991). Modulation of milk protein synthesis through alteration of the cytoskeleton in mouse mammary epithelial cells cultured on a reconstituted basement membrane. J. Cell Physiol. 146, 117–130.

Shatkin, A.J. (1985). mRNA cap binding proteins: essential factors in initiation translation. Cell 40, 223–224.

Shestakova, E.A., Motuz, L.P., Minin, A.A., Gelfand, V.I., & Gavrilova, L.P. (1991). Some of eukaryotic initiation factor 2 is colocalized with actin microfilament bundles in mouse embryo fibroblasts. Cell Biol. Int. Repts. 15, 75–84.

Shimotahno, K., Kodama, Y., Hashimoto, J., & Miura, K.I. (1977). Importance of 5′-terminal blocking structure to stabilize mRNA in eucaryotic protein synthesis. Proc. Natl. Acad. Sci. USA 74, 2734–2738.

Shore, G.C. & Tata, J.R. (1977). Two fractions of rough endoplasmic reticulum from rat liver. II. Cytoplasmic mRNA's which code for albumin and mitochondrial proteins are distributed differently between the two fractions. J. Cell Biol. 74, 315–321.

Singer, R.H., Langevin, G.L., & Lawrence, G.B. (1989). Ultrastructural visualization of cytoskeletal mRNAs and their associated proteins using double-label in situ hybridization. J. Cell Biol. 108, 2343–2353.

Snow, C.M., Senior, A., & Gerace, L. (1987). Monoclonal antibodies identify a group of nuclear pore complex glycoproteins. J. Cell. Biol. 104, 1143–1156.

Steinert, P., Zackoff, R., Aynardi-Whitman, M., & Goldman, R.D. (1982). Isolation and characterization of intermediate filaments. Methods Cell Biol. 24, 399–419.

Steward, O. & Levy, W.B. (1982). Preferential localization of polyribosomes under the base of dendritic spines in granule cells of the dentate gyrus. J. Neurosci. 2, 284–291.

Sundell, C.L. & Singer, R.H. (1991). Requirement of microfilaments in sorting of actin messenger RNA. Science 253, 1275–1277.

Svardal, A.M. & Pryme, I.F. (1980). Aspects of the role of the endoplasmic reticulum in protein synthesis. Subcellular Biochem. 7, 117–170.

Svardal, A.M., Pryme, I.F., & Dalen, H. (1981). Ultrastructure and polysome content of the microsomal subfractions of mouse plasmacytoma cells. Mol. Cell. Biochem. 34, 165–175.

Toh, B., Lolait, S., Mathy, J., & Baum, R. (1980). Association of mitochondria with intermediate filaments and of polyribosomes with cytoplasmic actin. Cell Tissue Res. 211, 163–169.

Trachsel, H., Sonenberg, N., & Shatkin, A.J. (1980). Purification of a factor that restores translation of vesicular stomatitis virus mRNA in extracts from polio-virus infected HeLa cells. Proc. Natl. Acad. Sci. USA 77, 770–774.

Traub, P. & Nelson, W. (1982). Polyribosomes are not associated with vimentin-type intermediate filaments in Ehrlich tumor cells. Hoppe-Seylers Z. Physiol. Chem. 363, 1177–1185.

van der Klei, I.J., Veenhuis, M., & Neupert, W. (1994). A morphological view of mitochondrial protein targeting. Micros. Res. Tech. 27, 284–293.

Van Eekelen, C.A.G. & Van Venrooij, W.J. (1981). HnRNA and its attachment to the nuclear matrix. J. Cell. Biol. 88, 554–563.

Van Venrooij, W.J., Sillekens, P.T.G., Van Eekelen, C.A.G., & Reinders, R.T. (1981). On the association of mRNA with the cytoskeleton in uninfected and adenovirus-infected human KB cells. Exp. Cell Res. 135, 79–91.

Vedeler, A., Pryme, I.F., & Hesketh, J.E. (1990). Insulin and step-up conditions cause redistribution of polysomes among free, cytoskeletal-bound and membrane-bound fractions in Krebs II ascites cells. Cell Biol. Int. Repts. 14, 211–218.

Vedeler, A., Pryme, I.F., & Hesketh, J.E. (1991a). The characterization of free, cytoskeletal and membrane-bound polysomes in Krebs II ascites and 3T3 cells. Mol. Cell. Biochem. 100, 183–193.

Vedeler, A., Pryme, I.F., & Hesketh, J.E. (1991b). The compartmentalization of polysomes into free, cytoskeletal-bound and membrane-bound populations. Biochem. Soc. Trans. 19, 1108–1111.

Vedeler, A., Pryme, I.F., & Hesketh, J.E. (1991c). Insulin induces changes in the subcellular distribution of actin and 5′-nucleotidase. Mol. Cell. Biochem. 108, 67–74.

Vignais, P.V., Huet, J. & André, J. (1969). Isolation and characterization of ribosomes from yeast mitochondria. FEBS Lett. 3, 177–181.

Vorbrodt, A. & Maul, G.G. (1980). Cytochemical studies on the relation of nucleoside triphosphatase activity to ribonucleoproteins in isolated rat liver nuclei. J. Histochem. Cytochem. 28, 27–35.

Watson, K. (1972). The organization of ribosome granules within mitochondrial structures of aerobic and anaerobic cells of Saccharomyces cerevisiae. J. Cell Biol. 55, 721–726.

Weeks, D.L. & Melton, D.A. (1987). A maternal mRNA localized to the vegetal hemisphere in Xenopus eggs codes for a growth factor related to TGF-beta. Cell 51, 861–867.

Welch, W.J. & Ferramisco, J.R. (1985). Disruption of the three cytoskeletal networks in mammalian cells does not affect transcription, translation or protein translocation changes induced by heat shock. Mol. Cell Biol. 5, 1571–1581.

Wharton, R.P. & Struhl, G. (1989). Structure of the Drosophila BicaudalD protein and its role in localizing the posterior determinant nanos. Cell 59, 881–892.

Wickens, M.P. & Gurdon, J.B. (1983). Post-transcriptional processing of SV40 late transcripts in injected frog oocytes. J. Mol. Biol. 163, 1–26.

Wolosewick, J.J. & Porter, K.R. (1976). Stereo high-voltage electron microscopy of whole cells of the human diploid line WI-38. Amer. J. Anat. 147, 303–324.

Wolosewick, J.J. & Porter, K.R. (1979). Microtrabecular lattice of the cytoplasmic ground substance. J. Cell Biol. 82, 114–139.

Worman, H.J., Yuan, J., & Blobel, G. (1988). A lamin B receptor in the nuclear envelope. Proc. Natl. Acad. Sci. USA 85, 8531–8534.

Yang, F., Demma, M., Warren, V., Dharmawardhane, S., & Condeelis, J. (1990). Identification of an actin-binding protein from Dictyostelium as elongation factor 1a. Nature 347, 494–496.

Yisraeli, J.K. & Melton, D.A. (1988). The maternal mRNA Vg1 is correctly localized following injection into Xenopus oocytes. Nature 336, 592–595.

Yisraeli, J.K., Sokol, S., & Melton, D.A. (1990). A two-step model for the localization of maternal mRNA in Xenopus oocytes: involvement of microtubules and microfilaments in the translocation and anchoring of Vg1 mRNA. Development 108, 289–298.

Zambetti, G., Schmidt, W., Stein, G., & Stein, J. (1985). Subcellular localization of histone messenger RNAs on cytoskeleton-associated free polysomes in HeLa S3 cells. J. Cell Physiol. 125, 345–353.

Zambetti, G., Stein, J., & Stein, G. (1990). Role of messenger RNA subcellular localization in the posttranscriptional regulation of human histone gene expression. J. Cell Physiol. 144, 175–182.

Zeevi, M., Nevins, J.R., & Darnell, J.E. (1982). Newly formed mRNA lacking poly(A) enters the cytoplasm and the polyribosomes but has a shorter half-life in the absence of poly(A). Mol. Cell. Biol. 2, 517–525.

Zumbe, A., Staehli, C., & Trachsel, H. (1982). Association of a 50K molecular weight cap binding protein with the cytoskeleton in baby hamster cells. Proc. Natl. Acad. Sci. USA 79, 2927–2931.

ROLES OF THE CYTOSKELETON AND MEMBRANE-CYTOSKELETON IN GENERATING AND MAINTAINING THE STRUCTURAL AND FUNCTIONAL ORGANIZATION OF POLARIZED EPITHELIAL CELLS

Kenneth A. Beck and W. James Nelson

Advances in Molecular and Cell Biology
Volume 12, pages 75–106.

ISBN: 1-55938-845-5

I. INTRODUCTION

The generation of nonrandom distributions of proteins in cells is a central feature in the development of cellular diversity in all organisms (Nelson, 1991). Analysis of muscle, neurons, and epithelia, for example, reveals that these diverse cells express both cell-type specific and common proteins, which have restricted distributions in the cytoplasm and at the cell surface. Mechanisms that generate and maintain these protein distributions are fundamental to the structural and functional organization of these cells.

The polarized organization of proteins in transporting epithelia has been extensively characterized, and there is detailed insight into the mechanisms involved (Simons and Fuller, 1985; Rodriguez-Boulan and Nelson, 1989). There is strong evidence for sorting of proteins in the Golgi complex of these cells, and subsequent delivery of sorted proteins in different populations of transport vesicles to different membrane domains (Simons and Wandinger-Ness, 1990; Hunziker and Mellman, 1992; Mostov et al., 1992; Wollner and Nelson, 1992). However, the process of protein sorting, *per se*, is unlikely to be solely responsible for the generation of the structural and functional organization of these cells (Wollner and Nelson, 1992). Other studies have shown that organelles of the secretory pathway (e.g., Golgi complex) have distinctive distributions (Rodriguez-Boulan and Nelson, 1989), that vesicles are targeted from the Golgi complex to specific membrane domains (Simons and Fuller, 1985; Rodriguez-Boulan and Nelson, 1989), and that some membrane proteins are specifically retained in the membrane (Nelson et al., 1990a).

This review will explore these latter aspects of the generation and maintenance of epithelial cell polarity. It is proposed that the cytoskeleton plays a role in the spatial organization of organelles in these cells, in the trafficking of vesicles between the Golgi complex and the cell surface, and the retention of proteins upon vesicle docking at the plasma membrane. However, to place these concepts in context, we will briefly review our current understanding of the mechanisms of protein sorting in these cells; subsequently, we will consider roles of the filamentous cytoskeleton and membrane-cytoskeleton in regulating protein distributions in polarized cells.

II. MECHANISMS FOR PROTEIN SORTING IN POLARIZED EPITHELIAL CELLS

Polarized epithelial cells form a barrier between two biological compartments, and regulate the ionic composition of these compartments by vectorial transport of ions, fluids, and solutes across the epithelium (Simons and Fuller, 1985; Rodriguez-Boulan and Nelson, 1989). The direction of vectorial transport is determined by the distribution of ion channels and transporters to domains of the cell surface that face these different biological compartments (Almers and Stirling, 1984). For example, the primary function of polarized epithelial cells that line the kidney nephron is the reabsorption of ions and solutes from the ultrafiltrate to the blood supply. Transport of Na^+ from the ultrafiltrate is a major driving force in this reabsorptive process in the proximal tubule. Na^+ enters the cell down its electrochemical gradient through Na^+ channels and co-transporters localized to the apical domain of the plasma membrane that faces the tubule lumen. Na^+ is rapidly pumped out of the cell by Na/K-ATPase localized in the basal-lateral membrane domain. The Na^+ gradient that is generated across the epithelium is used for the co-transport of other ions and solutes from the ultrafiltrate to the blood supply (Molitoris and Nelson, 1990).

Both physiological analysis of ion transport function and cell biological investigation of the subcellular location of specific ion transporter proteins has revealed restricted distributions of specific proteins to the apical and basal-lateral membranes (Simons and Fuller, 1985; Rodriguez-Boulan and Nelson, 1989). In almost all cases, each protein is restricted to one or other of the membrane domains, and reflects whether the cell is a secretory or reabsorptive epithelium. For example, as noted above, Na/K-ATPase is localized to the basal-lateral membrane in the kidney nephron, a reabsorptive epithelium (Simons and Fuller, 1985; Rodriguez-Boulan and Nelson, 1989) in the choroid plexus, a secretory epithelium, Na/K-ATPase is restricted to the apical membrane and regulates fluid secretion into the ventricles of the brain (Wright, 1972).

How are different protein distributions generated in these cells? Conceptually, there are two simple pathways for generating a restricted distribution of proteins in a membrane, termed *selective targeting* and *selective retention* (Nelson, 1992). Selective targeting involves protein sorting prior to delivery to the cell surface, and then direct targeting of proteins in vesicles to the appropriate membrane domain.

Selective retention involves random delivery of proteins to the cell surface, but retention there of only a subset of the proteins. In both cases, long term maintenance of proteins at the cell surface requires exclusion of proteins from nonspecific internalization.

Initial studies on protein trafficking in polarized epithelial cells used enveloped RNA viruses (Rodriguez-Boulan and Sabatini, 1978). These studies showed that the envelope glycoprotein of influenza virus, hemagglutinin, was delivered to the apical membrane of Madin-Darby canine kidney epithelial (MDCK) cells. In contrast, the envelope glycoprotein of VSV, G protein, was delivered to the basal-lateral membrane (Figure 1). Detailed microscopical analysis of transport

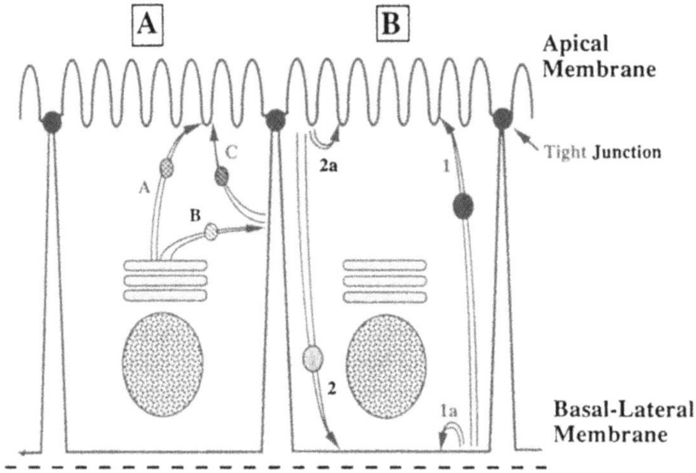

Figure 1. Vesicle trafficking in polarized epithelial cells. The targeting of vesicles between different membranes plays a critical role in the development and maintenance of cell surface polarity of epithelial cells. **A.** Newly-synthesized proteins are transported through the endoplasmic reticulum and Golgi complex to the trans Golgi complex (TGN). In the TGN, proteins are sorted into different vesicle populations based upon intrinsic sorting signals in the proteins (see text for details). In MDCK cells, and to a lesser extent in Caco-2 cells, apical (A) and basal-lateral membrane proteins (B) are sorted into distinct vesicle populations which are delivered vectorially to the appropriate membrane domain. In hepatocytes, and to a lesser extent in Caco-2 cells, proteins are delivered from the TGN to the basal-lateral membrane (B only), and apical proteins are retrieved by endocytosis and subsequently delivered to the apical membrane (C). **B.** A subset of proteins (e.g., poly IgA receptor; see text for details) that are delivered to either the apical or basal-lateral membrane are internalized, upon binding ligand, and either recycled to the same membrane (1a, 2a), or delivered vectorially to the opposite membrane domain (1, 2) by a process termed trancytosis.

pathways showed that proteins destined for different membrane domains in MDCK cells were sorted from each other in the trans-Golgi network (TGN) into different vesicle populations, which were then selectively targeted directly to either the apical or basal-lateral plasma membrane domain (Rodriguez-Boulan and Sabatini, 1978; Matlin and Simons, 1984; Rindler et al., 1984; Rindler et al., 1985; see Figure 1). Studies of other cells indicate a more complex pathway. In hepatocytes (Bartles et al., 1987; Bartles and Hubbard, 1988), and, to a lesser extent, in intestinal cells (Le Bivic, et al., 1990; Matter, et al., 1990; Gilbert et al., 1991), proteins are delivered directly from the TGN to the basal-lateral membrane (Figure 1). Upon arrival at the membrane, basal-lateral membrane proteins are selectively retained, while apical membrane proteins were internalized into vesicles which are then transported directly to the apical membrane (Figure 1). These results have raised the question of the nature of the sorting signals in proteins and the sorting machinery. Studies on endogenous proteins has revealed two putative signal-mediated sorting signals which may be recognized by sorting machinery in the TGN (Simons and Wandinger-Ness, 1990; Hunziker and Mellman, 1992; Mostov et al., 1992; Wollner and Nelson, 1992).

An apical sorting signal may involve affinity of proteins for glycosphingolipids, which have also been shown to be sorted predominantly to the apical membrane of renal and intestinal polarized epithelial cells (Simons and Wandinger-Ness, 1990). Detailed analysis of the sorting pathway of proteins anchored to the membrane by a glycosphingolipid moiety (GPI-proteins) demonstrated that they are delivered exclusively to the apical membrane of MDCK cells (Brown et al., 1989; Lisanti et al., 1989; Brown and Rose, 1992). Significantly, when a GPI-anchor was transferred to the ectodomain of a protein normally delivered to the basal-lateral membrane, the chimera was delivered to the apical membrane (Brown et al., 1989; Lisanti et al., 1989). It has been proposed that sorting of GPI-anchored proteins from other proteins in the TGN is mediated by clustering of glycosphingolipids which causes vesicle budding from the TGN membrane (Simons and Wandinger-Ness, 1990). Recently, Brown and Rose (1992) isolated a nonionic detergent insoluble fraction that contains a high concentration of glycosphingolipids and the apical protein hemagglutinin. Since hemagglutinin is a transmembrane protein, it has been proposed that it either interacts directly with glycosphingolipids, or indirectly through binding to an apical sorting receptor which binds to glycosphingolipids (Simons and Wandinger-Ness, 1990). It remains to be determined if this is the only mechanism for sorting apical membrane proteins, or whether glycosphingolipid-independent mechanisms exist.

The second putative sorting signal has been located on proteins destined for the basal-lateral membrane domain (Hopkins, 1991; Mostov et al., 1992). It is characterized by a short amino acid sequence (<14 amino acids) that is located close to the transmembrane domain. In some proteins, the signal includes a tyrosine residue that confers an increased rate of endocytosis from the cell surface (Brewer and Roth, 1991; Le Bivic et al., 1991); in other proteins the signal for basal-lateral sorting and endocytosis are distinct (Casanova et al., 1991; Hunziker et al., 1991). The structural characteristics of this basal-lateral sorting signal suggest that it may be recognized by cytosolic proteins related to the family of adaptor proteins that regulate signal mediated endocytosis of receptor proteins from the cell surface (Hunziker and Mellman, 1992; Mostov et al., 1992). The endocytosis sorting signal also comprises a short spacer sequence (>7 amino acids) that separates the transmembrane domain from a structurally conserved tetrapeptide (Pearce and Robinson, 1990). The tetrapeptide forms a tight β-turn (Bansal and Gerisch, 1991; Eberle et al., 1991), which is recognized by cytosolic adaptor proteins that are components of the clathrin-coated pit assembly (Pearce and Robinson, 1990). It is thought that binding of proteins through the basal-lateral sorting signal to a specific class of adaptor proteins that are localized to the Golgi complex could induce clustering of those proteins in the TGN membrane and initiate vesicle formation and budding (Hunziker and Mellman, 1992; Mostov et al., 1992).

Although sorting of apical and basal-lateral proteins occurs in the TGN of MDCK cells, protein sorting in hepatocytes and intestinal epithelial cells occurs following delivery of proteins to the basal-lateral membrane (sinusoidal membrane in hepatocytes) (Bartles et al., 1987; Bartles and Hubbard, 1988; Le Bivic et al., 1990; Matter et al., 1990; Gilbert et al., 1991; see Figure 1). At present it is not known whether mechanisms involved in protein sorting in the TGN of MDCK cells are the same as those on the basal-lateral membrane of other cells.

Vesicles are delivered with a high degree of fidelity either to the apical or basal-lateral membrane of MDCK cells, or to first the basal-lateral and then apical membrane in hepatocytes and intestinal epithelial cells (Simons and Fuller, 1985; Rodriguez-Boulan and Nelson, 1989; see Figure 1). This indicates that there are specific mechanisms for regulating vesicle traffic in the cytoplasm and/or docking of vesicles with the appropriate membrane domain. Little is known about how vesicles are targeted to, and recognized by, different membrane domains. Reconstitution studies with purified post-TGN transport vesicles and MDCK cell basal-lateral membranes showed that vesicle binding was dependent on temperature, energy, and components in the cytosol. In addition, vesicle binding was inhibited by GTPγs, indicating that both GTP-binding proteins and GTP hydrolysis are required (Gravotta et al., 1990; Mayer et al., 1992).

After delivery to the cell surface, proteins have several fates. For example, proteins may diffuse away from the site of delivery; be selectively retained at the site of delivery; or may be internalized (selectively or randomly) and either cycled back to the same membrane (retrieval), delivered to another membrane domain (transcytosis), or degraded (Wollner and Nelson, 1992). These different fates have distinct effects on the generation and maintenance of cell surface polarity of proteins.

III. DISTRIBUTION OF THE CYTOSKELETON IN POLARIZED EPITHELIAL CELLS

Polarized epithelial cells contain discrete spatial distributions of the three major components of the cytoskeleton: intermediate filaments, microtubules, and actin filaments (Nelson, 1991; see Figure 2). Keratin intermediate filaments interlink desmosomes at sites of cell–cell contact to form a structural continuum throughout the epithelium (Fey et al., 1984). Actin filaments have more complex distributions (Figure 2). In cells that have a well-formed brush border (e.g., enterocytes, proximal kidney tubule cells), actin is localized in bundles in the core of each microvillus; actin bundles penetrate the cytoplasm below the microvilli and connect into another actin formation, the terminal web, which circumscribes the cell at the level of the apical junction complex (Mooseker, 1985). The apical junctional complex comprises the tight junction and *zonula adherens*, and is indirectly linked to the actin network (Geiger, 1989; Geiger et al., 1990; Gumbiner, 1990; Takeichi, 1991;

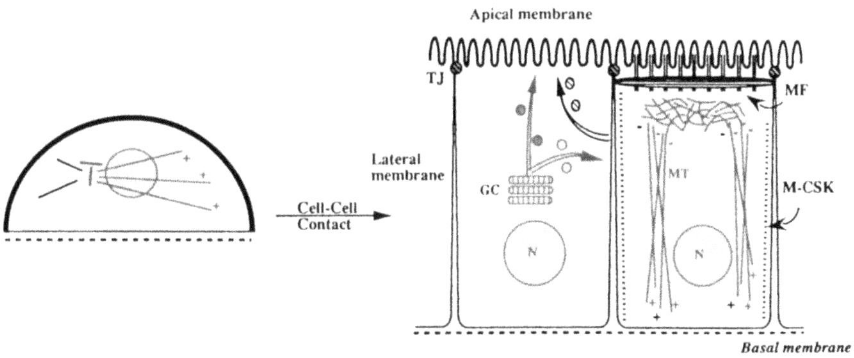

Figure 2. Cytoskeleton organization in MDCK cells. In the absence of cell–cell contact (left), MDCK cells exhibit a diffuse array of cytoskeletal components similar to that in fibroblasts. The microtubule network is nucleated from the centrosome located next to the nucleus (N). Following cell–cell contact, there is a dramatic reorganization of the cytoskeleton (see text for details). The microtubules (MT) become localized in a dense mat of short, randomly oriented filaments in the apical cytoplasm, and in long bundles of filaments of identical polarity (plus ends at the base of the cell) that traverse the cytoplasm from the apical cytoplasm to the base of the cell. The microtubule array is closely associated with the distribution of the Golgi complex (GC), and may play a role in the trafficking of transport vesicles from the Golgi complex to different membrane domains. The actin microfilament (MF) array becomes localized to the apical membrane in association with the apical microvilli, and in a cortical ring of filaments that is linked to the apical intercellular junctional complex. Actin is also associated with the membrane-cytoskeleton (M-CSK) which accumulates on the lateral membranes of cells in association with cell–cell contacts (see text for details).

Kemler, 1992). In the *zonula adherens*, actin filaments are linked indirectly to cell adhesion molecules through cytoplasmic proteins termed catenins (Nagafuchi and Takeichi, 1988; Nagafuchi and Takeichi, 1989; Ozawa et al., 1989; Ozawa et al., 1990). Actin filaments are also localized at the basal membrane of cells in the form of stress fibers associated with local adhesion plaques that contain integrins and associated cytoplasmic proteins (vinculin, talin, α-actinin) (Burridge et al., 1988).

Actin and actin-binding proteins are also located along the lateral membranes and sites of cell–cell contacts; actin filaments are linked to membrane proteins through complexes of ankyrin and fodrin (the membrane-cytoskeleton; see Figure 2) (Nelson et al., 1990a). It is thought that the membrane-cytoskeleton of non-erythroid cells may be assembled from component proteins similar to those that have been well described in the human erythrocyte (Bennett, 1990a,b). In the latter cells, spectrin tetramers form the predominant structural component of the membrane-cytoskeleton. Spectrin is comprised of a tetramer of two nonidentical subunits (α,β) (Shotton et al., 1979) and forms a lattice of hexagonal arrays joined at the vertices by complexes of protein 4.1, adducin and short actin oligomers (Bennett, 1990a,b). The spectrin lattice is attached to the membrane through ankyrin and protein 4.1. Ankyrin binds to the β-subunit of spectrin and the Cl^-/HCO_3 exchanger (Band 3 or AE1) with high affinity (Bennett and Stenbuck, 1979; Bennett and Stenbuck, 1980), while protein 4.1 binds directly to glycophorin (Anderson and Lovrien, 1984). In polarized MDCK epithelial cells interactions between ankyrin and other membrane proteins (e.g., Na/K-ATPase) have been described (see below). Also homologs of protein 4.1 and adducin have been localized in MDCK cells to the basal-lateral membrane, although direct interactions with ankyrin/fodrin complexes have not yet been demonstrated (Bennett, 1990a,b).

Microtubules also have a specialized distribution in polarized epithelial cells (Bacallao et al., 1989). At the apical pole of the cell, above the nucleus, the microtubule array comprises short, randomly-oriented filaments that form a dense meshwork (Figure 2). In addition, long bundles of microtubules extend down the length of the cell, parallel to cell–cell contacts; the "plus"-end of these microtubules are located in the basal portion of the cell (Bacallao et al., 1989; see Figure 2).

IV. SYNOPSIS: PROTEIN SORTING AND CYTOSKELETON DISTRIBUTION IN POLARIZED EPITHELIAL CELLS

The juxtaposition in the text of mechanisms for protein sorting and the distribution of the cytoskeleton is highlighted in several instances: formation of the basal-lateral membrane and the organization of intermediate filaments and actin filament network (membrane-cytoskeleton) in association with cell–cell contacts; spatial organization of the Golgi complex and trafficking of vesicles to the apical and basal-lateral membranes, and the spatial organization of microtubules and actin filaments (Figure 2); and localization of membrane proteins to the basal-lateral membrane that bind to the membrane-cytoskeleton. In the following sections, we discuss possible roles for the cytoskeleton in the positional organization of organelles, in the trafficking of vesicles from the TGN and apical and basal-lateral membranes, and in the recruitment and retention of proteins at the cell surface. This analysis reveals that the cytoskeleton plays a potentially significant role in both the generation and maintenance of epithelial cell polarity.

V. REGULATION OF VESICLE TRAFFICKING BETWEEN THE TGN AND PLASMA MEMBRANE: ROLE OF MICROTUBULES AND ACTIN FILAMENTS

As described earlier, both microtubules and actin filaments have non-random distributions in polarized epithelial cells (Figure 2). Microtubules are localized to the apical cytoplasm around the Golgi complex as a dense mat of short, randomly-oriented filaments, and as long bundles of tubules that extend down the length of the cell; significantly, the latter microtubule array has a distinctive unilateral polarity with plus ends located at the apical membrane in close proximity to the Golgi complex (Bacallao et al., 1989; see Figure 2). This polarized orientation of microtubules suggests that a plus-end-directed, kinesin-like motor protein (Vallee, 1990) could participate in vesicle transport from the vicinity of the Golgi complex towards the basal-lateral pole of the cell, and that a "minus"-end-directed, dynein-like motor protein (Vallee and Shpetner, 1990) could be involved in vesicle transport towards the apical pole.

In the case of actin filaments, the apical pole is a rich source of two types of filament alignment (Figure 2): a circumferential band of actin filaments associated with the apical junctional complex (tight junction, *zonula adherens*); and actin filaments that form the core of each microvillus and the terminal web below the microvilli (Mooseker, 1985; Geiger, 1989; Geiger et al., 1990). Both actin arrays are in the close vicinity of the Golgi complex, and span the space between the Golgi complex and the apical membrane. This organization raises the possibility that mechanoenzymes such as myosin I could participate in transport of vesicles along actin filaments to the apical membrane.

A direct test of the roles of microtubules or actin filaments in vesicle delivery is to determine whether disruption of either filament system affects the efficiency or fidelity of transport. Disruption of microtubules with nocodozole, under conditions in which filaments are disrupted completely (i.e., at 4°C), effected a subset of vesicle transport pathways (Nelson, 1991). The delivery of proteins from the TGN to the apical membrane was slowed (Salas et al., 1986; Rindler et al., 1987; Parczyk et al., 1989; Breitfeld et al., 1990; Hunziker et al., 1990; Van Zeijl and Matlin, 1990), and transcytotic transport from the basal-lateral to apical membrane (MDCK cells), and basal-lateral to apical membrane (enterocytes) was blocked (Breitfeld et al., 1990; Hunziker et al., 1990); vesicle transport between the TGN and basal-lateral membrane in MDCK and Caco-2 cells was not affected (Salas et al., 1986; Rindler et al., 1987; Parczyk et al., 1989; Breitfeld et al., 1990; Hunziker et al., 1990; Van Zeijl and Matlin, 1990). Disruption of actin filaments with cytochalasin D had little or no effect on any of the transport pathways (Salas et al., 1986; Rindler et al., 1987; Parczyk et al., 1989; Breitfeld et al., 1990; Hunziker et al., 1990; Van Zeijl and Matlin, 1990). The results of such inhibitor studies are equivocal and it is possible that either some filaments are resistant to disruption by these drugs, or vesicles can reach the membrane by secondary transport pathway (e.g., direct affinity for the membrane or transport along the filament system not disrupted) (Nelson, 1991).

At present, little is known about the expression, distribution, or function of the mechanoenzymes, kinesin, dynein, or myosin I in polarized epithelial cells. Recent results, however support the notion that these proteins are directly involved in vesicle transport in these cells. First, vesicle transport between apical and basal-lateral endosomes isolated from MDCK cells was shown to require intact microtubules and the mechanoenzymes dynein and kinesin (Bomsel et al., 1990). Second, post-TGN apical and basal-lateral transport vesicles from MDCK cells were both shown to bind directly to microtubules *in vitro* (Van Der Sluijs et al., 1990). Third, Golgi-derived transport vesicles (Fath and Burgess, 1993) and the apical membrane domain of intestinal epithelial cells (Mooseker and Coleman, 1989) have been shown to contain the mechanoenzyme myosin I. Fourth, the development of the polarized distributions of both actin filaments and microtubules is dependent upon induction of cell–cell and cell–substratum contacts (Bacallao et al., 1989; see Figure 2), and the formation of the complex microtubule array (~48–72 h) coincides temporally with the establishment of vesicle targeting from the TGN to the basal-lateral membrane domain in MDCK cells (D. A. Wollner and W. J. Nelson, in preparation).

In summary, despite results that inhibitors of microtubule and actin filament organization do not have a dramatic effect on patterns of vesicle transport to the cell surface, further studies are warranted to determine whether there are redundancies in the system that result in apparently normal vesicle delivery in the absence of filaments.

VI. A ROLE FOR THE MEMBRANE-CYTOSKELETON AS A SORTING MECHANISM

A critical step in the analysis of the mechanism of polarized targeting and accumulation of membrane proteins in epithelial cells is the identification of cellular factors that are responsible for the sorting of apical from basal-lateral membrane proteins. Although the composition of this putative "sorting machinery" remains virtually unknown, it is anticipated that such a machinery should possess two important features: (1) an ability to specifically recognize and bind membrane protein populations destined to reside in a given membrane domain; and (2) an ability to segregate proteins belonging to one domain from those destined to reside in the other.

As discussed above, studies that have examined the kinetics of delivery of newly synthesized membrane proteins to a given membrane domain have established that the site at which sorting takes place can be localized to either the TGN (MDCK cells), the basal-lateral membrane (hepatocytes), or both of these locations (Caco-2 cells) (Rodriguez-Boulan and Nelson, 1989). These results indicate that the putative sorting machinery is expected to reside and function at either of two locations, the basal-lateral membrane or the TGN. Whether or not these results demonstrate the existence of multiple types of sorting machinery which are expressed differentially in different cell types or a single sorting machinery capable of functioning at both locations is not clear at present.

Additional information regarding the nature of the putative sorting machinery can be derived from studies examining the polarized distribution of transfected membrane proteins (Hopkins, 1991; Mostov et al., 1992). These investigations have shown that membrane proteins destined to reside in the basal-lateral membrane contain structural determinants located within their cytoplasmic domains which serve as signals for basal-lateral delivery (Brewer and Roth, 1991; Casanova et al., 1991; Hunziker et al., 1991; Le Bivic et al., 1991). These observations indicate that a sorting machinery responsible for routing basal-lateral proteins to their appropriate destination should be able to interact with discrete cytoplasmic domain determinants (Figure 3). Thus, we can conclude that in evaluating the potential of a given cellular apparatus as a sorting mechanism the following properties should exist: 1) specific binding to the appropriate subset of membrane proteins; 2) an ability to sequester that subset of proteins away from other membrane proteins; 3) residence at either the plasma membrane or the TGN; and 4) an ability to interact with cytoplasmic domains of basal-lateral proteins. We will explore the possibility that the membrane-cytoskeleton is a candidate for a sorting machinery.

It is well established that the membrane-cytoskeleton is capable of binding specifically and with high affinity to a diverse set of integral membrane proteins, and that these interactions occur through the cytoplasmic domains of these proteins (Bennett, 1990a,b). Furthermore, the coupling of this membrane protein binding activity to the ability of membrane-cytoskeleton components to assemble into an extensive meshwork of filaments can lead to a restriction of diffusion and hence distribution of bound membranes proteins (Bennett, 1990a,b). Restriction of protein diffusion is an event that is capable of facilitating the segregation of one group of membrane proteins from another.

That the membrane-cytoskeleton is capable of selectively recognizing a subset of membrane proteins belonging to a specific membrane domain is implied by the observations that the membrane-cytoskeleton localizes to discrete cell surface domains in a variety of polarized cell types, including epithelial cells (Drenckhahn et al., 1985; Nelson and Veshnock, 1986; Koob et al., 1987; Morrow et al., 1989; Nelson and Hammerton, 1989; Nelson et al., 1990b), neurons (Lazarides et al.,

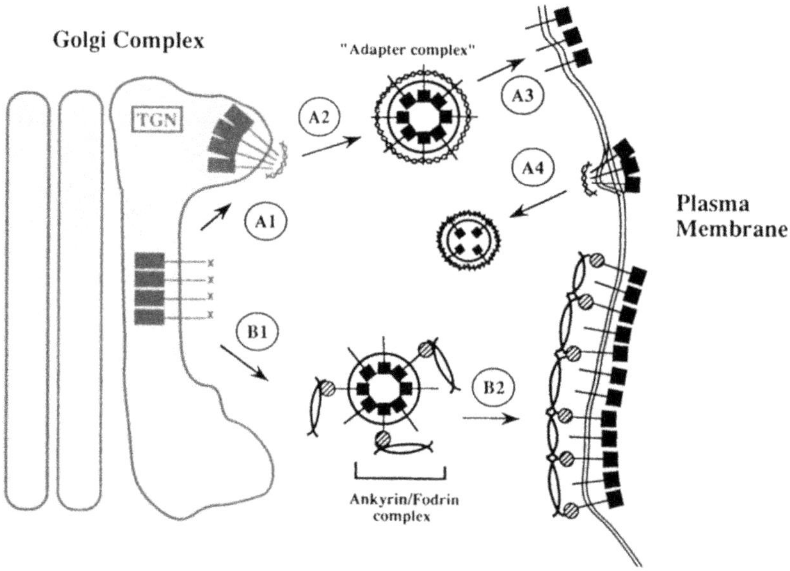

Figure 3. Possible sorting mechanisms of membrane proteins in the Golgi complex, and fates of proteins following arrival at the plasma membrane. Proteins that arrive in the trans Golgi network (TGN) are thought to be segregated from other (apical) membrane proteins by selective clustering in the plane of the lipid bilayer. Based upon analysis of basal-lateral targeting signals (see text), it has been proposed that basal-lateral membrane protein clustering is mediated by "adapter" proteins (A1) similar to those involved in endocytosis at the cell surface. Vesicles are delivered (A2) to, and fused (A3) with the plasma membrane. Unless proteins are retained in the membrane (see below), they will be internalized (A4). We propose that an additional mechanism may play a role in protein sorting and retention (B). Since a number of transport proteins have an affinity for the membrane-cytoskeletal protein complex of ankyrin/fodrin (see text for details), these proteins may interact with the complex in the TGN (B1). Transport vesicles may contain membrane-cytoskeletal complexes that would have an intrinsic affinity for the membrane-cytoskeleton assembled at the membrane and, hence, mediate docking of the vesicle with the appropriate membrane domain (B2). Subsequently, integration of the newly-delivered proteins with the membrane-cytoskeleton would retain the proteins in the membrane and exclude them from the internalization.

1984; Zagon et al., 1986; Bennett et al., 1991; Kordeli and Bennett, 1991) and muscle cells (Bloch and Morrow, 1989). More specifically, in MDCK cells, components of the membrane-cytoskeleton accumulate exclusively at the basal-lateral membrane, where they exist in the form of a stable cytoskeletal assembly (Nelson and Veshnock, 1986; Morrow et al., 1989; Nelson and Hammerton, 1989).

In addition, a number of studies have shown that two well established basal-lateral membrane proteins, Na/K-ATPase and the cell adhesion molecule E-cadherin, are capable of binding with high affinity to membrane-cytoskeleton components *in vitro* [Na/K-ATPase, (Koob et al., 1987; Nelson and Veshnock, 1987; Morrow et al., 1989)], and can be isolated from intact MDCK cells in the form of complexes with membrane-cytoskeleton components [Na/K-ATPase and E-cadherin (Nelson and Hammerton, 1989; Nelson et al., 1990b)]. Since it is well established that the membrane-cytoskeleton underlie the plasma membrane (Bennett, 1990a,b), it can be concluded that these interactions involve membrane protein cytoplasmic domains.

Thus, in MDCK cells the membrane-cytoskeleton appears to exhibit several important functional properties anticipated for a sorting machinery in these cells: (1) localization to the basal-lateral membrane; (2) an ability to bind domain-specific proteins (basal-lateral) through their cytoplasmic domains; and (3) an ability to form high oligomers which could potentially facilitate the segregation of apical and basal-lateral proteins.

VII. EVIDENCE FOR A DYNAMIC MEMBRANE-CYTOSKELETON CAPABLE OF INTERACTION WITH A VARIETY OF INTRACELLULAR SITES

In contrast to the relatively static condition of the erythrocyte plasma membrane, the cell surfaces of essentially all other cell types, including those of polarized cells, are continually in communication with a variety of intracellular membranes. This communication is manifested in the combined processes of endocytosis and secretion which together give rise to a steady-state turnover of plasma membrane that can occur at rates on the order of one cycle per hour (Cohn and Steinman, 1982; Balcarova et al., 1984). The establishment and maintenance of discrete membrane domains in polarized cells is integrally linked to specific, targeted secretory and endocytic processes. Hence, any model which attempts to describe the mechanism whereby the membrane-cytoskeleton participates in the domain-specific accumulation of membrane proteins must consider how the membrane-cytoskeleton responds to these dynamic membrane events.

In addition, there is a growing body of evidence which indicates that the localization of membrane-cytoskeleton components is not limited to the plasma membrane. Rather, membrane-cytoskeleton components are found to be associated with a variety of intracellular membranes including those that participate in plasma membrane formation (e.g., secretory organelles; see below). This raises the possibility that the membrane-cytoskeleton may function at the level of organellar membranes.

One possible way in which the membrane-cytoskeleton could facilitate the sorting of apical and basal-lateral proteins in polarized epithelial cells is through selective retention of membrane proteins, mediated by membrane protein–membrane-cytoskeleton interactions, following their delivery to the appropriate membrane domain (Figure 3). Such a model has been proposed for the domain specific accumulation of the Na/K-ATPase (Hammerton et al., 1991). This protein, unlike all other proteins thus far examined in MDCK cells, is initially targeted equally to both the apical and basal-lateral membrane domains. However, since the half-life of the protein is substantially longer at the basal-lateral membrane than at the apical membrane (Hammerton et al., 1991), it accumulates at steady state exclusively at the basal-lateral membrane. This differential stability has been attributed to the ability of the Na/K-ATPase to interact with the basal-laterally localized membrane-cytoskeleton (Nelson and Hammerton, 1989; Hammerton et al., 1991). Presumably, other membrane proteins which also interact with the membrane-cytoskeleton would be similarly affected. This model has the membrane-cytoskeleton acting solely at the plasma membrane where it serves to prevent a defined subset of membrane proteins from gaining access to the endocytic pathway subsequent to their delivery to the cell surface.

It is also possible that the membrane-cytoskeleton participates in events that precede the insertion of newly synthesized membrane protein into the plasma membrane (see Figure 3). For instance, in adrenal chromaffin cells the distribution of fodrin changes in response to stimulated secretion from a uniform plasma membrane ring to patchy cell surface staining (Perrin and Aunis, 1985) This was interpreted as a loss of fodrin from large portions of the cell surface. A similar event was observed in parotid acinar cells following cAMP- and Ca^{2+}-mediated secretion. Under those conditions, a complete loss of plasma membrane fodrin staining correlated with the stimulated secretory response (Perrin et al., 1992). These observations have led to the proposal that a necessary prerequisite to secretory vesicle fusion is a rapid, second messenger-mediated dissolution of the plasma membrane spectrin lattice (Perrin and Aunis, 1985). This implies that at steady state the fodrin membrane-cytoskeleton lattice serves as a barrier to the fusion of secretory vesicles (Figure 3). In support of this, deep-etch replicas of human erythrocyte membrane-cytoskeletons (Coleman et al., 1989) reveal that assembled spectrin and its associated proteins form a dense meshwork with pore sizes smaller than 100 nm. It can be argued that if such a compact arrangement of material also exists in non-erythroid cells it could be capable of preventing access to the plasma membrane of the relatively large secretory granules of chromaffin cells.

It is feasible that the dense matrix formed by the membrane-cytoskeleton may also prevent fusion of smaller transport vesicles that participate in constitutive secretion in polarized epithelial cells (Figure 3). Accordingly, the ability of the membrane-cytoskeleton to act as a barrier to secretory vesicle fusion could represent a critical event in the mechanism whereby polarized cells deliver discrete membrane protein populations to distinct membrane domains. Selective fusion of Golgi-derived transport vesicles with a given plasma membrane domain could be limited to only those vesicles which express an appropriate signal or machinery required for localized membrane-cytoskeleton dissociation (Figure 3). In this way the membrane-cytoskeleton could serve as a fidelity check for membrane protein targeting by preventing random, nonspecific fusions of secretory vesicles.

Introduction of antibodies to fodrin into permeabilized chromaffin cells results in a partial but specific inhibition of secretory vesicle fusion with the plasma membrane (Perrin et al., 1987), providing strong evidence that fodrin plays a critical role in this process. This result can be interpreted as being consistent with the proposed "barrier to fusion" model in that it suggests that the fodrin antibodies block the dissociation of the membrane-cytoskeleton by preventing the apparatus responsible for membrane-cytoskeleton dissolution from gaining access to its substrate. It is also possible that the antibodies act by binding to a spectrin-like molecule which has been found to be present on the chromaffin granule surface (Aunis and Perrin, 1984). Here it could be argued that membrane-cytoskeleton components present on the surface of peripheral secretory compartments participate in a critical interaction between the secretory vesicle membrane and the plasma membrane that is necessary for fusion to take place. It is interesting to note that Zagon et al. (1986) have found in neurons, by immunoelectron microscopy, that spectrin is associated with the surfaces of synaptic vesicles and with filamentous strands (possibly containing microtubules) that connect cytoplasmic vesicles to each other as well as to the plasma membrane (Zagon et al., 1986). Similar structures have been seen in the terminal web region in chicken intestinal brush borders (Hirokawa et al., 1983). Perhaps these spectrin-decorated strands represent such an intermediate in the fusion process.

If spectrin dependent connections are also important in the fusion of constitutive secretory vesicles with the plasma membrane they could provide a means for the selective and targeted delivery of newly synthesized membrane proteins in polarized cells (Figure 3). Since fodrin isoforms localize to a given plasma membrane domain and interact with membrane proteins which ultimately localize to the same cell surface domain (Koob et al., 1987; Nelson and Veshnock, 1987; Morrow et al., 1989; Nelson and Hammerton, 1989), it is possible that Golgi-derived transport vesicles containing a given subset of membrane proteins may be tagged with fodrin molecules of the appropriate subtype. These vesicles could then be targeted to a specific membrane domain by the interaction of the fodrin tag with a receptor located on the cytoplasmic surface of the target membrane. While the nature of such a receptor is highly speculative at this time, it could simply be membrane-cytoskeleton proteins that are localized to the plasma membrane. That is, membrane-cytoskeleton molecules residing on the surface of an approaching vesicle may interact with the membrane-cytoskeleton of the plasma membrane. This interaction would form a continuum of assembled cytoskeleton that encompassed both the plasma membrane and the recently arrived transport vesicle. It is easy to imagine that an intermediate of such a reaction would resemble the spectrin-decorated, filamentous strands described above. Since in polarized cells specific spectrin isoforms are differentially localized to distinct plasma membrane domains (see above), the specificity of this interaction could come about through isoform-specific co-assembly of plasma membrane and transport vesicle membrane-cytoskeleton proteins.

The association of membrane-cytoskeleton components with intracellular membranes is not limited to the highly specialized chromaffin granules and synaptic vesicles. In neurons and glial cells fodrin isoforms have been found associated with a variety of organelles including endoplasmic reticulum, nuclear membrane, mitochondria, as well as numerous unidentified vesicles (Zagon et al., 1986). More interestingly, observations made in lymphocytes (Black et al., 1988) and rat sperm precursor cells (DeCesaris et al., 1989) suggest that membrane cytoskeletal components associate with proximal regions of the secretory apparatus, including Golgi membranes and the TGN. Indirect immunofluorescence localization of fodrin isoforms in rat spermatids have shown that an erythroid isoform of spectrin is predominantly localized to the perinuclear region in a staining pattern reminiscent of the Golgi complex (DeCesaris et al., 1989). Unfortunately, no double labeling experiments were performed to more definitively establish co-localization of spectrin with Golgi markers.

In lymphocytes, fodrin exists as a large cytoplasmic aggregate that is located in close proximity to the TGN (Black et al., 1988). Mitogen-stimulated secretion of interleukin-2 in these cells is accompanied by a protein kinase C-dependent rearrangement of the cytoplasmic pool of fodrin, which ultimately results in a dissolution of the aggregate followed by translocation of fodrin to plasma membrane patches (Lee et al., 1988; Gregorio et al., 1992). The translocation of fodrin to the plasma membrane occurs in parallel to stimulated secretion, implying a direct role of membrane-cytoskeleton components in secretory processes. At the ultra structural level, the fodrin aggregate shows no obvious association with morphologically distinct Golgi cisternae. However, numerous membranous structures, perhaps representing elements of the TGN, are found imbedded within the fodrin aggregate. It is tentative, at present, to conclude from these results that the membrane-cytoskeleton is associated with Golgi cisternae and the TGN. The possibility that the membrane-cytoskeleton functions within the secretory apparatus at the level at which membrane protein sorting takes place is particularly tantalizing when one considers, as we have discussed above, that the membrane-cytoskeleton has the potential to act as a membrane protein sorting machinery (Figure 3).

In summary, an examination of documented events involving an interplay between the membrane-cytoskeleton and dynamic membrane systems has revealed a number of possible subcellular sites at which the membrane-cytoskeleton could serve to facilitate or potentiate polarized protein targeting. That is, there is a potential for the membrane-cytoskeleton to function as a sorting machinery at a number of the intermediate steps which define the secretory pathway.

Studies examining the targeting and accumulation of the Na/K-ATPase in polarized membrane domains of MDCK cells have indicated that the basal-lateral plasma membrane localized membrane-cytoskeleton serves as an immobilized lattice which binds to a specific membrane protein and prevents them from gaining access to the endocytic pathway (Figure 3). Here, then, is an example of the membrane-cytoskeleton acting as a sorting apparatus at the level of the plasma membrane by directly facilitating the accumulation of a specific membrane protein species subsequent to its delivery to the appropriate plasma membrane domain.

While this system fits well with the known behavior of the Na/K-ATPase, it fails to account for the fact that virtually all other membrane proteins thus far studied in MDCK cells are sorted properly before they reach the plasma membrane (see above). In this context, the observations made in lymphocytes and rat germ cells (see above) suggesting that membrane-cytoskeleton components associate with Golgi membranes become significant. If the membrane-cytoskeleton could be definitively shown to be present on the surfaces of Golgi membranes then it could be easily imagined how it could act as a sorting machinery at this site. A restricted assembly of a membrane domain-specific isoform of fodrin could cause a localized accumulation of newly synthesized membrane proteins. This accumulation of proteins would form a microdomain within the organellar membrane composed of proteins destined to reside in a given plasma membrane region. Subsequent budding of vesicles from this microdomain would give rise to transport vesicles containing only proteins belonging to the cell surface domain specified by the membrane-cytoskeleton proteins involved (e.g., the basal-lateral membrane in MDCK cells). These vesicles would then be transported to the cell periphery via a process that involves microtubules and microtubule motor proteins (see above). It is interesting to note that both fodrin and ankyrin have been found to bind microtubules (Bennett, 1990a,b), suggesting that membrane-cytoskeleton proteins may be involved in microtubule based vesicle transport. Alternatively, these binding sites may be important in orienting and/or anchoring cytoskeletal filaments to the lateral membrane domain.

Upon reaching the plasma membrane the vesicle-associated fodrin may serve as a signal that assesses whether the appropriate domain has been reached (see above and Figure 3). A localized dissolution or rearrangement of the membrane-cytoskeleton at the plasma membrane similar to that seen in stimulated chromaffin cells may then occur to promote vesicle fusion with the plasma membrane. Finally, the newly-inserted membrane proteins would be stabilized by membrane-cytoskeleton interactions (Figure 3), whereas proteins mistakenly included during transport vesicle formation would be rapidly internalized and degraded due to their inability to interact with the membrane-cytoskeleton (Figure 3). We conclude that the membrane-cytoskeleton has the potential to participate in a number of steps along the secretory pathway.The advantage of this multiple participation of the membrane-cytoskeleton lies in the possibility that it could serve to markedly increase the overall fidelity of membrane protein trafficking in polarized cells.

If the membrane-cytoskeleton acts as a sorting machinery in polarized cells it is necessary to consider what types of membrane proteins are involved. Membrane-cytoskeleton components have been shown to directly interact with a number of membrane proteins including E-cadherin (Nelson et al., 1990b) Na/K-ATPase (Koob et al., 1987; Nelson and Veshnock, 1987; Morrow et al., 1989), the voltage gated Na^+ channel (Srinivasan et al., 1988; Smith et al., 1991), and the anion transporter (Bennett, 1978; Bennett and Stenbuck, 1980). Additionally, the abundance and diversity of uncharacterized high affinity membrane binding sites for isoforms of membrane-cytoskeletal proteins identified in crude membrane preparations from kidneys (Davis et al., 1989) reveal a potential for many more specific membrane protein-membrane-cytoskeleton interactions. It is probably not a trivial coincidence that a significant number of the proteins shown to interact with the membrane-cytoskeleton are involved in ion transport processes. As mentioned above the distribution of the membrane-cytoskeleton (as well as the Na/K-ATPase) differs in kidney-derived MDCK cells (where fodrin is basal-lateral) and choroid plexus epithelial cells (where fodrin is apical; J. A. Marrs and W. J. Nelson, in preparation). It is unreasonable to expect that all of the basal-lateral proteins of MDCK cells are inserted into the apical domain of the choroid plexus epithelium. It is not, however, unreasonable to expect that the distribution of proteins involved in ion transport processes would conform to the differential distribution of membrane-cytoskeleton components seen in these cells. In fact, such a redistribution is expected for the Na/K-ATPase because the direction in which kidney tubule epithelial cells transport Na^+ is opposite to that of the choroid plexus (Wright, 1972). This leads us to conclude that if the membrane-cytoskeleton serves to facilitate the sorting of membrane proteins in polarized epithelial cells, it is most likely that this function will be primarily involved in the sorting of proteins involved in ion transport processes. It is significant to note that membrane-cytoskeleton isoforms have been shown to be differentially distributed in neurons (Lazarides and Nelson, 1983; Lazarides et al., 1984; Zagon et al., 1986; Bennett et al., 1991; Kordeli and Bennett, 1991). Hence, in these cells there also exists the potential for a membrane-cytoskeleton dependent sorting of ion channel proteins.

VIII. REGULATION OF MEMBRANE-CYTOSKELETON ASSEMBLY BY CELL–CELL ADHESION

For the membrane-cytoskeleton to function directly in regulating protein sorting, vesicle docking, and protein retention at the cell surface, it must be assembled at specific membrane domains. Below, we discuss experimental evidence for the differential assembly of the membrane-cytoskeleton on the basal-lateral membrane of MDCK cells.

How is the assembly of the membrane-cytoskeleton in erythrocytes regulated, and can we apply this information to polarized epithelial cells? Although spectrin subunits will spontaneously form dimers, tetramers, and higher order oligomers, assembly of the spectrin lattice appears to be regulated by several post-translational modifications of ankyrin, protein 4.1, and adducin (Bennett, 1990a,b). Ankyrin binds to spectrin tetramers with higher affinity than to dimers (Weaver et al., 1984), which may favor formation of higher ordered spectrin oligomers at the plasma membrane. Phosphorylation of ankyrin lowers its affinity for spectrin tetramers, but does not affect its affinity for dimers (Lu et al., 1985; Lu and Tao, 1986); phosphorylation of ankyrin, therefore, may limit formation of higher ordered oligomers and assembly of the membrane-cytoskeleton. Ankyrin also may be subject to post-translational modification by the addition of palmitic acid (Staufenbiel, 1988; Staufenbiel and Lazarides, 1986), which may provide a mechanism for direct linkage of ankyrin and the plasma membrane lipid bilayer. Phosphorylation of 4.1 also decreases its affinity for spectrin and reduces its ability to potentiate spectrin-actin interactions (Eder et al., 1986; Ling et al., 1988). Adducin is a substrate for phosphorylation by protein kinase C (Palfrey and Waseem, 1985) and cAMP-dependent protein kinase (Ling et al., 1988), but the effects of phosphorylation on adducin function(s) are unknown. Adducin binds calmodulin, which inhibits its interaction with spectrin and actin (Gardner and Bennett, 1987; Mische et al., 1987). In summary, many of the components of the erythrocyte membrane-cytoskeleton are subject to phosphorylation, which, in general, decreases interactions between cytoskeletal proteins.

Little is known about the regulation of membrane-cytoskeleton assembly in nonerythroid cells. However, based upon information described above on the regulation of protein interactions in erythrocytes, it is possible that regulation of both phosphorylation and Ca^{2+} levels could directly affect assembly of the membrane-cytoskeleton in epithelial cells in which similar proteins have been identified. As we discussed earlier, cell–cell interactions have been shown to lead to activation of protein kinases, phosphatases and second messengers that could mediate changes in interactions between these proteins (Doherty et al., 1991; Frei et al., 1992).

IX. CONSEQUENCES OF MEMBRANE-CYTOSKELETON ASSEMBLY ON CELL SURFACE POLARITY OF MEMBRANE PROTEINS

The restriction of the membrane-cytoskeleton to the basal-lateral membrane in MDCK cells raises two important questions: 1) is the membrane-cytoskeleton linked to cell adhesion proteins; and 2) which other membrane proteins are bound to the membrane-cytoskeleton?

Aside from the fact that ankyrin and fodrin colocalize with cell adhesion proteins in polarized epithelial cells, there are two pieces of experimental evidence to indicate that there is a direct association. First, expression of exogenous E-cadherin in fibroblasts induces co-localization of fodrin and E-cadherin at cell–cell contacts; deletion of portions of the cytoplasmic domain of E-cadherin that bind to cytoplasmic proteins results in a nonpolarized distribution of fodrin in transfected cells (McNeill et al., 1990). Second, analysis of membrane-cytoskeleton complexes isolated from whole extracts of MDCK cells demonstrated that E-cadherin is a component protein (Nelson et al., 1990a). Although these results do not prove a direct interaction between the membrane-cytoskeleton and E-cadherin, they strongly suggest that they form a complex (see also Itoh et al., 1991). These results raise the interesting possibility that homotypic interactions between cadherins on adjacent cells nucleate assembly and patching of membrane-cytoskeleton at contact sites on the cell surface. Under these conditions, other membrane proteins directly linked, or with affinity for, the membrane-cytoskeleton would also be recruited to those contact sites on the membrane.

Detailed analysis of the subcellular distribution of Na/K-ATPase, ankyrin, and fodrin in a variety of polarized epithelia have demonstrated that these proteins co-localize on the plasma membrane (Nelson and Veshnock, 1986; Koob et al., 1987; Morrow et al., 1989; Nelson et al., 1990a). More importantly, however, these proteins have been shown to bind directly. Reconstitution of the complex with purified membrane-bound Na/K-ATPase and ankyrin demonstrated high affinity binding (Koob et al., 1987; Nelson and Veshnock, 1987; Morrow et al., 1989; Davis and Bennett, 1990). Significantly, a complex comprising Na/K-ATPase, ankyrin and fodrin tetramers was identified from MDCK cell following fractionation of a cell extract in sucrose gradients and non-denaturing polyacrylamide gels (Nelson and Hammerton, 1989).

The identification of the Na/K-ATPase/membrane-cytoskeleton complex raised the possibility that localized assembly of the membrane-cytoskeleton would result in the concomitant localized recruitment of Na/K-ATPase. Two observations support this possibility. First, induction of cell–cell contact in cultures of MDCK cells results in the gradual recruitment and accumulation of Na/K-ATPase, ankyrin and fodrin at sites of cell–cell contact (Nelson and Veshnock, 1986); the co-localization of these proteins also coincides with an increase in their resistance to extraction in buffers containing non-ionic detergents, which has been interpreted as an indication of cytoskeleton association (Nelson and Veshnock, 1986; Hammerton et al., 1991).

Second, fibroblasts transfected with E-cadherin exhibit a different distribution of Na/K-ATPase compared to untransfected cells (McNeill et al., 1990). In untransfected cells, Na/K-ATpase is localized over the entire cell surface. However, in cells expressing E-cadherin, Na/K-ATPase appears to be restricted to sites of E-cadherin-mediated cell–cell contacts; this is similar to the distribution of fodrin in these cells. Significantly, deletion of the portion of the cytoplasmic domain of E-cadherin that binds cytoplasmic proteins resulted in little change in Na/K-ATPase distribution from that found in untransfected cells (see above). Thus, loss of binding between E-cadherin and cytoplasmic proteins (catenins, fodrin) uncoupled Na/K-ATPase from cell–cell adhesion resulting in maintenance of a nonpolarized distribution of Na/K-ATPase over the entire cell surface (McNeill et al., 1990).

X. INTERRELATIONSHIP BETWEEN CYTOSKELETAL PROTEINS IN GENERATING EPITHELIAL CELL POLARITY

Polarized transporting epithelial cells are derived from non-polarized precursor cells (Ekblom et al. 1986; Ekblom, 1989). Previous studies have clearly demonstrated that specific cell–cell and cell–substratum contacts are required to induce the development of cellular polarity. This dramatic cellular reorganization requires coordinate changes in the spatial distribution of the secretory pathway, and the generation of structural and functional membrane domains based on the distinction between free and bounded cell surfaces.

Studies of how polarized epithelial cells arise from nonpolarized precursor cells have provided insight into the roles of cell–cell and cell–substratum adhesion. For example, single MDCK cells in suspension culture exhibit a partial polarized distribution of apical, but not basal-lateral, membranes (Ojakian and Schwimmer, 1988). Induction of cell–cell contacts in these cultures results in the formation of a bounded surface (cell–cell contacts) and a free surface; marker proteins of the basal-lateral membrane domain (e.g., Na/K-ATPase, E-cadherin) become restricted to the bounded surface, and apical membrane proteins become restricted to the free surface membrane (Wang et al., 1990a,b). This result suggests two general principles. First, the distinction between a free and a bounded surface (cell–cell or cell–substratum) is a criterion for formation of different membrane domains. Second, cell–cell contact specifically induces formation of the (basal)-lateral membrane domain (Wang et al., 1990a,b). In conclusion, the free cell surface in these cells is designated as the apical membrane, however formation of a bounded cell surface itself is not sufficient to result in specific localization of proteins such as Na/K-ATPase; cell–cell contacts are required for the formation of the (basal)-lateral membrane domain.

At present it is unknown how these extracellular contacts initiate the formation of cell surface domains. One important process may be to induce recruitment of subsets of vesicles to different domains. This is supported by the observation that cell polarity at the first cleavage of the fertilized *Xenopus* egg is apparently generated in part by specific recruitment of a subset of stored vesicles to the cleavage membranes (Roberts et al., 1992). In addition, under certain conditions, formation of the apical membrane in MDCK cells may be facilitated by direct fusion of storage vesicles containing apical marker proteins with the free cell surface following induction of cell-cell contacts (Rodriguez-Boulan and Nelson, 1989).

There are several possible mechanisms for transducing signals from extracellular contacts that result in vesicle recruitment to the cell surface. For example, occupation of cell adhesion proteins induces a classical second messenger response involving an increase of IP_3 and activation of Ca^{2+} channels, that lead to an elevation of intracellular Ca^{2+} that could trigger localized vesicle fusion with the membrane (Doherty et al., 1991; Frei et al., 1992). Also, activity of tyrosine kinases and phosphatases modulate protein interactions involved in assembly of junctional complexes and the cytoskeleton (Shattil and Brugge, 1991) (see below). Finally, cell–cell adhesion induces the assembly of the actin-based membrane-cytoskeleton at sites of cell–cell contacts and reorganization of the microtubule cytoskeleton (Hirano et al., 1987; Bacallao et al., 1989).

As discussed above, cell–cell adhesion induces specific changes in the organization of all components of the cytoskeleton. In summary, formation of the junction complex results in: 1) attachment of the keratin intermediate filament network to desmosomes that line the lateral membrane from the apex to the base of the cell (Pasdar and Nelson, 1988a,b); 2) the actin-based network becomes associated with the apical junctional complex (tight junction, *zonula adherens*) in the form of a circumferential band of actin filaments (Geiger, 1989; Geiger et al., 1990; Hirano et al., 1987; Geiger et al., 1990); 3) the actin-based network becomes associated with other areas of cell–cell adhesion along the lateral membrane where cadherins are linked (directly or indirectly?) to the ankyrin/fodrin lattice (Nelson et al., 1990a; Nelson, 1991); and 4) microtubules become localized to both the cytoplasm in proximity to the Golgi complex, and as long unipolarized filament bundles parallel to the lateral membranes (Bacallao et al., 1989).

XI. THE DEVELOPMENT OF PROTEIN SORTING PATHWAYS

Little is known about how this dynamic reorganization of organelles, cytoskeleton, and plasma membrane domains is set in motion by cell–cell contact. As noted at the beginning, many studies have focused on sorting pathways in the secretory pathway, and how they play a role in generating cell surface polarity. As cells generate cell surface polarity from the nonpolarized precursor state, it can be imagined that either the sorting pathways are induced *de novo*, or the pathways for sorting apical and basal-lateral membrane proteins pre-existed in the nonpolarized cells but delivery and retention of proteins between the TGN and cell surface domains needed to be established (Wollner et al., 1992; Wollner and Nelson, 1992).

There is little evidence to distinguish whether protein sorting of apical and basal-lateral membrane proteins in the TGN is a constitutive process, or one that is induced upon cell–cell adhesion (Wollner et al., 1992; Wollner and Nelson, 1992). As noted above, marker proteins of the apical membrane are localized on the free surface of single, substratum-attached MDCK cells (Ojakian and Schwimmer, 1988); however, in these cases, endogenous basal-lateral membrane proteins are localized over the entire cell surface. This supports the notion that there is constitutive sorting of some membrane proteins in nonpolarized single MDCK cells (Wollner et al., 1992; Wollner and Nelson, 1992), but they do not rule out the possibility that proteins are randomly delivered to the cell surface and then diffuse to another site on the membrane (there are no tight junctions in these cells), or exhibit differential stability on a free versus substratum-bounded membrane (i.e., sorting after delivery to the cell surface). These possibilities, however, can be excluded when a secreted protein is analyzed (Wollner et al., 1992). In polarized MDCK cells, gp84 is secreted extremely efficiently from the apical membrane domain (Parczyk et al., 1989; Wollner et al., 1992). By rapid sampling of media from cultures of MDCK cells grown in the absence of cell contact (i.e., nonpolarized cells), Wollner and Nelson (1992) showed that gp84 was still secreted exclusively from the free cell surface that in polarized cells constitutes the apical membrane domain. Since different proteins were also secreted from the bounded cell surface, the possibility that all secretion was to the free cell surface domain could be excluded.

Taken together, these results suggest that sorting of proteins in the TGN is a constitutive process that is independent of the status of cell surface polarity. To take this a step further, it could be proposed that establishment of the sorting pathways requires directed vesicle trafficking to different membrane domains, and the establishment of recognition and docking mechanisms specific for vesicle subpopulations and distinct membrane domains. Both of these processes could be dependent on the coordinate reorganization of different components of the cytoskeleton (Nelson, 1991).

We suggest that an important initial process in the generation of cell polarity is the reorganization of the cytoskeleton in response in cell–cell contacts mediated by cadherins. Formation of these specific contacts is a rapid process and occurs within a few minutes of a random cell contact between MDCK cells. High resolution analysis of E-cadherin at cell contacts during these times shows that there is a lag phase of 10 minutes following formation of a stable contact after which time E-cadhedrin becomes insoluble in buffers containing Triton X-100 at the contact site (McNeill et al., 1993). Significantly, E-cadherin insolubility in Triton X-100 occurs only within the boundary of the contact site, but does not immediately coincide with any changes in the distribution of other cytoskeletal proteins, actin, and fodrin. At present it is not known how E-cadherin becomes insoluble in Triton X-100 buffers under these conditions. However, since E-cadherin insolubility precedes reorganization of other components of the cytoskeleton, it is possible that it acts as a nucleating center for cytoskeleton assembly. Indeed, previous studies have shown that actin filaments and components of the membrane-cytoskeleton become localized to cadherin-mediated cell contacts on the order of hours and days after induction of cell–cell contacts (Nelson and Veshnock, 1986; Gumbiner et al., 1988; Hirano et al., 1987). Cadherin-mediated cell adhesion has also been shown to be required for assembly of other components of the junctional complex (tight junction, desmosomes) (Gumbiner et al., 1988). Hence, cadherin-mediated cell adhesion leads also to assembly of desmosome and, therefore, keratin intermediate filament redistribution to cell–cell contacts. Note that actin filaments are partitioned to a circumferential band of filaments at the cell periphery in a single nonpolarized MDCK cells, and then as a circumferential band of filaments at the apex of the lateral membrane in association with the apical junctional complex following cell–cell contact (Nelson, 1991). At present, it is not known whether these two distinctive actin structures are the same, or whether there is a reorganization of actin to form the cell adhesion-associated complex.

The identification of preassembled unit complexes of membrane-cytoskeletal complexes (see above) has led to the proposal that cell–cell adhesion induces polymerization of the individual units to $[unit]_N$ (Nelson et al., 1990a). At present, it is unknown how this polymerization reaction is initiated, but since it appears to occur specifically at cell–cell contacts it has been proposed that cadherin binding is involved. As noted earlier, cadherin-mediated adhesion results in the generation of second messages including changes in the intracellular levels of IP_3 and Ca^{2+}, and in localization of both tyrosine kinases and phosphatases to cell–cell contacts (for references, see above). Furthermore, protein–protein interactions in the erythrocyte membrane-cytoskeleton have been shown to be modulated by the state of phosphorylation of ankyrin and protein band 4.1, and calmodulin/Ca^{2+} binding to adducin (see above). It is interesting to note that in single MDCK cells, in which membrane-cytoskeleton components are unassembled (unit), protein 4.1 homologs and adducin do not co-sediment with ankyrin- fodrin- membrane protein complexes in sucrose gradients (W. J. Nelson, unpublished results). One interpretation of this result is that proteins that may perform the spot-welding of these unit complexes are not associated with the units until after induction of cell–cell contacts. It is possible, therefore, that cell–cell adhesion induces some form of signal that results in initiation of assembly of units of the membrane-cytoskeleton, specifically in the vicinity of the cell–cell contact. Subsequent membrane-cytoskeleton assembly could be generated by mass action following nucleation of the initial complexes at the contact site (Nelson et al., 1990a).

Detailed confocal microscope studies of microtubule distribution following induction of cell–cell contacts in culture of MDCK cells have shown a dramatic, gradual reorganization (Bacallao et al., 1989). Initiation of cell–cell contacts results in the splitting of the centrosome and the appearance of multiple microtubule organizing centers in the cytoplasm at the apical pole; at the same time there appears a large number of short microtubules in this region of the cell, and long bundles of microtubules begin to extend down the length of the cell towards the base (plus ends). The significance of this change in microtubule organization on vesicle transport and targeting has not been determined in detail. However, preliminary studies indicate that the establishment of polarized delivery of vesicles to the basal-lateral membrane domain in MDCK cells occurs at ~48–72 hours after cell–cell adhesion and coincides with the time course required to form the microtubule network in these cells (D.A. Wollner and W.J. Nelson, unpublished results). This raises the possibility that vesicle delivery to the basal-lateral membrane may require establishment of microtubule tracks for trafficking from the TGN to the cell surface, and the establishment of a vesicle recognition/docking mechanism at the cell surface. These two processes are not mutually exclusive. If the recognition/docking mechanism comprises membrane proteins (GTP-binding protein receptors?), then these need to be accumulated at the membrane in sufficient concentration to recruit vesicles. Perhaps the generation of the microtubule array initially drives vesicles carrying the recognition/docking mechanism delivery to the basal-lateral membrane where the recognition/docking mechanism is established and accumulates to the extent that, in the absence of microtubules (by incubation with nocodozole; see above), vesicles still retain high binding affinity with the basal-lateral membrane due to the accumulated recognition/docking mechanism.

As discussed earlier, the assembly of the membrane-cytoskeleton on the basal-lateral membrane in MDCK cells also provides a possible affinity site for a subset of transport vesicles and proteins. Given that interactions between membrane-cytoskeletal proteins are generally of high affinity (see above), it is possible that ankyrin/fodrin association with membrane proteins in transport vesicles could interact with the assembled membrane-cytoskeleton on the basal-lateral membrane and, hence, mediate vesicle recognition/docking with that membrane.

XII. CONCLUSIONS

The generation and maintenance of epithelial cell polarity has, to a great extent, focused on the sorting of proteins in the TGN. In MDCK cells, and to a lesser extent in Caco-2 cells, sorting of proteins in the TGN is an important process in maintaining cell surface polarity, but also in Caco-2 cells and hepatocytes protein sorting at the plasma membrane is important. Note that the generation of cell surface polarity in MDCK cells also appears to require protein sorting at the level of the cell surface after induction of cell–cell contacts. However, it has become clear that other aspects of the sorting pathway are critical. Upon sorting of apical and basal-lateral membrane proteins in the TGN of MDCK cells, transport vesicles are delivered to the appropriate membrane domain and therein proteins are retained. The obvious regulatory mechanism for vesicle trafficking in the cytoplasm is along cytoskeletal filaments. There is a variety of experimental evidence that both actin filaments and microtubules play roles in vesicle delivery between the Golgi complex and either the apical or basal-lateral membrane domain; it should be noted, however, that there is not definitive evidence of a *requirement* for these filaments for vesicle trafficking. It is also possible, but not shown, that the membrane-cytoskeleton plays a role in vesicle recruitment to the basal-lateral membrane of MDCK cells. Upon arrival at the cell surface, proteins may have a number of fates, and specific attachment to the membrane-cytoskeleton may be important in retention of the protein and maintenance of polarized cell surface distributions.

Our understanding of how these pathways and interactions are generated in a developmental context is not complete. Clearly extracellular contacts, and in particular cadherin-mediated cell–cell adhesion, are critical to inducing the development of the sorting pathway. Future attempts to directly correlate the reorganization of the cytoskeleton and assembly of the membrane-cytoskeleton, with the generation of targeted delivery pathways for vesicle transport between the TGN and different membrane domains, should provide further insight into the role(s) of the cytoskeleton in the generation and maintenance of epithelial cell polarity.

NOTE ADDED IN PROOF

Recently our laboratory has found evidence for the existence of isoforms of both spectrin (Beck, K. A. et al. (1993). J. Cell Biol., 127, 707–723) and ankyrin (Beck, K. A. and Nelson, W. J., manuscript in preparation) which localizes exclusively to the Golgi complex.

ACKNOWLEDGMENTS

Work from the authors laboratory was supported by grants from the National Institutes of Health, March of Dimes Foundation, and American Cancer Society. Ken Beck was also supported by a postdoctoral fellowship from the Program in Cancer Biology, Stanford University, and James Nelson is the recipient of an Established Investigator Award from the American Heart Association.

REFERENCES

Almers, W. & Stirling, C. (1984). Distribution of transport proteins over animal cell membranes. J. Membr. Biol. 77, 169–186.

Anderson, R.A. & Lovrien, R.E. (1984). Glycophorin is linked by band 4.1 protein to the human erythrocyte membrane skeleton. Nature (Lond.) 307, 655–658.

Aunis, D. & Perrin, D. (1984). Chromaffin granule membrane-F-actin interactions and spectrin-like protein of subcellular organelles: A possible relationship. J. Neurochem. 42, 1558–1569.

Bacallao, R., Antony, C., Karsenti, E., Stelzer, E., & Simons, K. (1989). The subcellular organization of Madin-Darby canine kidney cells during the formation of a polarized epithelium. J. Cell Biol. 109, 2817–2832.

Balcarova, S.J., Pfeiffer, S.E., Fuller, S.D., & Simons, K. (1984). Development of cell surface polarity in the epithelial Madin-Darby canine kidney (MDCK) cell line. EMBO J. 3, 2687–2694.

Bansal, A. & Gerisch, L.M. (1991). The NPXY internalization signal of the LDL receptor adopts a reverse-turn conformation. Cell 67, 1195–1201.

Bartles, J.R. & Hubbard, A.L. (1988). Plasma membrane protein sorting in epithelial cells. Do secretory pathways hold the key? Trends biochem. Sci. 13, 181–184.

Bartles, R.T., Feracci, H.M., Steiger, B., & Hubbard, A.L. (1987). Biogenesis of the rat hepatocyte plasma membrane in vivo: Comparison of the pathways taken by apical and basolateral proteins using subcellular fractionation. J. Cell Biol. 105, 1241–1251.

Bennett, V. (1978). Purification of an active proteolytic fragment of the membrane attachment site for human erythrocyte spectrin. J. Biol. Chem. 253, 2292–2299.

Bennett, V. (1990a). Spectrin-based membrane skeleton: A multipotential adaptor between plasma membrane and cytoplasm. Physiol. Rev. 70, 1029–1065.

Bennett, V. (1990b). Spectrin: A structural mediator between diverse plasma membrane proteins and the cytoplasm. Curr. Op. Cell Biol. 2, 51–56.

Bennett, V., Otto, E., Kunimoto, M., Kordeli, E., & Lambert, S. (1991). Diversity of ankyrins in the brain. Biochem. Soc. Trans. 19, 1034–1039.

Bennett, V. & Stenbuck, P.J. (1979). Identification and partial purification of ankyrin, the high affinity membrane attachment site for human erythrocyte spectrin. J. Biol. Chem. 254, 2533–2541.

Bennett, V. & Stenbuck, P.J. (1980). Association between ankyrin and the cytoplasmic domain of band 3 isolated from human erythrocyte membranes. J. Biol. Chem. 255, 6424–6432.

Black, J.D., Koury, S.K., Bankert, R.B., & Repasky, E.A. (1988). Heterogeneity in lymphocyte spectrin distribution: Ultra-structural identification of a new spectrin-rich cytoplasmic structure. J. Cell Biol. 106, 97–109.

Bloch, R.J. &. Morrow, J.S. (1989). An unusual β-spectrin associated with clustered acetylcholine receptors. J. Cell Biol. 108, 481–493.

Bomsel, M., Parton, R., Kuznetsov, S.E., Schroer, T.A., & Gruenberg, J. (1990). Microtubule- and motor-dependent fusion in vitro between apical and basolateral endocytic vesicles from MDCK cells. Cell 62, 719–731.

Breitfeld, P.P., McKinnon, W.C., & Mostov, K.E. (1990). Effect of nocodazole on vesicular traffic to the apical and basolateral surfaces of polarized MDCK cells. J. Cell Biol. 111, 2365–2373.

Brewer, C.B. & Roth, M.B. (1991). A single amino acid change in the cytoplasmic domain alters the polarized delivery of influenza virus hemagglutinin. J. Cell Biol. 114, 413–421.

Brown, D.A., Crise, B., & Rose, J.K. (1989). Mechanism of membrane anchoring affects polarized expression of two proteins in MDCK cells. Science (Wash.) 245, 1499–1501.

Brown, D.A. & Rose, J.K. (1992). Sorting of GPI-anchored proteins to glycolipid-enriched membrane subdomains during transport to the apical cell surface. Cell 68, 533–544.

Burridge, K., Fath, K., Kelly, T., Nuckolls, G., & Turner, C. (1988). Focal adhesion: Transmembrane junctions between the extracellular matrix and the cytoskeleton. Annu. Rev. Cell Biol. 4, 487–525.

Casanova, J.E., Apodaca, G., & Mostov, K. (1991). An autonomous signal for basolateral sorting in the cytoplasmic domain of the polymeric immunoglobulin receptor. Cell 66, 65–75.

Cohn, Z.A. & Steinman, R.M. (1982). Phagocytosis and fluid-phase pinocytosis in membrane recycling. In: Ciba Foundation Symposium 92, (Evered, D. & Collims, G., eds) pp. 15–33, Pitman Book Ltd., London.

Coleman, J.R., Fishkind, D.J., Mooseker, M.S., & Morrow, J.S. (1989). Functional diversity among spectrin isoforms. Cell. Motil. Cytoskel. 12, 225–247.

Davis, J. & Bennett, V. (1990). The anion exchanger and Na^+,K^+,ATPase interact with distinct sites on ankyrin in in vitro assays. J. Biol. Chem. 265, 17252–17256.

Davis, J., Davis, L., & Bennett, V. (1989). Diversity in membrane binding sites of ankyrins. J. Biol. Chem. 264, 6417–6426.

DeCesaris, P., Filipini, A., Stefanini, M., & Ziparo, E. (1989). Spectrin, fodrin and protein 4.1-like proteins in differentiating rat germ cells. Differentiation. 41, 216–222.

Doherty, P., Rowett, L.H., Moore, S.E., Mann, D.A., & Walsh, F.S. (1991). Neurite outgrowth in response to transfected N-CAM and N-cadherin reveals fundamental differences in neuronal responsiveness to CAMs. Neuron 6, 247–58.

Drenckhahn, D., Schulter, K., Allen, D., & Bennett, V. (1985). Colocalization of band 3 with ankyrin and spectrin at the basal membrane of intercalated cells in the rat kidney. Science (Wash.) 230, 1287–1289.

Eberle, W., Sander, C., Klaus, W., Schmidt, B., von Figura, K., & Peters, C. (1991). The essential tyrosine of the internalization signal in lysosomal acid phosphatase is part of a β turn. Cell 67, 1203–1209.

Eder, P., Soong, C., & Tao, M. (1986). Phosphorylation reduces the affinity of protein 4.1 for spectrin. Biochemistry 25, 1764–1770.

Ekblom, P. (1989). Developmentally regulated conversion of mesenchyme to epithelium. FASEB J. 3, 2141–2150.

Ekblom, P., Vestweber, D., & Kemler, R. (1986). Cell-matrix interactions and cell adhesion during development. Annu. Rev. Cell Biol. 2, 27–47.

Fath, K.R. & Burgess, D.R. (1993). Golgi-derived vesicles from developing epithelial cells bind actin filaments and possess myosin-I as a cytoplasmically oriented peripheral membrane protein. J. Cell Biol. 120, 117–127.

Fey, E.G., Wan, K., & Penman, S. (1984). The epithelial cytoskeletal framework and nucleus matrix-intermediate filament scaffold: Three-dimensional organization and protein composition. J. Cell Biol. 91, 1973–1984.

Frei, T., von Bohlen und Halbach, F., Wille, W., & Schachner, M. (1992). Different extracellular domains of the neural cell adhesion molecule (N-CAM) are involved in different functions. J. Cell Biol. 118, 177–194.

Gardner, K. & Bennett, V. (1987). Modulation of spectrin-actin assembly by erythrocyte adducin. Nature (Lond.) 328, 359–362.

Geiger, B. (1989). Cytoskeleton-associated cell contacts. Curr. Opin. Cell Biol. 1, 103–109.

Geiger, B., Ginsberg, D., Salomon, D., & Volberg, T. (1990). The molecular basis for the assembly and modulation of adherens-type junctions. Cell Diff. & Dev. 32, 343–353.

Gilbert, T., Lebivic, A., Quaroni, A., & Rodriguez-Boulan, E. (1991). Microtubular organization and its involvement in the biogenetic pathways of plasma membrane proteins in Caco-2 intestinal epithelial cells. J. Cell Biol. 113, 275–284.

Gravotta, D., Adesnik, M., & Sabatini, D.D. (1990). Transport of influenza virus HA from the trans Golgi network to the apical surface of MDCK cells permeabilized in their basal-lateral membrane: Energy dependence and involvement of GTP-binding proteins. J. Cell Biol. 111, 2893–2908.

Gregorio, C.C., Kubo, R.T., Bankert, R.B., & Repasky, E.Z. (1992). Translocation of spectrin and protein kinase C to a cytoplasmic aggregate upon lymphocyte activation. Proc. Natl. Acad. Sci. USA. 89, 4947–4951.

Gumbiner, B. (1990). Generation and maintenance of epithelial polarity. Curr. Opp. Cell Biol. 2, 881–887.

Gumbiner, B., Stevenson, B., & Grimaldi, A. (1988). The role of the cell adhesion molecule uvomorulin in the formation and maintenance of the epithelial junctional complex. J. Cell Biol. 107, 1575–1587.

Hammerton, R.W., Krzeminski, K.A., Mays, R.W., Ryan, T.A., Wollner, D.A., & Nelson, W.J. (1991). Mechanism for regulating cell surface distribution of Na^+,K^+-ATPase in polarized epithelial cells. Science (Wash.) 254, 847–850.

Hirano, S., Nose, A., Hatta, K., Kawakami, A., & Takeichi, M. (1987). Calcium-dependent cell-cell adhesion molecules (cadherins): Subclass specificities and possible involvement of actin bundles. J. Cell Biol. 105, 2501–2510.

Hirokawa, N., Cheney, R.W., & Willard, M. (1983). Location of a protein of the fodrin-spectrin-TW260/240 family in the mouse intestinal brush border. Cell 32, 953–965.

Hopkins, C.R. (1991). Polarity signals. Cell 66, 827–829.

Hunziker, W., Harter, C., Matter, K., & Mellman, I. (1991). Basolateral sorting in MDCK cells requires a distinct cytoplasmic domain determinant. Cell 66, 907–920.

Hunziker, W. & Mellman, I. (1992). Relationships between sorting in the exocytic and endocytic pathways of MDCK cells. Sem. Cell Biol. 2, 397–410.

Hunziker, W., Vale, P., & Mellman, I. (1990). Differential microtubule requirements for transcytosis in MDCK cells. EMBO J. 9, 3515–3525.

Itoh, M., Yonemura, S., Nagafuchi, A., Tsukita, S., & Tsukita, S. (1991). A 220-kD undercoat-constitutive protein: Its specific localization at cadherin-based cell-cell adhesion sites. J. Cell Biol. 115, 1449–1462.

Kemler, R. (1992). Classical cadherins. Sem. Cell Biol. 3, 149–155.

Koob, R., Zimmerman, M., Schoner, W., & Drenkhahn, D. (1987). Co-localization and co-precipitation of ankyrin and Na^+,K^+-ATPase in kidney epithelial cells. Eur. J. Cell Biol. 45, 230–237.

Kordeli, E. & Bennett, V. (1991). Distinct ankyrin isoforms of neuronal cell bodies and nodes of Ranvier resolved using erythrocyte ankyrin-deficient mice. J. Cell Biol. 114, 1243–1259.

Lazarides, E., Nelson, W.J., & Kasamatsu, T. (1984). Segregation of two spectrin forms in the chicken optic system: A mechanism for establishing restricted membrane-cytoskeletal domains in neurons. Cell 36, 269–278.

Lazarides, E. & Nelson, W.J. (1983). Expression of the erythrocyte form of spectrin in the cerebellum: Existence of distinct membrane-cytoskeleton domains in neurons. Science (Wash.) 220, 1295–1296.

Le Bivic, A., Sambuy, Y., Patak, A., Patil, M., Chao, M., & Rodriguez-Boulan, R. (1991). An internal deletion in the cytoplasmic tail reverses the apical localization of human NGF receptor in transfected MDCK cells. J. Cell Biol. 115, 607–618.

Le Bivic, A., Quaroni, A., Nichols, B., & Rodriguez-Boulan, E. (1990). Biogenetic pathways of plasma membrane proteins in Caco-2, a human intestinal line. J. Cell Biol. 111, 1351–1361.

Lee, J.D., Black, J.D., Repasky, E.A., Kubo, R.T., & Bankert, R.B. (1988). Activation induces a rapid reorganization of spectrin in lymphocytes. Cell 55, 807–816.

Ling, E., Danilov, Y., & Cohen, C. (1988). Modulation of red cell band 4.1 function by cAMP-dependent kinase and protein kinase C phosphorylation. J. Biol. Chem. 263, 2209–2216.

Lisanti, M.P., Caras, I.W., Davitz, M.A., & Rodriguez, B.E. (1989). A glycophospholipid membrane anchor acts as an apical targeting signal in polarized epithelial cells. J. Cell Biol. 109, 2145–2156.

Lu, P.W., Soong, C.J., & Tao, M. (1985). Phosphorylation of ankyrin decreases its affinity for spectrin tetramer. J. Biol. Chem. 260, 14958–14964.

Lu, P.W. & Tao, M. (1986). Phosphorylation of protein tyrosine by human erythrocyte casein kinase A. Biochem. Biophys. Res. Commun. 139, 855–860.

Matlin, K.S. & Simons, K. (1984). Sorting of an apical plasma membrane glycoprotein occurs before it reaches the cell surface in cultured epithelial cells. J. Cell Biol. 99, 2131–2139.

Matter, K., Brauchbar, M., Bucher, K., & Hauri, H.-P. (1990). Sorting of endogenous plasma membrane proteins occurs from two sites in cultured human intestinal epithelial cells (Caco-2). Cell 60, 429–437.

Mayer, A., Ivanov, I., Adesnik, M., & Sabatini, D.D. (1992). Cell-free reconstitution of the polarized transport of a viral glycoprotein from the TGN to the basal-lateral membrane in MDCK cells. Mol. Biol. Cell 3, 307a.

McNeill, H., Ozawa, M., Kemler, R., & Nelson, W.J. (1990). Novel function of the cell adhesion molecule uvomorulin as an inducer of cell surface polarity. Cell 62, 309–316.

McNeill, H., Ryan, T., Smith, S.J., & Nelson, W.J. (1993). Spatial and temporal dissection of immediate and early events following cadherin-mediated epithelial cell adhesion. J. Cell Biol. (in press).

Mische, S., Mooseker, M.S., & Morrow, J.S. (1987). Erythrocyte adducin: A calmodulin-regulated actin-bundling protein that stimulates spectrin-actin binding. J. Cell Biol. 105, 2837–2849.

Molitoris, B. & Nelson, W.J. (1990). Alterations in the establishment and maintenance of epithelial cell polarity as a basis for disease processes. J. Clin. Invest. 85, 1–11.

Mooseker, M.S. (1985). Organization, chemistry, and assembly of the cytoskeletal apparatus of the intestinal brush border. Annu. Rev. Cell Biol. 1, 209–242.

Mooseker, M.S. & Coleman, T.R. (1989). The 100-kd protein-calmodulin complex of the intestinal microvillus (brush border myosin I) is a mechanoenzyme. J. Cell Biol. 108, 2395–2400.

Morrow, J.S., Cianci, C.D., Ardito, T., Mann, A.S., & Kashgarian, M. (1989). Ankyrin links fodrin to the alpha subunit of Na^+,K^+,ATPase in Madin-Darby canine kidney cells and in intact renal tubule cells. J. Cell Biol. 108, 455–465.

Mostov, K., Apodaca, G., Aroeti, B., & Okamoto, C. (1992). Plasma membrane protein sorting in polarized epithelial cells. J. Cell Biol. 116, 577–583.

Nagafuchi, A., Shirayoshi, Y., Okazaki, K., Yasuda, K., & Takeichi, M. (1987). Transformation of cell adhesion properties by exogenously introduced E-cadherin cDNA. Nature (Lond.) 329, 340–343.

Nagafuchi, A. & Takeichi, M. (1988). Cell binding function of E-cadherin is regulated by the cytoplasmic domain. EMBO J. 7, 3679–3684.

Nagafuchi, A. & Takeichi, M. (1989). Transmembrane control of cadherin-mediated cell adhesion: A 94kd protein functionally associated with a specific region of the cytoplasmic domain of E-cadherin. Cell Regulation 1, 37–44.

Nelson, W.J. (1991). Cytoskeleton functions in membrane traffic in polarized epithelial cells. Sem. Cell Biol. 2, 375–386.

Nelson, W.J. (1992). Regulation of cell surface polarity from bacteria to mammals. Science (Wash.) 258, 948–955.

Nelson, W.J. & Hammerton, R.W. (1989). A membrane-cytoskeletal complex containing Na^+,K^+-ATPase, ankyrin, and fodrin in Madin-Darby canine kidney (MDCK) cells: Implications for the biogenesis of epithelial cell polarity. J. Cell Biol. 108, 893–902.

Nelson, W.J., Hammerton, R.W., Wang, A.Z., & Shore, E.M. (1990a). Involvement of the membrane-cytoskeleton in the development of epithelial cell polarity. Sem. Cell Biol. 1, 359–371.

Nelson, W.J., Shore, E.M., Wang, A.Z., & Hammerton, R.W. (1990b). Identification of a membrane-cytoskeletal complex containing the cell adhesion molecule uvomorulin (E-cadherin), ankyrin, and fodrin in Madin-Darby canine kidney epithelial cells. J. Cell Biol. 110, 349–57.

Nelson, W.J. & Veshnock, P.J. (1986). Dynamics of membrane-skeleton (fodrin) organization during development of polarity in Madin-Darby canine kidney epithelial cells. J. Cell Biol. 103, 1751–65.

Nelson, W.J. & Veshnock, P.J. (1987). Ankyrin binding to Na^+,K^+-ATPase and implications for the organization of membrane domains in polarized cells. Nature (Lond.) 328, 533–536.

Ojakian, G.K. & Schwimmer, R. (1988). The polarized distribution of an apical cell surface glycoprotein is maintained by interactions with the cytoskeleton of Madin-Darby canine kidney cells. J. Cell Biol. 107, 2377–2388.

Ozawa, M., Baribault, H., & Kemler, R. (1989). The cytoplasmic domain of the cell adhesion molecule uvomorulin associates with three independent proteins structurally related in different species. EMBO J. 8, 1711–1717.

Ozawa, M., Ringwald, M., & Kemler, R. (1990). Uvomorulin-catenin complex formation is regulated by a specific domain in the cytoplasmic region of the cell adhesion molecule. Proc. Natl. Acad. Sci. USA 87, 4246–50.

Palfrey, H.C. & Waseem, A. (1985). Protein kinase C in the human erythrocyte. Translocation to the plasma membrane and phosphorylation of bands 4.1 and 4.9 and other membrane proteins. J. Biol. Chem. 260, 16021–16029.

Parczyk, K., Hasse, W., & Kondor, K.C. (1989). Microtubules are involved in the secretion of proteins at the apical cell surface of the polarized epithelial cell, MDCK. J. Biol. Chem. 264, 16837–16846.

Pasdar, M. & Nelson, W.J. (1988a). Kinetics of desmosome assembly in Madin-Darby canine kidney epithelial cells: Temporal and spatial regulation of desmoplakin organization and stabilization upon cell-cell contact. I. Biochemical Analysis. J. Cell Biol. 106, 677–685.

Pasdar, M. & Nelson, W.J. (1988b). Kinetics of desmosome assembly in Madin-Darby canine kidney epithelial cells: Temporal and spatial regulation of desmoplakin organization and stabilization upon cell-cell contact. II Morphological analysis. J. Cell Biol. 106, 687–695.

Pearce, B.M.F. & Robinson, M. (1990). Clathrin, adaptors, and sorting. Annu. Rev. Cell Biol. 6, 151–171.

Perrin, D., Langley, O.K., & Aunis, D. (1987). Anti-α-fodrin inhibits secretion from permeabilized chromaffin cells. Nature (Lond.) 326, 498–501.

Perrin, D., Moller, K., Hanke, K., & Soling, H.-D. (1992). cAMP and Ca^{2+}-mediated secretion in parotid acinar cells is associated with reversible changes in the organization of the cytoskeleton. J. Cell Biol. 116, 127–134.

Perrin, D. &. Aunis, D. (1985). Reorganization of α-fodrin induced by stimulation in secretory cells. Nature (Lond.) 315, 589–591.

Rindler, M.J., Ivanov, I.E., & Sabatini, D.D. (1987). Microtubule-acting drugs lead to the nonpolarized delivery of the influenza hemagglutinin to the cell surface of polarized Madin-Darby canine kidney cells. J. Cell Biol. 104, 231–241.

Rindler, M.J., Ivanov, I.E., Plesken, H., & Sabatini, D.D. (1985). Polarized delivery of viral glycoproteins to the apical and basolateral membranes of Madin-Darby canine kidney cells infected with temperature sensitive viruses. J. Cell Biol. 104, 231–252.

Rindler, M.J., Ivanov, I.E., Plesken, M., Rodriguez-Boulan, E., & Sabatini, D.D. (1984). Viral glycoproteins destined for apical or basolateral plasma membrane domains traverse the same Golgi apparatus during their intracellular transport in doubly infected Madin-Darby canine kidney cells. J. Cell Biol. 98, 1304–1319.

Roberts, S.J., Leaf, D.S., Moore, H.-P., & Gerhart, J.C. (1992). The establishment of polarized membrane traffic in *Xenopus laevis* embryos. J. Cell Biol. 118, 1359–1370.

Rodriguez-Boulan, E. & Nelson, W.J. (1989). Morphogenesis of the polarized epithelial cell phenotype. Science (Wash.) 245, 718–725.

Rodriguez-Boulan, E. & Sabatini, D.D. (1978). Asymmetric budding of viruses in epithelial monolayers: A model system for the study of epithelial cell polarity. Proc. Natl. Acad. Sci. USA 75, 5071–5075.

Salas, P.J., Misek, D.E., Vega, S.D., Gundersen, D., Cereijido, M., & Rodriguez, E. (1986). Microtubules and actin filaments are not critically involved in the biogenesis of epithelial cell surface polarity. J. Cell Biol. 102, 1853–1867.

Shattil, S.J. & Brugge, J.S. (1991). Protein tyrosine phosphorylation and the adhesive functions of platelets. Curr. Opp. Cell Biol. 3, 869–879.

Shotton, D.M., Burke, B.E., & Branton, D. (1979). The molecular structure of erythrocyte spectrin. J. Mol. Biol. 131, 303–329.

Simons, K. & Fuller, S.D. (1985). Cell surface polarity in epithelia. Ann. Rev. Cell Biol. 1, 243–288.

Simons, K. & Wandinger-Ness, N.A. (1990). Polarized sorting in epithelia. Cell 62, 207–210.

Smith, P.R., Saccomani, G., Joe, E.H., Angelides, K.J., & Benos, D.J. (1991). Amiloride-sensitive sodium channel is linked to the cytoskeleton in renal epithelial cells. Proc. Natl. Acad. Sci. USA 88, 6971–6975.

Srinivasan, Y., Elmer, L., Davis, J., Bennett, V., & Angelides, K. (1988). Ankyrin and spectrin associate with voltage-dependent sodium channels in brain. Nature (Lond.) 333, 177–180.

Staufenbiel, M. (1988). Fatty acids covalently bound to erythrocyte proteins undergo a differential turnover in vivo. J. Biol. Chem. 263, 13615–13622.

Staufenbiel, M. & Lazarides, E. (1986). Ankyrin is fatty acid acylated in erythrocytes. Proc. Natl. Acad. Sci. USA 83, 318–322.

Takeichi, M. (1991). Cadherin cell adhesion receptors as a morphogenetic regulator. Science (Wash.) 251, 1451–5.

Vallee, R.B. & Shpetner, H.S. (1990). Motor proteins of cytoplasmic microtubules. Annu. Rev. Biochem. 59, 909–932.

Van Der Sluijs, P., Bennett, M., Antony, C., Simons, K., & Kreis, T.E. (1990). Binding of exocytic vesicles from MDCK cells to microtubules in vitro. J. Cell Sci. 95, 545–553.

Van Zeijl, M.J.A.H. & Matlin, K.S. (1990). Microtubule perturbation inhibits intracellular transport of an apical membrane glycoprotein in a substrate-dependent manner in polarized Madin-Darby canine kidney epithelial cells. Cell Regulation 1, 921–936.

Wang, A.Z., Ojakian, G.K., & Nelson, W.J. (1990a). Steps in the morphogenesis of a polarized epithelium. I. Uncoupling the roles of cell-cell and cell substratum contact in establishing plasma membrane polarity in multicellular epithelial (MDCK) cysts. J. Cell Sci. 95, 137-151.

Wang, A.Z., Ojakian, G.K., & Nelson, W.J. (1990b). Steps in the morphogenesis of a polarized epithelium. II. Disassembly and assembly of plasma membrane domains during reversal of epithelial cell polarity in multicellular epithelial (MDCK) cysts. J. Cell Sci. 95, 153-165.

Weaver, D.C., Pasternack, G.R., & Marchesi, V.T. (1984). The structural basis of ankyrin function. J. Biol. Chem. 259, 6170–6175.

Wollner, D.A., Krzeminski, K.A., & Nelson, W.J. (1992). Remodelling the cell surface distribution of membrane proteins during the development of epithelial cell polarity. J. Cell Biol. 116, 889–899.

Wollner, D.A. & Nelson, W.J. (1992). Establishing and maintaining epithelial cell polarity: Roles of protein sorting, delivery and retention. J. Cell Sci. 102, 185–190.
Wright, E.M. (1972). Mechanisms of ion transport across the choroid plexus. J. Physiol. 226, 545–571.
Zagon, I.S., Higbee, R., Riederer B.M., & Goodman, S.R. (1986). Spectrin subtypes in mammalian brain: An immunoelectron microscope study. J. Neurosci. 6, 2977–2986.

DEVELOPMENT OF THE AXONAL AND DENDRITIC CYTOSKELETON

Beat M. Riederer

Advances in Molecular and Cell Biology
Volume 12, pages 107–142.

ISBN: 1-55938-845-5

I. HISTORY

In the 18th and 19th centuries, neuroscientists such as Remak, Purkinje, and Valentin had already observed that a nerve cell consisted of a cell body and several emerging neurites, of which the axon was morphologically distinct from the dendrites (for review see Van der Loos, 1967). Purkinje (1838) was the first to postulate that the soma is the site of generation of energy, and that the nerve fibers are conductors of energy; in a sense this was the first notion of neuronal polarity. In 1865, Deiters pointed out that axons must be different from the cell body and dendrites. These early studies were based solely on morphological observations, but they all suggested a difference between axons and the somato-dendritic part of neurons. However, the physiological and molecular basis for such differences has only been discovered in the last 20 years. It is clear that the cytoskeleton is among the structures responsible for the differences in cellular form, structure and function that occur between axons and dendrites. Indeed, the existence of a cellular skeleton was first mentioned by Schultze (1878) who observed an organized and fibrous array in the cytoplasm of ganglion cells. With the use of modern immunocytochemical and molecular techniques, a more precise analysis of the axons and dendrites, and of their subcellular regions, has been possible, and it has become obvious that the different parts within a neuron have even distinct micro-domains which differ in their cytoskeletal composition (Matus et al., 1983; Fath and Lasek, 1988; Nixon, 1991).

II. THE CYTOSKELETAL STRUCTURES

The cytoskeleton is essential for the organization and maintenance of neuronal shape and function. Its name does not do justice to the dynamic functions it is involved in, including structural growth and plasticity, transport functions, regional specialization, intercellular communication, and cell division. It is beyond the scope of this essay to discuss this matter in depth and the reader is referred to the many reviews available for further information (Nunez, 1986; Olmsted, 1986; Cronly-Dillon and Nona, 1988; Filiatreau, 1988; Goodman et al., 1988; Matus, 1988a; Mitchison and Kirschner, 1988; Robinson and Anderson, 1988; Riederer, 1990; Tucker, 1990; Burgoyne, 1991; Portier, 1992). The neuronal cytoskeleton consists of several morphologically well-defined structures: In the cytoplasm, one finds microtubules, neurofilaments, and microfilaments; and at the cytoplasmic surface of plasma and organelle membranes a dense filamentous meshwork of spectrin and actin oligofilaments. Furthermore, several associated proteins act as a link between membranes and the cytoplasmic cytoskeleton (Goodman et al., 1988, Bennett, 1990). In the following sections, typical features of the axonal or dendritic cytoskeleton, their molecular and functional differences, and their potential role during dendritic or axonal development will be discussed. Proteins of interest are summarized in Table 1 (p. 110). It is essential to mention that several mechanisms may be involved in the establishment of neuronal polarity, especially post-translational modifications that could influence targeting of cytoskeletal proteins.

III. THE AXONAL CYTOSKELETON

The cytoskeletal composition of a growing axon is quite different from that of a mature one. During early development, an axon undergoes considerable growth, and depends on its motile elements. Once synaptic contacts have been established, the cytoskeletal scenario changes, and several elements involved in the juvenile plastic process are replaced by proteins that are essential for the stabilization of the axon, and are therefore considered adult-type proteins. Both types of cytoskeletal proteins will be discussed. Although no clear separation between juvenile and adult-type cytoskeletal proteins is made in the following sections, typical juvenile or adult-type axonal proteins will be pointed out. The transition from a juvenile to an adult set of cytoskeletal proteins occurs within a precise time schedule, e.g., in the rat between postnatal days 15 and 20 (Binder et al., 1984; Burgoyne and Cumming, 1984; Riederer and Matus, 1985), and in the cat between postnatal days 20 and 40 (Riederer et al., 1990; Riederer and Innocenti, 1991, 1992). One may also postulate that the cytoskeleton changes from a plastic to a more rigid structure; this may be an oversimplification, but it serves as a working hypothesis to explain the various changes in the cytoskeleton composition during the critical period of brain development. A change to a more adult-type cytoskeleton must favor the formation of synaptic structures and the maintenance of axonal diameter and function (for discussion see Matus, 1988a,b; Riederer, 1990).

A. Axon Growth and Axonal Transport

Over the last 20 years it has been shown that microtubules are important for the outgrowth of neurites (Seeds et al., 1970; Yamada et al., 1970). A growing nerve fiber consists of an axoplasm, a growth cone shaft, and a growth cone (Mitchison and Kirschner, 1988). An elementary question, which preoccupied researchers for years, was how and in what form cytoskeletal elements are transported in the growing neurite and how they participate in elongation and stabilization of an axon. It was first believed that constituents of the axonal cytoskeleton were synthesized and assembled in the cell body, and then pushed into the axoplasm of the axon shaft

Table 1. Cytoskeletal Proteins in Neurons: Their Distribution in Axons and Dendrites, and Their Molecular Differences

Protein	*kDa*	*Molecular Difference, Localization & Development*
α- & β- tubulin	54/56	Are ubiquitous proteins, which originate from different genes. They exist in several post-translational modifications. Axons contain detyrosinatied and acetylated tubulin.
MAP1b, MAP5, MAP1x, MAP1.2	320	Expressed by one gene. Are present as several isoforms early in development; in neuroblasts, axons, cell bodies, and dendrites. The axonal form is phosphorylated, but must contain another type of modification (MAP1x). Is also present in glia.
MAP2a, b	280/260	Encoded from one gene. Good marker for neuronal cell bodies and their dendrites. MAP2b is early present, while MAP2a appears late. In phosphorylated form they are less effective in microtubule assembly.
MAP2c	65	Early presence, in axons and in glia. It is a splice variant of MAP2b, and lacks a 200-kDa sequence. It exists in phosphorylated form. The glial form was termed MAP2d and contains a fourth repeat in the microtubule-binding domain.
tau	45–70	Encoded from one gene. Exists in juvenile and adult splice variants. Axonal and dendritic forms are differentially phosphorylated. In phosphorylated form they are less effective in microtubule assembly.
HMW-tau	110	A unique splice variant is found in DRG and peripheral axons.
actin	46	Is an ubiquitous protein, which is encoded by different genes. It is post-translationally modified by acetylation, methylation, and phosphorylation.
α- & β-brain spectrin	240/235	Encoded by different genes. The erythrocyte-related SpIΣ2 appears late, in somata and dendrites. The axonal form, SpIIΣ1, is present early. The role of phosphorylation is unknown.
NF-H	200	Encoded by one gene. It is the latest NF to appear. The axonal form is highly phosphorylated. It is essential for the formation of crossbridges between NFs.
NF-M & NF-L	160/68	Both are expressed early and are distributed in dendrites, cell bodies and perikarya. They form the core of NF. NF-M is able to crossbridge to other NFs. Both subunits exist in phosphorylated form.
peripherin	57	It is expressed by one gene, and typically found in peripheral axons and ganglia. It is phosphorylated near the N-terminal end.

Notes: For further details and references, see text.

and transported distally at a slow velocity of 0.2–4 mm/day, termed slow components a (SCa) and b (SCb) (Lasek and Hoffman, 1983, Lasek et al., 1984). It turned out not to be that simple. The majority of cytoskeletal elements are assembled soon after synthesis (Black et al., 1986), but some remained soluble or particulate and have the ability to move at higher velocity in the presence of ATP (Weisenberg et al., 1987). Tubulin and actin may also move at a faster than average transport rate in the axon, and integrate into the cytoskeleton at different points along the axon or in the growth cone neck (Bamburg et al., 1986; Okabe and Hirokawa, 1988, 1990). It has also been suggested that microtubules do not move in an extending neurite, and that tubulin exchanges locally between polymerized and soluble pools so adding to axonal growth. However, recent experiments using photobleaching techniques (Reinsch et al., 1991), have demonstrated that the movement of axonal microtubules is mediated by microtubule polymer sliding. In addition, it was postulated that tau proteins may stabilize axons by promoting microtubule bundling, and thus could mediate axon elongation and branching of the growth cone neck (Cleveland and Hoffman, 1991). It is interesting that transport of cytoskeletal components such as actin, tubulin, and neurofilaments occurs at higher velocity in immature axons than in mature ones (Willard and Simon, 1983; McQuarrie et al., 1989; Watson et al., 1989). A reasonable explanation for this phenomenon may be that an increase in structural cross-linking of cytoskeletal elements with maturation may be the cause for a slowing down of the axonal transport (for reviews of cytoskeletal interactions see Nixon, 1991; Riederer, 1990; Hirokawa, 1991). As already pointed out by Hollenbeck (1989), it remains to be defined how cytoskeletal proteins are modified and reduced in their ability to assemble into the cytoskeleton, and thus move faster in the axon. In consequence, there must be a local difference in enzymes between proximal and distal, as well as central and cortical regions which may remodify cytoskeletal proteins along the axon to favor their incorporation into the cytoskeleton.

B. The Versatility of Tubulins

Alpha- and β-tubulins are the major constituents of microtubules and are found in all eukaryotic cells. They arise from many genes (Cleveland, 1987), and furthermore undergo post-translational modifications such as acetylation, detyrosination, phosphorylation, and glutamylation, resulting in more than 20 tubulin isoforms (Binet and Meininger, 1988; Edde et al., 1990). Such modifications may modulate the behavior of tubulin to polymerize into microtubules (Hollenbeck, 1989), and post-translational modifications seem to be the basis for the versatility of tubulin isoform function. Interestingly, phosphorylation of tubulin enhances its interaction with membranes (Hargreaves et al., 1986), but it remains to be seen to which membrane protein this tubulin is binding, and perhaps is linking microtubules. Several monoclonal antibodies have been used to identify a subset of tubulin isoforms in axons; it has been shown that acetylation and detyrosination may play a role in the maintenance of a stable microtubule population in the axon (Cumming et al., 1983; Cumming and Burgoyne, 1984a,b; Cambray-Deakin and Burgoyne, 1987), while tyrosination in the growth cone may destabilize microtubules (Robson and Burgoyne, 1989). These studies demonstrated that tubulins undergo development-dependent modifications in a selected set of axons. Thus, post-translational modifications may provide a tool for modulating microtubules in growth cone motility and axon extension (Hollenbeck, 1989). One could speculate that enzymes involved in the generation of modified tubulins must have a local distribution in the axons too, but we lack detailed information on their temporal occurrence and subcellular distribution during development.

C. Is Tau an Axonal Marker?

Other elements which may influence the stability, transport, and cross-linking of microtubules are microtubule-associated proteins (MAPs). In rodents, a single tau protein of 45 kDa was found in juvenile brain tissue while the number of isoforms, with a molecular weight between 48 and 70 kDa, increased up to 6 in adult tissue (Cleveland et al., 1977). Juvenile tau protein differs from adult forms in its ability to polymerize microtubules (Nunez, 1988; Goedert and Jakes, 1990). Tau proteins are expressed by only one gene, and the cause for the tau protein diversity is alternate splicing of its mRNA (for reviews see Lee, 1990; Goedert et al., 1991). Juvenile tau protein contains three imperfect repeat sequences in the microtubule-binding domain, while adult tau proteins have four (Lee et al., 1989; Kosik et al., 1989), which could be a potential reason for the difference between juvenile and adult tau proteins to induce microtubule assembly. Furthermore, post-translational modifications such as phosphorylation reduced the ability of tau proteins to induce microtubule polymerization (Lindwal and Cole, 1984). Initial immunocytochemical localizations of tau proteins suggested that in the central nervous system these proteins occur exclusively in axons (Binder et al., 1985), and are prominent during the phase of axonal outgrowth (Brion et al., 1988). In axons of the peripheral nervous system tau occurred in two variants in immature dorsal root ganglia (DRG) as low and high molecular weight form (the latter form was called high M weight HMW or big tau), while only the HMW tau of approximately 110 kDa was present in adult DRG and peripheral nerves (Georgieff et al., 1991; Oblinger et al., 1991; Goedert et al., 1992). This HMW tau, sequenced in N115 murine neuroblastoma cell lines, has an additional sequence of 237 amino acids between exons 4 and 5, as well as a domain of 66 residues homologous to exon 6 of the bovine gene that is absent in bovine rat and mouse tau cDNAs (Couchie et al., 1992).

On first glance, tau proteins seem to be good axonal markers. However, several studies indicate that tau is not located exclusively in the axon. When brain tissue was dephosphorylated with alkaline phosphatase and immunostained with TAU-1 monoclonal antibody, Papasozomenos and Binder (1987) observed immunostaining in perikarya and dendrites, as well as in glial cells. This result suggested that there are phosphorylation-dependent differences between axonal and dendritic tau proteins, and that phosphorylation hides the TAU-1 epitope in dendrites. In visual cat cortex, two juvenile and two adult tau forms were identified with the same antibody that showed the same phosphorylation-dependent distribution (Riederer and Innocenti, 1991). The dendritic form was differentially phosphorylated and detectable with TAU-1 antibody only after dephosphorylation of cortical tissue. In addition, strong evidence was presented that the somato-dendritic tau proteins are abundant in early brain tissue and decrease with progressing maturation. Other results had already indicated that only a phosphorylated subclass of tau proteins was transported axonally in retinal ganglion cells (Tytell et al., 1984), and that tau may exist in differentially phosphorylated forms. It is not clear if phosphorylation is the cause for the difference in the axonal or dendritic distribution of tau proteins and it remains to be studied if phosphorylation of tau occurs in restricted regions.

Phosphorylation of tau proteins is one possible mechanism to reduce the potential of tau proteins to polymerize microtubules (Yamamoto et al., 1983; Lindwall and Cole, 1984), and thus to influence the stability of microtubules, or eventually the interaction of tau in crosslinking to other structural components. Two novel kinases have been described which phosphorylate tau and the heavy neurofilament subunits at the KSP sites (defined by lysine-serine-proline amino acid sequence) in the heavy phosphorylated C-terminal tail regions, and which are distinct from common kinases such as cAMP-dependent kinases, protein kinase C, or second messenger activated mechanisms (Roder and Ingram, 1991). This indicates that tau proteins, NF-H and NF-M, may underlie the same phosphorylation mechanisms; however, neither kinase is associated with the cytoskeleton.

The use of tau antisense oligonuclotides on neurite formation of cultured cerebellar neurons suggested that tau is important for the establishment of axonal growth but is of less influence for minor neurite or dendrite-like processes (Caceres et al., 1991). Sf9 nonneuronal cells of the moth ovary overexpressed tau proteins after an infection with a baculovirus containing a tau cDNA (Knops et al., 1991), and the transfection resulted in the appearance of long cellular processes. As to the mechanism of these effects, elegant experiments of Drubin and Kirschner (1986) pointed to tau having a microtubule-stabilizing effect. Tau proteins have also the potential to link and bundle microtubules (Hirokawa, 1991).

It is important to point out that, due to the diversity of tau expression and post-translational modifications, antibodies may recognize specific variants occurring in axons only, or in the cell body and dendrites as well, and tau proteins are by far and exclusively axonal markers. Furthermore, one can assume that tau is not only involved in the regulation of microtubule sliding or stability in neurons since tau proteins have been localized in glial cells by the use of antibodies (Couchie et al., 1985; Papsozomenos and Binder, 1987; Riederer and Innocenti, 1991). In addition, tau mRNA was found in glial cells by the use of specific ribosomal RNA (Kosik et al., 1989) confirming results on the protein level. Furthermore, tau has been demonstrated in nuclei of a human neuroblastoma cell line (Loomis et al.,

1990), and its presence was recently discovered in a monocyte lineage (Kim et al., 1991). In conclusion tau proteins may have a much broader distribution than initially thought.

D. Is MAP2c an Axonal and Dendritic Marker?

MAP2c (apparent molecular weight (M_r) of 65–70 kDa) is an isoform of MAP2. It was first known under the name "tau slow" because it migrated slower than tau proteins on SDS-electrophoresis gels (Francon et al., 1982). Later, it was found to be immunologically related to the high molecular weight MAP2 (Riederer and Matus, 1985; Garner et al., 1988a). MAP2c and the high Mr MAP2b form are transcribed possibly from a single gene and arise via alternative splicing of a primary MAP2 transcript (Garner and Matus, 1988). The MAP2c lacks the middle 1363 amino acid sequence, while N- and C-terminal ends are identical to MAP2b. The carboxy-terminus of MAP2 shows a high degree of similarity with the C-terminal end of tau, with three tandem repeats (Lewis et al., 1988; Papandrikopoulou et al., 1989; Kindler et al., 1990). Thus, MAP2c must have similar microtubule-binding properties as tau proteins. MAP2c was expressed during early rat and quail brain development (Riederer and Matus, 1985; Tucker et al., 1988b), but disappears before maturation of axons. It is enormously abundant at the time of axon formation, e.g., in the motor neurons of quail spinal cord at postnatal day P3 (Tucker et al., 1988a). The localization of MAP2c was defined by a subtraction method (for review see Tucker, 1990) using two different antibodies, one that recognized all three MAP2 forms (monoclone C, Huber and Matus, 1984), and another that reacted only with the high-M_r MAP2a and 2b, but not with MAP2c (monoclone AP14, Binder et al., 1986).

MAP2c is frequently found in optic nerve axons, in parallel fibers of the cerebellum, and in peripheral axons. In addition it has been found in some glial cells. The glial form was termed MAP2d and contains four tandem repeats in the microtubule binding domain (Doll et al., 1993). In quail cerebellum, MAP2c occurred as a phosphorylated and non-phosphorylated form (Tucker et al., 1988b), both of which disappeared by the ninth day after hatching. In cat brain, MAP2 was present in several phosphorylated forms (Riederer and Innocenti, 1992). During callosal and visual cortical development in the cat, MAP2c was initially present in several phosphorylated forms, but during postnatal maturation the phosphorylated forms disappeared earlier than the less or non-phosphorylated forms of MAP2c. Using a myc-tagged MAP2c its occurrence was shown also in cell bodies and dendrites (Meichsner et al., 1993), but it is not clear whether MAP2c phosphorylation differs between axons and dendrites.

Puzzling questions concern the role of MAP2c and the means by which phosphorylation changes its function. An interesting proposition was made by Kindler et al. (1990); since MAP2c lacks a considerable part of the middle sequence, one could predict that MAP2c prevents microtubules from interacting with the cytoskeleton, by occupying potential MAP2b or tau microtubule-binding sites. However, MAP2c does behave similarly to MAP2b in subcellular fractionation and MAP2c was found in detergent-insoluble cytoskeleton fractions, suggesting that it behaves like MAP2b in binding to microtubules and interacting with other structures (Riederer and Innocenti, 1992). Moreover, the restriction of MAP2b to the dendrites remains to be explained. A local difference in MAP2c and MAP2b could depend on differences in distribution of their encoding mRNA (Garner et al., 1988b; Tucker et al., 1989). Another explanation could be that MAP2c depends on different post-translational modifications, but because MAP2c has the same sequences as MAP2b, one may expect a similar phosphorylation pattern as MAP2b. Phosphorylation of MAP2c could add to the dynamic plasticity of immature axons. MAP2c has been studied with respect to its influence on kinesin-dependent microtubule-motility (Heins et al., 1991); on a kinesin-coated glass surface microtubules, together with MAP2, were repelled more than those microtubules containing MAP2c or tau proteins. Because kinesin is involved in anterograde axonal transport the presence of MAP2b (but not MAP2c) may regulate kinesin interaction between microtubules and organelles. This certainly adds to the ongoing debate as to whether MAPs act as cross-linkers or as spacers. Furthermore, one could speculate that only the full length MAP2, but not MAP2c, could have an influence on the kinesin-dependent vesicle transport along microtubules.

E. Phosphorylated MAP5 Is Essential for Axonal Growth

Another prominent MAP during brain development is MAP5 (Riederer et al., 1986a), an important cytoskeletal protein of 320 kDa, also known under the names MAP1b (Bloom et al., 1985), MAP1x (Calvert and Anderton, 1985), or MAP1.2 (Greene et al., 1983). It is often referred to that these proteins are identical, however, it has not been shown that they all have the same primary sequence (Table 1). Furthermore, biochemical and immunocytochemical results suggest that it may be a group of isoforms, differing by post-translational modifications, as will be discussed below. Therefore, the names of each isoform will be maintained throughout this review.

MAP5 has been found to be abundant in embryonic brain tissue, and to be present in neuroblasts (Riederer et al., 1986a); it is an essential protein during axonal growth, and one of the first MAPs to appear during axonal outgrowth in motor neurons (Tucker et al., 1988a), in cat cerebellum (Figure 1), or in cultured PC12 (Brugg and Matus, 1988). MAP1b behaved similarly in neuroblastoma cells during neurite formation (Avila et al., 1988). Furthermore, antisense oligonucleotides directed against MAP5 inhibited process formation in nerve growth factor induced PC12 cells (Reddy et al., 1991), which corroborates the importance of MAP5 for axonal growth. This protein is probably also involved in cellular plasticity during differentiation, because in olfactory receptor neurons, which still divide and differentiate in adult animals, MAP5 is expressed at high levels (Viereck et al., 1989).

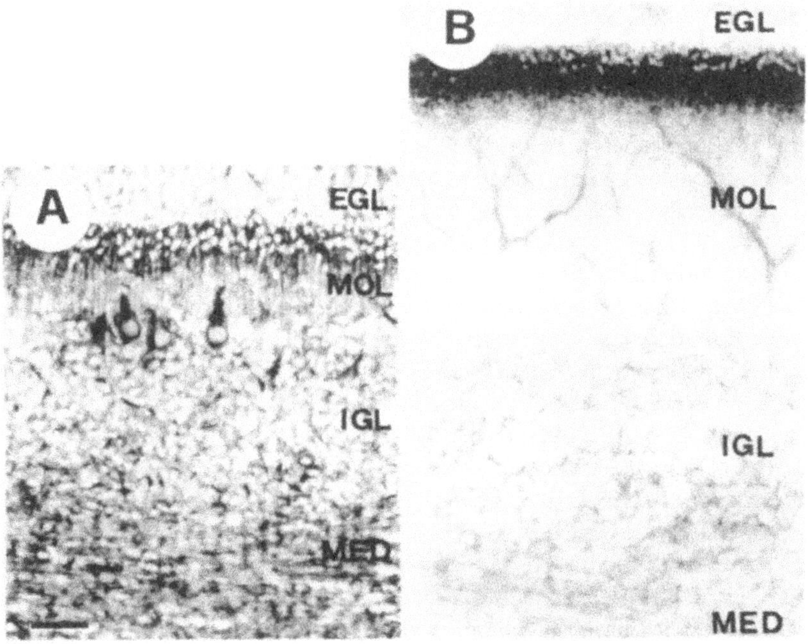

Figure 1. Immunocytochemical localization of MAP5 in cat P3 (**A**) and P28 cerebellum (**B**). Brain tissue was fixed with 4% paraformaldehyde and postfixed as described by Riederer et al. (1990). Cerebellum was cut with a vibratome in 50μm thick sections and immunostained with monoclonal antibody AA6, peroxidase conjugated secondary antibody, and chloronaphthol. Note that AA6 stains MAP5 in axons and dendrites at both ages. At P28 intense axonal staining is seen beneath the EGL. EGL = external germinal layer; MOL = molecular layer; IGL = internal granular layer; MED = medullary layer. Bar = 50 μm.

MAP1b consists of 2464 amino acids (Noble et al., 1989), and contains a repeated sequence motif in the N-terminal binding domain, unrelated to tau or MAP2. It is encoded as polyprotein and associates with a light chain to form a complex of proteins which is possibly involved in the binding to microtubules (Hammarback et al., 1991). MAP1b and MAP5 are present in different phosphorylated forms, the phosphorylated form being typically found in immature axons (Sato-Yoshitake et al., 1989; Viereck and Matus, 1989; Fischer and Romano-Clarke, 1990; Riederer et al., 1990). During postnatal development the degree of phosphorylation decreases with maturation. On immunoblots, MAP5 of cat visual cortex was identified as two readily distinguishable proteins; MAP5a, the phosphorylated form, occurred in axons of the corpus callosum and cortex, while the dendritic and non-phosphorylated form, MAP5b, was prominent only in the visual cortex but absent from corpus callosum (Riederer et al., 1990). MAP5a was found associated with structural components while MAP5b was present in the soluble fraction after high speed centrifugation (Riederer et al., 1990). Thus, phosphorylation of MAP5 may be an important modification to change its function, especially for the early neuronal plasticity during axonal outgrowth and elongation. It has been suggested that MAP1b is involved in cross-bridging of microtubules (Sato-Yoshitake et al., 1989). It remains to be seen if phosphorylation of MAP1b changes its ability to induce polymerization of microtubules, or its ability to cross-link or space microtubules and organelles. A casein kinase II-like activity could be responsible for MAP1b phosphorylation because identical phosphorylation patterns have been found *in vitro* and *in vivo* (Avila et al., 1988). In cultured PC12 cells, nerve growth factor (NGF) was found essential to induce phosphorylation of MAP1.2 with incorporation of up to 4.5 moles phosphate per mole protein (Aletta et al., 1988). Furthermore a monoclonal antibody against a phosphorylated epitope reacted with MAP1a, MAP1b, NF-H, and NF-M (Luca et al., 1986). This may point to a common phosphorylation mechanism, at least for one epitope in these proteins, and may indicate some structural homology as well as a common phosphorylation-dependent function.

It is interesting that MAP1x was found exclusively in developing axons (Calvert et al., 1987), but not in regenerating ones (Woodhams et al., 1989). This is distinct from neurofilaments and tubulins, which recapitulate the developmental program during axonal regeneration (Hoffman and Cleveland, 1988). MAP5 and MAP1x were shown to be immunologically related (Garner et al., 1989), however, the difference in immunocytochemical distribution of MAP1x and MAP5 is not due to phosphorylation. It must be the result of another modification, or of specific conformational differences. Recent results on rat recombinant MAP1b demonstrated that the MAP5 antibody (clone AA6) reacted with the one microtubule binding site closer to the N-terminal end, while MAP1b and MAP1x antibodies detected a sequence between the two microtubule binding sites (Zauner et al., 1992). One can assume that MAP1b consists of a group of highly related proteins that differ in their post-translational modifications. This may explain why some antibodies react with MAP1b in glial cells (Bloom et al., 1985), while others do not (Riederer et al., 1986a, 1990; Calvert et al., 1987), or why MAP1x is only found in axons (Calvert et al., 1987). It should be pointed out that MAP1b is structurally and functionally related to MAP1a (Schoenfeld et al., 1989). In contrast to MAP1b, MAP1a is a protein that appears late during development (Riederer and Matus, 1985). In addition, MAP1a is expressed by another gene than MAP1b (Garner et al., 1990). It still remains to be defined what the reasons are for the differences in the immunohistochemical staining between the different monoclonal antibodies. This requires a molecular identification of the sites of modification, the type of modification, the cell specificity, and the functional role of these modifications.

F. Actin and Microfilaments: Is There a Differential Distribution Within Axons?

Actin is an ubiquitous protein found in all eukaryotic cells (Table 1). It is one of the prominent proteins found in slow components a and b of axonal transport. Localization studies of actin initially suggested that actin was localized in the cortical region within an axon rather than in the central region (Spooner and Holladay, 1981; Letourneau, 1983). However, it turned out that much more actin is present in the central regions of axons, and the regional differences reported previously are due to the solubility of actin (Brady et al., 1984; Heriot et al., 1985). Nevertheless, it was shown that actin exists either as monomers or as filaments (Fath and Lasek, 1988). The role of axonal actin is unclear, but, given its motile character, one can assume roles in local movement, shape changes, and organellar transport (Brady et al., 1984).

The growth cone itself is characterized by a massive concentration of actin and soluble tubulin, and by high motility of cone tip and filopodia (Sobue and Kanda, 1989; Gordon-Weeks and Mansfied, 1992). During growth cone movement and changes in shape, actin undergoes dynamic reorganization (Okabe and Hirokawa, 1991; Bamburg et al., 1992). Furthermore, an essential requirement for locomotion and stability of the cone are the actin-binding proteins of the cortical cytoskeleton (Cunningham et al., 1992). These proteins may be involved in the cross-linking of filaments (for reviews see Bamburg and Bernstein, 1991; Hartwig and Kwiatkowski, 1991; Matsudaira, 1991). An interesting observation suggests that actin microfilaments in filopodia are capable of capturing microtubules (Gordon-Weeks, 1991). In consequence, it seems interesting to speculate whether MAPs such as MAP1b, MAP2, or tau proteins may be involved in such processes, and whether they participate in the directional growth of filopodia.

G. Brain Spectrin Isoforms and Their Nomenclature

Brain spectrin is the major component of the membrane-related cytoskeleton and links the cytoplasmic cytoskeleton to the plasma membrane. Brain spectrin is composed of two subunits of 240 and 235 kDa, which form a heterotetramer and link to actin (for review see Goodman et al., 1988; Bennett, 1990). Typical features of all spectrins are their 107 amino acid repeat sequences and the actin-binding sequence (Speicher and Marchesi, 1984; Speicher, 1986; for review see Dhermy, 1991). Interestingly, these features have also been found in other proteins such as α-actinin and dystrophin; therefore, spectrin is thought to be one member of a superfamily consisting of many proteins.

Brain spectrin, also named fodrin, was first demonstrated as a component of slow axonally transported proteins (Levine and Willard, 1981). Several studies have shown that spectrins exist in several isoforms in mammalian and avian brain and differ in their temporal and subcellular expression during development (Lazarides and Nelson, 1983a,b; Riederer et al., 1986b, 1987). In mammalian neurons, one isoform, brain spectrin (240/235), is expressed early in germinal cells and undifferentiated neurons, and located prominently in mature axons. The other form, brain spectrin (240/235E), appears during neuronal differentiation in perikarya and dendrites. It is likely that the presence of brain spectrin (240/235) in immature cells is essential for the formation of an early membrane-related cytoskeleton (Barakat-Walter and Riederer, 1991). The 235E subunit, which is immunologically related to red blood cell (rbc) spectrin, was found to be 59% and 66% homologous at the protein and nucleotide level respectively to the rbc β spectrin (Zimmer et al., 1992; Ma et al., 1993). A large number of names like brain-actin-binding protein, calmodulin-binding protein I, fodrin, or brain spectrin (240/235) and (240/235E) were given to this group of proteins (Goodman et al., 1988). This was useful for the structural, morphological, and functional characterization, but based on the molecular properties, a new nomenclature was recently proposed (for details see Zimmer et al., 1992; Winkelmann and Forget, 1993): erythroid α-spectrin whose gene is found on mouse and human chromosome 1 is named α-SpI; nonerythroid α-spectrin with its gene on human chromosomes 2 and 9 is named α-SpII; erythroid β spectrin whose gene is located on mouse chromosome 12 and human chromosome 14 will be referred to as β-SpI; and nonerythroid β-spectrin with genes found on mouse chromosome 11 and human chromosome 2 will be called β-SpII. The multiplicity of spliced transcripts from single genes is designated by Arabic numbers after the symbol Σ to denote subtypes, or by an asterisk * for unknown subtypes. Thus, erythrocyte β-spectrin is β-SpIΣ1, the β-subunit of brain spectrin (240/235E) is referred to as β-SpIΣ2, while the α- and β-subunits of brain spectrin (240/235) are called α-SpIIΣ1 and β-SpIIΣ1 respectively. This nomenclature should simplify the confusing current terminology.

Observations at the electron microscopic level indicate that both isoforms are situated at the plasma membrane, and are associated with the different organellar membranes, or microtubules, in axons and in dendrites (Zagon et al., 1986; Shimo-Oka and Atsumi, 1986; and Figure 2). The differential subcellular distribution also suggests that the function of the spectrin isoforms is more diverse and compartment specific. For one, spectrins may be involved in membrane plasticity and in controlling lateral mobility of membrane-bound proteins. Furthermore SpIIΣ1 may be involved in synaptic transmission while SpIΣ2 may influence receptors in dendritic spines (for review see Goodman et al., 1988).

Recent findings demonstrated that immunoreactivity of the axonal brain spectrin isoform α-SpIIΣ1 is changed in Alzheimer's disease (Masliah et al., 1990), and breakdown products increased with age in the telencephalon of mouse (Bahr et al., 1991). This suggests that brain spectrin metabolism undergoes age- and disease-dependent changes. Brain spectrin was a good marker to define the "well being" of cultured cells (Riederer et al., 1992) since fragile cells had less intact spectrin and a higher amount of breakdown products. It is known that the 240-kDa subunit

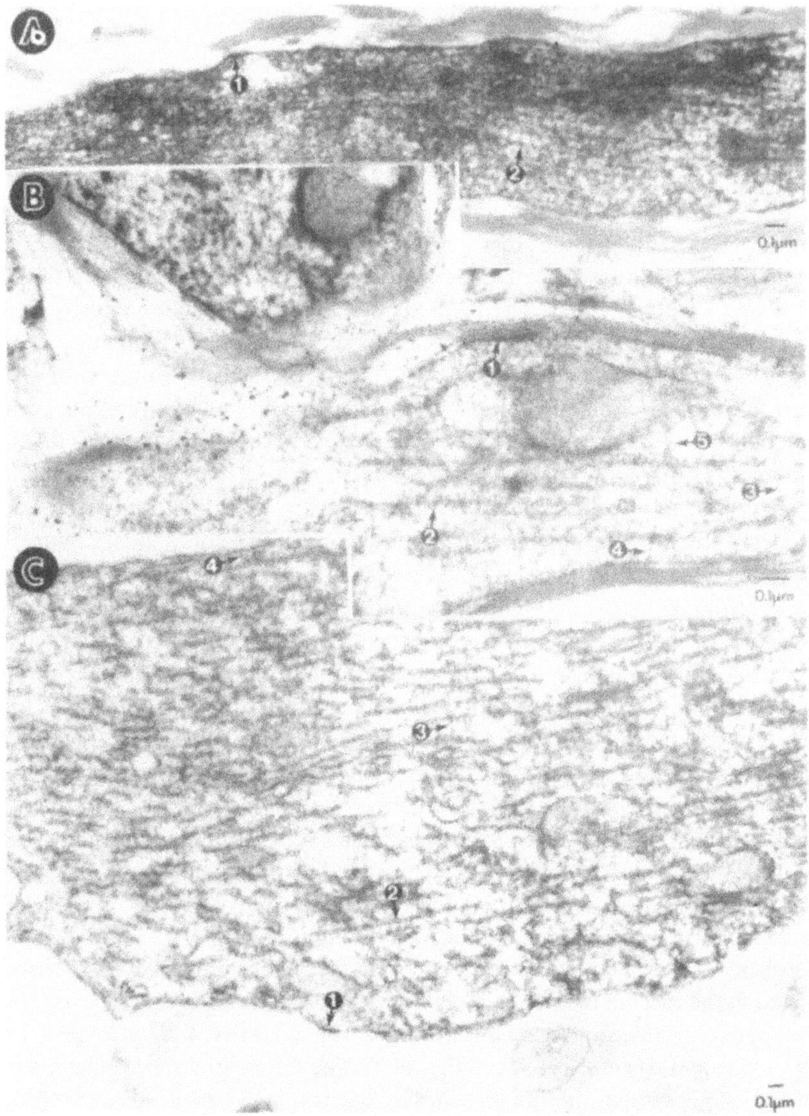

Figure 2. Spectrin isoforms were immunostained in mouse cerebellum at electron microscopic level with antibodies for brain spectrin (240/235), α-SpIIΣ1, (panels **A** and **B**) in myelinated axons of the medullary layer, and for brain spectrin(240/235E), α- and β-SpIΣ*/2, in a dendrite of a Purkinje cell (**C**). A peroxidase-conjugated secondary antibody and diaminobenzidine were used to visualize the specific staining. Note the decoration of the plasma membrane (1), of microtubules (2), of microtubule-crosslinking elements (3), of elements which link microtubules to the plasma membrane cytoskeleton (4), or structures which surround or link microtubules to mitochondria (5). A detailed description of the subcellular localization brain spectrin isoforms was given by Zagon and colleagues (1986). This micrograph is courtesy of Dr. I. Zagon.

α-SpIIΣ1 has in the middle of the amino acid sequence a sensitive cleavage site for proteases (Harris and Morrow, 1988).

Little is known about brain spectrin phosphorylation. The 235-kDa subunit can be phosphorylated by a cAMP-independent protein kinase (Goodman et al., 1984), or by an insulin receptor tyrosine kinase (Kadowaki et al., 1985). However, these studies were performed before the presence of brain spectrin isoforms was discovered. The significance of spectrin phosphorylation is unclear. The location of spectrin at the plasma membrane makes it a primary target for phosphorylation and signal transduction. This has been shown for other cytoskeletal proteins; e.g., MAP2 was phosphorylated after stimulation of insulin receptor mediated tyrosine kinase (Almas et al., 1992). It remains to be studied how the structural organization and plasticity of the spectrin cytoskeleton may be affected by phosphorylation.

Furthermore, it is interesting that ankyrin, a protein that anchors spectrin to the membrane (Nelson and Lazarides, 1984), and other spectrin-binding proteins such as synapsin I and amelin, which are immunologically related to 4.1 protein of red blood cells, exist in isoforms with axonal or dendritic location respectively (Krebs et al., 1987a,b). Thus, a whole group of proteins seems to be involved in the formation of an axon-typical or dendrite-specific, membrane-bound cytoskeleton.

H. Phosphorylated NF-H as an Axonal Marker

Mammalian neurofilament proteins consist of three polypeptide subunits of approximately 200, 160, and 70 kDa, termed NF-H, NF-M, and NF-L, respectively. They belong to the class IV intermediate filament proteins (Steinert and Roop, 1988; Robinson and Anderton, 1988; Shaw, 1991) and are exclusively found in nervous tissue. Intermediate filaments are characterized by a central rod domain which forms a coiled-coil structure; four molecules bundle together and align to a filamentous structure (Hirokawa, 1991). Neurofilament proteins, together with actin and tubulin, are part of the cytoskeleton that is transported at slow velocity in the axon, called slow components SCa and SCb (for review see Lasek et al., 1984; Nixon, 1991). Of the three neurofilament subunits, the NF-M and NF-L proteins form the core of the neurofilaments, and are distributed in both axons and dendrites (Hirokawa et al., 1984). The C-terminal tail of NF-H subunit extends from neurofilaments and is responsible for the formation of cross-bridges to neighboring filaments (Hirokawa et al., 1984). It was soon realized, by the use of monoclonal antibodies, that the NF-H in its phosphorylated form is typically found in axons while a non-phosphorylated form localizes in dendrites (Sternberger and Sternberger, 1983; Shaw et al., 1986). The NF-H subunit is the most abundantly phosphorylated of the three subunits, with 21 phosphates per protein (Julien and Mushinsky, 1982). The major phosphorylation sites of NF-H are found in the C-terminal tail region, characterized by numerous serine residues arranged in the typical KSP (lysine-serine-proline) sequence, which repeats several times and forms a multi-repeat sequence (Shaw, 1991). NF-M and NF-L may also be phosphorylated in the N-terminal region (Nixon and Sihag, 1991; Sihag and Nixon, 1991).

However, the role of these phosphorylations is not known. The tail regions of NF-H and NF-M extend out of the filaments and may be responsible for the cross-bridge formation between filaments, and it is not clear if such spacing could be due to attraction or repulsion of the heavily phosphorylated regions (as discussed by Matus, 1988b). Neurofilament assembly after NF-subunit synthesis does not depend on phosphorylation (Oblinger, 1987; Nixon et al., 1989); nor does dephosphorylation of neurofilament proteins induce neurofilament disassembly (Hisanaga and Hirokawa, 1989). However, it was demonstrated *in vitro* that phosphorylated NF-L could not reassemble into filaments (Hisanaga et al., 1990), and 1–2 phosphates per protein were sufficient to inhibit assembly. One can assume that phosphorylation results in numerous phosphate groups of functional heterogeneity (Eyer and Leterrier, 1988). Phosphorylation seems an essential modification to change the mobility of neurofilaments during axonal transport, making filaments less mobile by increasing their structural stability, as well as protecting them against turnover (Goldstein et al., 1987). Dephosphorylation was shown to induce a binding of neurofilaments with microtubules, and phosphorylation dissociated this association (Hisanaga and Hirokawa, 1990; Hisanaga et al., 1991). During slow axonal transport, NF-H and NF-M become increasingly phosphorylated (Nixon, 1991), and two populations of neurofilaments seem to exist which differ in their transport rates; one is stable while the other is mobile. Thus, phosphorylation may not only be an important step in the stabilization of the axonal neurofilaments and essential for axonal morphology, but may also regulate the interaction between neurofilaments and other structures such as microtubules, actin, or spectrin (for review see Hirokawa, 1991; Shaw, 1991). It is interesting to mention that neurofilaments may have the possibility of bidirectional transport (Glass and Griffin, 1991). However, this was observed in transected nerves, and it is not clear if retrograde transport occurs in undisturbed axons.

During rat brain development, M- and L-NF are already detectable in neurons at birth, whereas the NF-H appears later (Shaw and Weber, 1982). The NF-H becomes phosphorylated several days after its synthesis, which may indicate axonal maturation (Dahl et al., 1986; Dahl and Bignami, 1986). In the cat, the temporal expression of the NF-H coincides with the loss of callosal axons, a major phenomenon in normal brain development (Figlewicz et al., 1988). Phosphorylated NF-H was present in callosal axons prior to the appearance of non-phosphorylated NF-H (Guadano-Ferraz et al., 1990), suggesting that phosphorylation is important early in development, and perhaps essential for the stabilization of the axonal cytoskeleton.

There is evidence that neurofilament proteins are not homogeneously distributed. This is based mainly on the use of monoclonal antibodies specific to either phosphorylated or phosphorylation-independent epitopes (Sternberger and Sternberger, 1983; Shaw et al., 1986; Marc et al., 1986; Campbell and Morrison, 1989;

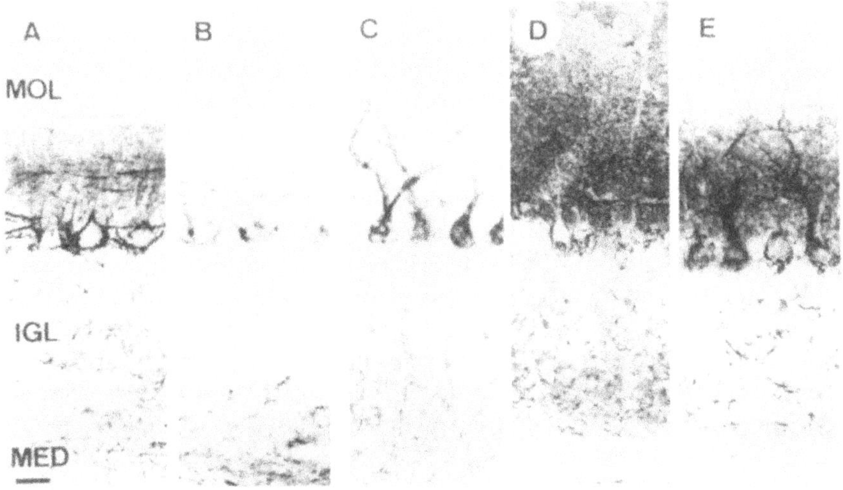

Figure 3. Immunocytochemical localization of neurofilament NF-H isoforms in adult cat cerebellum with a panel of antibodies, as was demonstrated in a recent study (Riederer et al., 1995). (**A**) A fixation insensitive polyclonal antibody stains axons in the various cerebellar layers of paraformaldehyde fixed tissue (Riederer, 1993). (**B**) An antibody against phosphorylated NF-H (Boehringer Mannheim, monoclone NE14) stains axons in cryostat sections after ethanol and acetone fixation, while SMI-32 antibody stains prominently Purkinje cell bodies and their axon and dendrites under the same fixation conditions (**C**). A novel monoclonal antibody, M9 (Marugg and Baier-Kusterman, 1988), stains in paraformaldehyde-fixed tissue exclusively axons (**D**). The same antibody stained in additional dendrites (**E**) in adjacent vibratome sections which were dephosphorylated with alkaline phosphatase 50U/section, at 37°C for 20 hours. Interestingly, this staining was less intense in perikarya, suggesting that dephosphorylation unmasks mostly assembled neurofilaments in the dendritic tree of Purkinje cells but not in the cell body. MOL = molecular layer; IGL = internal granular layer; MED = medullary layer. Bar = 50 µm.

Fisher and Shea, 1991; Lee et al., 1986; Szaro et al., 1990). For instance, in the cerebellum different NF-H staining patterns were observed (Marc et al., 1986; Kondo et al., 1991; Riederer et al., 1995). In Figure 3 a difference in NF-H staining between four different antibodies is shown in cat cerebellum. Two antibodies against phosphorylated NF-H stained axons differently, in that one reacted less with parallel fibers (Figure 3B). The appearance of neurofilaments in these fibers is a relatively late event, occurring during the third postnatal month. Transient expression of neurofilament proteins without filament formation has been observed during Purkinje cell development (Bignami et al., 1985). Cerebellar granule cells were either devoid of NF-H (Vitadello and Denis-Donini, 1990), or they showed a transient expression of neurofilament-like proteins (Cambray-Deakin and Burgoyne, 1986). Szaro and colleagues (1990) have demonstrated that phosphorylation-dependent epitopes of NF-H differ between corticospinal and primary sensory fibers. This indicates again that differential phosphorylation between different fiber systems may exist. Injection of an antibody against hyperphosphorylated neurofilament proteins caused collapse of the neurofilament network in motor neurons but not in dorsal root ganglion cells (Durham, 1992). This indicates that phosphorylation is essential for neurofilament organization in some but not all cell types.

It has been postulated that the number of neurofilaments may control axon diameter (Hoffman et al., 1984). This is of functional importance because the caliber of an axon determines its conduction velocity. Results from transgenic mice suggest that an increase in NF-L expression is not sufficient to increase axonal diameter (Cleveland et al., 1991), but axonal diameter may possibly depend on the presence of NF-M, NF-H, or on all three subunits. Interestingly, in cultured rat brain cell aggregates NF-H expression and phosphorylation were increased after stimulation by elevated potassium in the culture medium (Riederer et al., 1992). This may be of importance during development since membrane activity could influence the expression of NF-H and enhance the stabilization of the cytoskeleton and of the axon itself.

Over the past decade several novel intermediate filament proteins have been discovered in neurons, thus making the study of neurofilament function even more complex (for review see Greene, 1989; Portier, 1992). Another intermediate filament protein of type IV, α-internexin, was found in the developing nervous system, and was expressed even earlier than NF-L (Pachter and Liem, 1985; Kaplan et al., 1990). Peripherin, a protein of type III intermediate filaments, was shown to be a neurofilament protein found prominently in axons of the peripheral nervous system (Portier et al., 1984; Escurat et al., 1988). This seems interesting because of the analogy to tau proteins, where the peripheral nervous system HMW tau form is distinct from the central ones.

IV. THE DENDRITIC CYTOSKELETON

A. Is There Dendritic Transport?

One might expect that in dendrites a transport mechanism exists to target proteins and structural elements to their correct localization as is known to occur in axons. Unfortunately, studying the cytoskeleton is not as simple in dendrites as in axons because there is not a typical dendritic projection free of axons, and dendrites are usually surrounded by a large number of synapses. Possibly, a new cell culture system may permit one to study protein synthesis and transport in growing processes with cytoskeleton composition characteristic of axons or dendrites (Torre and Steward, 1992). One should stress that axons and dendrites differ not only with respect to the cytoskeleton but in their organelle composition as well (Steward and Reeves, 1988; Black and Baas, 1989). A typical difference is the presence of mRNA

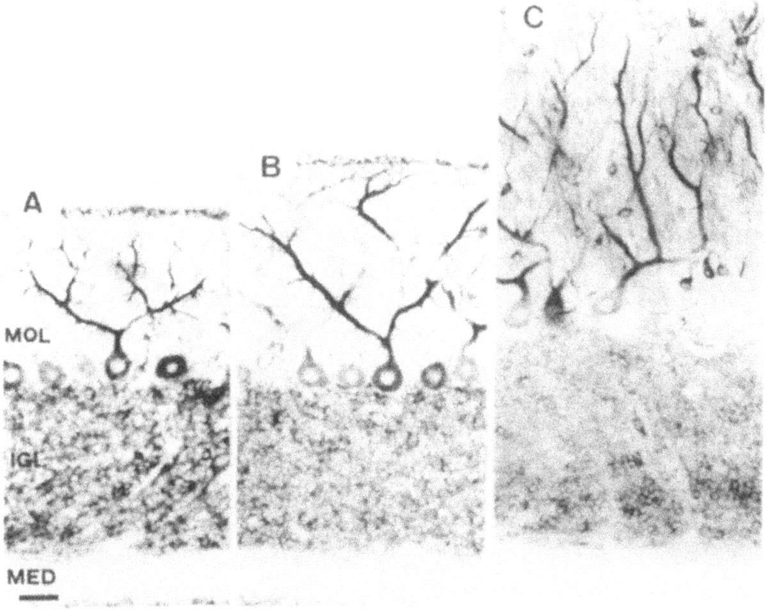

Figure 4. Immunocytochemical localization of MAP2 in P19 (**A**), P39 (**B**) and adult (**C**) cerebellum of the cat. Brain tissue was fixed with 4% paraformaldehyde and postfixed. Cerebellum was cut with a vibratome in 50μm thick sections and immunostained with monoclonal antibody AP14, peroxidase conjugated secondary antibody, and chloronaphthol. Note that dendrites and perikarya of Purkinje and granule cells are stained at all ages, while axonal staining is lacking in MED. Basket and stellate cells are stained in MED at P39 and adult stages. MOL = molecular layer; IGL = internal granular layer; MED = medullary layer. Bar = 50 μm

in dendrites (Garner et al., 1988b; Papandrikopoulou et al., 1989). The presence of specific mRNA and ribosomes in dendrites carries the advantage that proteins can be locally synthesized, and may permit a quick response to signals from the environment. This may also require different regulation mechanisms as compared to axons, where signals must first reach the perikaryon to induce de novo synthesis of proteins destined for the axon. Recently synthesized RNA was transported at a velocity of 0.55 mm/day in dendrites (Davies et al., 1990) which is slightly slower than determined for the slow axonal transport.

B. MAP2a and b are Dendritic Proteins

Typical MAP2 isoforms found in the somato-dendritic part of neurons are MAP2a and b (Izant and McIntosh, 1980; Bernhardt and Matus, 1982). The largest molecule (MAP2a, 280kDa) appears late during development, while MAP2b (slightly smaller than 2a) is present in considerable amounts already at birth, and remains throughout postnatal development (Binder et al., 1984; Burgoyne and Cumming, 1984; Bernhardt et al., 1985; Riederer and Matus, 1985). It may be involved in the outgrowth of dendrites (Bernhardt and Matus, 1982). The typical dendritic location is shown in developing cat cerebellum in Figure 4. In contrast to the axons, MAP2 mRNA is transported into dendrites, suggesting that in dendrites MAP2a and b may be synthesized locally. Furthermore, MAP2 expression may depend on local signals from the environment, possibly glial factors (Chamak et al., 1987). It has been shown that phosphorylation of MAP2 (the high M_r MAP2 forms) reduces its ability to induce microtubule polymerization (Hoshi et al., 1988; Murthy and Flavin, 1983) and modifies its binding to neurofilaments and microfilaments (Leterrier et al., 1982; Sattilaro, 1986; Selden and Pollard, 1983). Results from Brugg and Matus (1991) demonstrated that the site of phosphorylation in MAP2 rather than the amount of phosphate incorporation is the critical factor in determining microtubule binding activity of MAP2. This implies that phosphorylation of an individual amino acid may result in a functional change. The multitude of kinases which have been reported to phosphorylate MAP2 suggests that MAP2 phosphorylation is rather complex, and one may expect different functions of MAP2 dependent on phosphorylation at the C-terminal microtubule-binding site or at its opposite end (Theurkauf and Vallee, 1983; Kyriakis and Avruch, 1990; Sano et al., 1990; Seger et al., 1991). Evidence has been presented that light stimulation decreased the phosphorylation of MAP2 in the visual cortex, during a critical period of ocular dominance development (Aoki and Siekevitz, 1985). Another report indicated MAP2 being dephosphorylated upon activation of NMDA receptors (Halpain and Greengard, 1990). Given the fact that unphosphorylated MAP2 polymerizes microtubules better, one could expect that a dephosphorylation of MAP2 may increase microtubule stability, and therefore may also increase dendritic stability and morphology.

It is still unclear if MAP2a is just a phosphorylated form of MAP2b. When adult rat and cat brain tissue was dephosphorylated by alkaline phosphatase treatment, MAP2a persisted on immunoblots (unpublished observations). However, MAP2a may be different from MAP2b, and possibly differs by another type of post-translational modification. It seems that MAP2a is predominantly found in rodents, while in quail or cat brain little or no MAP2a was found (Tucker et al., 1988a,b; Riederer and Innocenti, 1992). Yet, the function of MAP2a remains to be defined.

C. Somato-Dendritic Tau is Differentially Phosphorylated

Antibodies specific for different epitopes in tau proteins suggest that phosphorylation of tau proteins determines their axonal or dendritic location (Papasozomenos and Binder, 1987). In cat brain, TAU-1 reacted with two adult tau proteins of higher M_r (60–70kDa), while in juvenile tissue several tau forms of lower M_r (50–60 kDa) were recognized (Riederer and Innocenti, 1991). The transition from juvenile to adult tau proteins occurred during the third and fourth week after birth. When brain tissue was dephosphorylated with alkaline phosphatase, an increase in immunoreactivity of juvenile tau was observed on blots, and the immunocytochemical location pointed to the presence of tau proteins in perikarya and dendrites, especially at younger stages. This suggests that the TAU-1 epitope in perikarya and dendrites must be masked by phosphorylation. The abundant presence of differentially phosphorylated tau proteins in dendrites in juvenile tissue could explain why such tau proteins are less effective in the polymerization of microtubules (Mareck et al., 1980; Francon et al., 1982; Goedert and Jakes, 1990), because phosphorylated tau proteins have a reduced potential to polymerize microtubules (Yamamoto et al., 1983; Lindwall and Cole, 1984). Phosphorylation has also been reported to render tau protein stiffer (Hagestedt et al., 1989). High speed centrifugation of juvenile and adult cortex tissue showed that juvenile tau proteins were more associated with the pellet fraction, while adult tau proteins were more soluble (Riederer and Innocenti, 1991). This may indicate that juvenile tau proteins associate in a different way to structural components than adult tau. Again, phosphorylation may be an important factor to modulate this association. Candidates for tau-binding sites in other structures than microtubules are the NF-L in neurofilaments (Heimann et al., 1985) and spectrin, as part of the membrane-attached cytoskeleton (Carlier et al., 1984). In the latter study erythrocyte spectrin was used to study tau binding, and one could speculate that in neurons the target protein for tau may be the brain spectrin isoform that is immunologically related to erythroid spectrin, SpIΣ2. The development-dependent distribution of protein kinase C in kitten visual cortex (Jia et al., 1990) exhibits a similar decrease as the somato-dendritic tau proteins. Ca^{2+}-calmodulin-dependent kinase (Steiner et al., 1990) and Ca^{2+}/phospholipid-dependent protein kinase C (Baudier et al., 1987) have been identified to phosphorylate tau proteins. It remains to be determined which sites are phosphorylated in dendritic or axonal tau, the functional implication of such phosphorylation, and which enzymes are responsible for the domain specific modification *in vivo*.

D. MAP5 in Dendrites Is not Phosphorylated

MAP5, MAP1b, and MAP1x are abundant proteins in the early cytoskeleton, and during brain development decrease in concentration (Bloom et al., 1985; Riederer et al., 1986a; Calvert et al., 1987). MAP5 has the ability to promote tubulin polymerization (Riederer et al., 1986a), and it is one of the first MAPs to appear in neuroblasts (Riederer et al., 1986a) or spinal cord neurons (Tucker et al., 1988a). The non- or less-phosphorylated form of MAP1b, also named MAP5b, is present in the perikarya and dendrites (Sato-Yoshitake, 1989; Riederer et al., 1990). MAP5b, at least at early stages, is restricted to somata and dendrites. In developing cerebellum, MAP5 was present in cells of the external germinal layer and was abundant in ingrowing axons of the medullary layer (Riederer et al., 1986a; Riederer, 1992; this is also shown in Figure 1). At subsequent stages MAP5 decreased in concentration in cells of the germinal layer, but it persisted into adulthood in granule cells and Purkinje cells. Biochemical experiments indicate that phosphorylation of MAP5 influences its state of association with cellular components (Riederer et al., 1990), and that MAP5b is more soluble. It is not clear if the state of phosphorylation influences the ability of MAP5 to induce microtubule polymerization, although phosphorylation appears to be an important factor in the modulation of MAP5 function in axons but not in dendrites. In cerebral cat cortex MAP5b is present in equimolar concentration to MAP5a. Thus, MAP5b must have an equal important function in dendrites as MAP5a has in axons, but less dependent on phosphorylation. Indeed MAP5 (MAP16) is subject to a different type of phosphorylation mediated by a casein-kinase throughout neuronal maturation, and such phosphorylated MAP5 (MAP16) is found in cell bodies, axons and dendrites, as discussed by Riederer (1995).

E. Dendritic Brain Spectrin Is Related to Erythrocyte Spectrin

The 240-kDa α-subunit of non-erythroid spectrin, α-SpII (McMahon et al., 1987; Leto et al., 1988) is thought to be present as a single form in mammalian neurons. Specific antibodies against brain spectrin (fodrin) identified the 240-kDa subunit and localized this protein in axons and dendrites (Levine and Willard, 1981), and in axons and some perikarya, respectively (Riederer et al., 1986b). In contrast, an antibody raised against red blood cell spectrin reacted mainly with the 235E brain spectrin (β-SpIΣ2) and little with the 240-kDa subunit, and labeled perikarya and dendrites, but not axons (Riederer et al., 1986b). This isoform, like the axonal one, was found attached to the plasma membrane and surrounding or cross-linking several organelles (Zagon et al., 1986). This is also shown in Figure 2. This spectrin type may play a crucial role in dendritic spines, and could be the spectrin type described as major calmodulin-binding protein of postsynaptic densities (Carlin et al., 1983). The essential role of spectrin to interact and cross-link cellular structures has been extensively discussed previously (Riederer, 1990) and is also illustrated at the electron-microscopic level in Figure 2 (1–5). In a comparative study, Riederer and colleagues (1988) have described an antibody raised against human red blood cell spectrin which reacted with the α brain spectrin (α-SpIΣ*) on blots, and which labeled perikarya and dendrites in tissue. Usually antibodies against the α brain spectrin subunit label axons, but not this antibody. Therefore, one has to consider that a dendrite-specific α-SpIΣ2 form may exist, despite the molecular data pointing to the presence of only one α-non-erythroid spectrin form, α-SpIIΣ1 (McMahon et al., 1987). Furthermore, recent data in an abstract hint at the possibility that splice variants of α-fodrin variants may exist (Wolda and Moon, 1990). It remains to be determined if the somato-dendritic α-SpIΣ2 is a post-translational modification or is generated, like the β-SpIIΣ1 and β-SpIIΣ2 isoforms, from different mRNAs (Ma et al., 1993).

F. Dendritic NF-H: Does Phosphorylation Make a Difference?

There is overwhelming evidence that all three neurofilament subunits are located in axons, perikarya, and dendrites, and that the exclusively axonal location of NF-H is due to its phosphorylation at the multiphosphorylation site in the C-terminal tail domain (for review see Shaw, 1991). Phosphorylation of NF-H seems an important modification in axons and its role in spacing neurofilaments has been discussed (Matus, 1988b). Phosphorylation of neurofilament proteins is more complex and heterogeneous than anticipated (Nixon and Sihag, 1991) and may play a crucial role during axonal transport. It is well established that antibodies against phosphorylation-independent epitopes on NF-H localize this protein also in dendrites (Lee et al., 1986; Sternberger and Sternberger, 1986; Shaw et al., 1986). These results all suggest that the dendritic NF-H is not phosphorylated, certainly not at the highly antigenic multiphosphorylation site. Two studies suggest that NF-H may also be expressed transiently during cerebellar development in some neurons such as granule cells (Cambray-Deakin and Burgoyne, 1986), and in Purkinje cells (Bignami et al., 1985). Furthermore the latter group has shown that NF-H is present in Purkinje cell bodies and dendrites but does not assemble into filaments. It is not clear if the disability of NF-H to assemble into filaments is due to a lack of phosphorylation. A monoclonal antibody (M9) described by Marugg and Baier-Kusterman (1988) reacted with a non-phosphorylated phosphorylation site in NF-H, which is masked in dendrites by phosphorylation (Riederer et al., 1995). M9 recognized NF-H *in vivo* only in axons (Figure 3D). Upon dephosphorylation with alkaline phosphatase the phosphorylated NF-H is unmasked and becomes detectable for this antibody (Figure 3E). These observations would suggest that NF-H in dendrites may be present in a phosphorylated form, but differently phosphorylated than axonal NF-H. The data add to the general view that post-translational modifications of NF-H may differ between different types of axons and cell types (Lee et al., 1986) and furthermore may differ between axons and dendrites. It remains to be investigated whether phosphorylation is an essential modification for the targeting of the molecule to the axon or towards dendrites, or whether it is required for neurofilament assembly and stabilization in addition to the spacing of neurofilaments.

V. NEURONAL POLARITY

The mechanisms involved in the formation of dendrites and axons, and the targeting of different cytoskeletal proteins selectively to the axon or to dendrites, are largely unknown. There are several leads which indicate that these may be complex and depend on various factors (Black and Baas, 1989; Craig and Banker, 1994; Ginzburg, 1991; Singer, 1992). One important role may be played by the microtubule polarity in directing transport of organelles. In axons, tubulin assembly into microtubules occurs preferentially at the distal (positive) end, while in dendrites the orientation of microtubules is mixed. This has been discussed as the basis for differential distribution of organelles such as the Golgi apparatus and ribosomes (Black and Baas, 1989). Motor proteins such as kinesin and dynein (MAP1c) may use this microtubule polarity for anterograde and retrograde transport, respectively (for reviews see Vale, 1987; Filiatreau, 1988; Vallee et al., 1989). Hirokawa and colleagues (1991) have shown that kinesin is responsible for the fast axonal anterograde organelle transport *in vivo.*

The dogma that axons lack ribosomes, the protein synthesizing machinery, may be true for most axons (Steward and Reeves, 1988) although some axons may have a limited potential for protein synthesis since some polyribosomes have been found in the initial segment of axons from pyramidal cells of the sensorimotor cortex of monkey (Steward and Ribak, 1986). Furthermore, it has been shown that the location of the Golgi complex does not play a major role in determining which of the processes initially extending from a neuron becomes an axon (Dotti and Banker, 1991).

Recent evidence indicates that subcellular sorting of tubulin isotypes is also based on the interaction with MAPs, such as MAP1b (Falconer et al., 1992). Many cytoskeletal proteins are synthesized in the cell body, assembled into cytoskeletal structures, and transported into neuronal processes (Black et al., 1986; Hollenbeck, 1989). In addition, mRNA for MAP2, but not for GAP43 or α tubulin (Bruckenstein et al., 1990), is transported in the dendrite. Notably, mRNA encoding MAP2b is found in dendrites (Garner et al., 1988b), in contrast to mRNA of MAP2c, which is located only in perikarya (Papandrikopoulou et al., 1989). This indicates that in dendrites a transport of specific mRNA and the presence of ribosomes provide the necessary tools for local protein synthesis and quick response to signals from the environment. But what is the signal that makes a molecule move into an axon or a dendrite? It has been postulated that mRNA targeting involves proteins which recognize specific localizing sequences at the 3′ end of the mRNA and then chaperone the mRNA, presumably along microfilaments, to its destination (Hesketh and Pryme, 1991; Singer, 1992; also see Pryme et al., this volume). For cytoskeleton such as tau, MAP5, MAP2, or NF-H, one has to take into consideration that a difference between axonal and dendritic forms is their state of phosphorylation. Post-translational modifications have to be considered as a mean to modify a protein in its function, association to other structures and eventually in directing it towards subcellular areas. This implies that protein kinases or phosphatases must have an adequate subcellular distribution to control such modifications. An essential point for targeting proteins may be their three dimensional conformation. A number of proteins, collectively known as chaperones, are involved in catalyzing protein folding (for review see Gething and Sambrook, 1992). They are a highly conserved protein family whose members are widely distributed in pro- and eukaryotic cells. Among many functions they are involved in protein translocation. Recent evidence was presented that in the cytosol a chaperone is essential in the tubulin biogenesis (Yaffe et al., 1992). How much chaperones participate in directing cytoskeletal proteins into axons and dendrites remains to be studied.

VI. PERSPECTIVES

In this review, some differences between the axonal and dendritic cytoskeleton have been pointed out, and the essential question of sorting cytoskeletal protein to axons and dendrites was addressed. Post-translational modifications appear to be essential in targeting proteins either directly or by changing their binding behavior to other domain-bound proteins or structures in order to guide them towards the axon or dendrite. Therefore, to study post-translational modifications and their influences on protein function it seems necessary to understand how the dendritic and axonal cytoskeletons are assembled. As mentioned, phosphorylation seems one of the possibilities. This immediately raises the question whether limited phosphorylation (e.g., of a single amino acid) is sufficient to target the protein or whether extensive phosphorylation is required. Furthermore, enzymes that seem involved in modulating the cytoskeletal proteins are not yet well defined, and have been mostly studied *in vitro*. In the future it is important that the subcellular location of such enzymes is studied so as to define if targeting of cytoskeletal proteins is related to the distribution of such enzymes and to protein modification. Another point to be addressed in more detail is the effects of protein modification and the resulting changes in function during neuronal development. It is obvious that molecular changes and post-translational modifications of cytoskeletal proteins condition the interaction between the different filament systems, and thus modify the structural stability of the cytoskeleton, which in consequence influences axonal and dendritic plasticity.

ACKNOWLEDGMENTS

The author would like to thank Drs. P.G.H. Clarke and P. Honegger for helpful comments, and Ms. R. Porchet for excellent technical help. This work was supported by the National Science Foundation of Switzerland grants 31.26624-89 and 31.33447-92.

REFERENCES

Aletta, J.M., Lewis, S.A., Cowan, N.J., & Greene, L.A. (1988). Nerve growth factor regulates the phosphorylation and steady-state levels of microtubule-associated protein 1.2 (MAP1.2). J. Cell Biol. 106, 1573–1581.

Almas, B., Pryme, I.F., Vedeler, V., & Hesketh, J.E. (1992). Insulin: Signal transmission and short-term effects on the cytoskeleton and protein synthesis. Int. J. Biochem. 24, 183–191.

Aoki, C. & Siekevitz, P. (1985). Ontogenetic changes in the cyclic adenosine 3′,5′-monophosphate-stimulatable phosphorylation of cat visual cortex proteins, particularly of microtubule-associated protein 2 (MAP2): Effects of normal and dark rearing and of the exposure to light. J. Neurosci. 5, 2465–2483.

Avila, J., Serrano, L., Hernandez, M.A., & Diaz-Nido, J. (1988). Phosphorylation of neuronal microtubule proteins. Protoplasma 145, 82–88.

Bahr, B.A., Vanderklish, P.W., Ha, L.T., Tin, M.T., & Lynch, G. (1991). Spectrin breakdown products increase with age in telencephalon of mouse. Neurosci. Lett. 131, 237–240.

Bamburg, J.R., Bray, D., & Chapman, K. (1986). Assembly of microtubules at the tip of growing axons. Nature 321: 788–790.

Bamburg, J.R. & Bernstein, B.W. (1991). Actin and actin-binding proteins in neurons. In: The Neuronal Cytoskeleton (Burgoyne, R.D., ed.), pp. 121–160, Wiley-Liss, Inc., New York.

Bamburg, J.R., Lockerbie, R.O., Bernstein, B.W., Morgan, T.E., Whitney, M., & Adams, M.E. (1992). Actin dynamics in neuronal growth cones. In: The Nerve Growth Cone (Letourneau, P.C., Kater, S.B., & Macagno, R., eds.), pp. 27–38, Raven Press Ltd., New York.

Barakat-Walter, I. & Riederer, B.M. (1991). Brain spectrin 240/235 and 240/235E: Differential expression during development of chicken dorsal root ganglia in vivo and in vitro. Eur. J. Neurosci. 3, 431–440.

Baudier, J., Lee, S.-H., & Cole, R.D. (1987). Separation of the different microtubule-associated Tau protein species from bovine brain and their mode II phosphorylation by Ca^{2+}/phospholipid-dependent protein kinase C. J. Biol. Chem. 262, 17584–17590.

Bennett, V. (1990). Spectrin-based membrane skeleton: A multipotential adaptor between plasma membrane and cytoplasm. Physiol. Rev. 70, 1029–1065.

Bernhardt, R. & Matus, A. (1982). Initial phase of dendrite growth: Evidence for the involvement of high molecular weight microtubule-associated proteins (HMWP) before the appearance of tubulin. J. Cell Biol. 92, 589–593.

Bernhardt, R., Huber, G., & Matus, A. (1985). Difference in the developmental patterns of three microtubule-associated proteins in the rat cerebellum. J. Neurosci. 5, 977–991.

Bignami, A., Grossi M., & Dahl, D. (1985). Transient expression of neurofilament protein without filament formation in purkinje cell development. Immunohistological and electron microscopic study of chicken cerebellum. Int. J. Devl. Neurosci. 3, 365–377.

Binder, L.I., Frankfurter, A., Kim, H., Caceres, A., Payne, M.R., & Rebhun, L.I. (1984). Heterogeneity of microtubule-associated protein 2 during rat brain development. Proc. Natl. Acad. Sci. USA 81, 5613–5617.

Binder, L.I., Frankfurter, A., & Rebhun, L.I. (1985). The distribution of tau in the mammalian neurons in situ. J. Cell Biol. 101, 545–557.

Binder, L.I., Frankfurter, A., & Rebhun, L.I. (1986). Differential localization of MAP2 and tau in mammalian neurons in situ. Ann. N.Y. Acad. Sci. 466, 145–166.

Binet, S. & Meininger, V. (1988). Modifications of tubulin heterogeneity during axonal growth in the embryonic nervous system. Dev. Brain Res. 38, 296–299.

Black, M.M., Keyser, P., & Sobel, E. (1986). Interval between the synthesis and assembly of cytoskeletal proteins in cultured neurons. J. Neurosci. 6, 1004–1012.

Black, M.M. & Baas, P.W. (1989). The basis of polarity in neurons. Trends in Neurosci. 6, 211–214.

Bloom, G.S., Luca, F.C., & Vallee, R.B. (1985). Microtubule-associated protein 1B: Identification of a major component of the neuronal cytoskeleton. Proc. Natl. Acad. Sci. USA 82, 5404–5408.

Brady, S.R., Lasek, R., Allen, R., Lin, H., & Stossel, T. (1984). Gelsolin inhibition of fast axonal transport indicates a requirement for actin microfilaments. Nature 310, 56–58.

Brion, J.P., Guilleminot, J., Couchie, D., Flament-Durand, J., & Nunez, J. (1988). Both adult and juvenile tau microtubule-associated proteins are axon specific in the developing and adult rat cerebellum. Neurosci. 25, 139–146.

Bruckenstein, D.A., Lein, P.J., Higgins, D., & Fremeau, R.T., Jr. (1990). Distinct spatial localization of specific mRNAs in cultured sympathetic neurons. Neuron 5, 809–819.

Brugg, B. & Matus, A. (1988). PC12 cells express juvenile microtubule-associated proteins during nerve growth factor-induced neurite outgrowth. J. Cell Biol. 107, 643–650.
Brugg, B. & Matus, A. (1991). Phosphorylation determines the binding of MAP2 to microtubules in living cells. J. Cell Biol. 114, 735–743.
Burgoyne, R.D. & Cumming, R. (1984). Ontogeny of microtubule-associated protein 2 in rat cerebellum: Differential expression of the doublet polypeptides. Neurosci. 11, 156–167.
Burgoyne, R.D. (1991). The Neuronal Cytoskeleton. Wiley-Liss, Inc., New York.
Caceres, A., Pontrebic, S., & Kosik, K.S. (1991). The effect of tau antisense oligonucleotides on neurite formation of cultured cerebellar macroneurons. J. Neurosci. 11, 1515–1523.
Calvert, R. & Anderton, B.H. (1985). A microtubule-associated protein (MAP1) which is expressed at elevated levels during development of the rat cerebellum. EMBO J. 4, 1171–1176.
Calvert, R., Woodhams, P.L., & Anderton, B.H. (1987). Localization of an epitope of a microtubule-associated protein 1x in outgrowing axons of the developing rat central nervous system. Neurosci. 23, 131–141.
Cambray-Deakin, M.A. & Burgoyne, R.D. (1986). Transient expression of neurofilament-like (RT97) immunoreactivity in cerebellar granule cells. Dev. Brain Res. 28, 282–286.
Cambray-Deakin, M.A. & Burgoyne, R.D. (1987). Posttranslational modifications of α-tubulin: Acetylated and detyrosinated forms in axons of rat cerebellum. J. Cell Biol. 104, 1569–1574.
Campbell, M.J. & Morrison, J.H. (1989). Monoclonal antibody to neurofilament protein (SMI-32) labels a subpopulation of pyramidal neurons in the human and monkey neocortex. J. Comp. Neurol. 282, 191–205.
Carlin, R.K., Bartelt, D.C., & Siekevitz, P. (1983). Identification of fodrin as major calmodulin-binding protein in postsynaptic density preparations. J. Cell Biol. 96, 443–448.
Carlier, M.F., Simon, C., Cassoly, R., & Pradel, F.A. (1984). Interaction between microtubule-associated protein tau and spectrin. Biochim. 66, 305–311.
Chamak, B., Fellous, A., Glowinski, J., & Prochiantz, A. (1987). MAP2 expression and neuritic outgrowth and branching are coregulated through region-specific neuro-astroglial interactions. J. Neurosci. 7, 3163–3170.
Cleveland, S.W., Hwo, H.Y., & Kirschner, M.W. (1977). Purification of tau, a microtubule-associated protein induces assembly of tubulin from purified tubulin. J. Mol. Biol. 116, 207–226.
Cleveland, D.W. (1987). The multitubulin hypothesis revisited: What have we learned? J. Cell Biol. 104, 381–383.
Cleveland, D.W. & Hoffman, P.N. (1991). Slow axonal transport models come full circle: Evidence that microtubule sliding mediates axon elongation and tubulin transport. Cell 67: 453–456.
Cleveland, D.W., Monteiro, M.J., Wong, P.C., Gill, S.R., Gerhart, J.D., & Hoffman, P.N. (1991). Involvement of neurofilaments in the radial growth of axons. J. Cell Sci., Suppl. 15, 85–95.
Couchie, D., Fags, C., Bridoux, A.M., Rolland, B., Tardy, M., & Nunez, J. (1985). Microtubule-associated proteins and "in vitro" astrocyte differentiation. J. Cell Biol. 101, 2095–2103.
Couchie, D., Mavilia, D., Georgieff, I., Liem, R.K.H., Shelanski, M.L., & Nunez, J. (1992). Primary structure of high molecular weight tau present in the peripheral nervous system. Proc. Natl. Acad. Sci. USA 89, 4378–4381.
Craig, A.M. & Banker, G. (1994). Neuronal polarity. Annu. Rev. Neurosci. 17, 267–310.
Cronly-Dillon, J. & Nona, S.N. (1988). Changes in cytoskeletal elements during postnatal development of cerebral cortex. Cerebral Cortex 7, 273–308.
Cumming, R., Burgoyne, R.D., & Lytton, N.A. (1983). Axonal subpopulation in the central nervous system demonstrated using monoclonal antibodies against α-tubulin. Eur. J. Cell Biol. 98, 347–351.
Cumming, R., Burgoyne, R.D., & Lytton, N.A. (1984a). Immuno-fluorescence distribution of α-tubulin, β-tubulin and microtubule-associated protein 2 during in vitro maturation of cerebellar granule cell neurons. Neurosci. 12, 775–782.

Cumming, R., Burgoyne, R.D., & Lytton, N.A. (1984b). Immunocytochemical demonstration of α-tubulin modification during axonal maturation in the cerebellar cortex. J. Cell Biol. 98, 347–351.

Cunningham, C.C., Gorlin, J.B., Kwiatkowski, D.J., Hartwig, J.H., Janmey, P.A., Byers, H.R., & Stossel, T.P. (1992). Actin-binding protein requirement for cortical stability and efficient locomotion. Science 325, 325–327.

Dahl, D., Crosby, C.J., Gardner, E.E., & Bignami, A. (1986). Delayed phosphorylation of the largest neurofilament protein in rat optic nerve development. J. Neurosci. Res. 15, 513–519.

Dahl, D. & Bignami, A. (1986). Neurofilament phosphorylation in development. A sign of axonal maturation? Exp. Cell Res. 162, 220–230.

Davis, L., Burger, B., Banker, G.A., & Steward, O. (1990). Dendritic transport: Quantitative analysis of the time course of somatodendritic transport of recently synthesized RNA. J. Neurosci. 10, 3056–3068.

Deiters, O. (1865). Untersuchungen über Gehirn und Rückenmark des Menschen und der Säugerthiere. Ed. M. Schultze, Vieweg.

Dhermy, D. (1991). The spectrin super-family. Biol. Cell 71, 249–254.

Doll, T., Meichsner, M., Riederer B.M., Honegger, P., & Matas, A. (1993). An isoform of microtubule-associated protein (MAP2) containing four repeats of the tubulin-binding motif. J. Cell Sci. 106, 633–640.

Dotti, C.G. & Banker, G. (1991). Intracellular organization of hippocampal neurons during the development of neuronal polarity. J. Cell Sci. 15, 75–84.

Drubin, D.G. & Kirschner, M. (1986). Tau protein function in living cells. J. Cell Biol. 103, 2739–2746.

Durham, H.D. (1992). An antibody against hyperphosphorylated neurofilament proteins collapses the neurofilament network in motor neurons but not in dorsal root ganglion cells. J. Neuropath. Exp. Neurol. 51, 287–297.

Edde, B., Rossier, J., LeCaer, J.-P., Desbruyères, E., Gros, F., & Denoulet, P. (1990). Posttranslational glutamylation of α-tubulin. Science 247, 83–85.

Escurat, M., Landon, F., Gumpel, M., Gros F., & Portier, M.-M. (1988). Peripherin, an intermediate filament protein is a developmental marker of specific neuronal populations. In: Structure and Functions of the Cytoskeleton (Rousset, B.A.F., ed.), pp. 169–177, Colloque INSERM, Paris.

Eyer, J. & Leterrier, J.-F. (1988). Influence of the phosphorylation state of neurofilament proteins on the interactions between purified filaments in vitro. Biochem. J. 252, 655–660.

Falconer, M.M., Echeverri, C.J., & Brown, D.L. (1992). Differential sorting of β tubulin isotypes into colchicine-stable microtubules during neuronal and muscle differentiation of embryonal carcinoma cells. Cell Mot. Cytoskel. 21, 313–325.

Fath, K.R. & Lasek, R.J. (1988). Two classes of actin microfilaments are associated with the inner cytoskeleton of axons. J. Cell Biol. 107, 613–621.

Figlewicz, D.A., Gremo, F., & Innocenti, G.M. (1988). Differential expression of neurofilament subunits in the developing corpus callosum. Dev. Brain Res. 42, 181–189.

Filiatreau, G. (1988). Les moteurs moléculaires du transport axonal. M/S médecine/science 6, 373–377.

Fischer, I. & Romano-Clarke, G. (1990). Changes in microtubule-associated protein MAP1b phosphorylation during rat brain development. J. Neurochem. 55, 328–333.

Fischer, I. & Shea, T.B. (1991). Differential appearance of extensively phosphorylated forms of the high molecular weight neurofilament protein in regions of mouse brain during postnatal development. J. Neuroimmunol. 31, 73–81.

Francon, J., Lennon, A.M., Fellous, A., Mareck, A., Pierre, M., & Nunez, J. (1982). Heterogeneity of microtubule-associated proteins and brain development. Eur. J. Biochem. 129, 465–472.

Garner, C.C., Brugg, B., & Matus, A. (1988a). A 70-kilodalton microtubule-associated protein (MAP2c), related to MAP2. J. Neurochem 50, 609–615.

Garner, C.C., Tucker, R.P., Matus, A. (1988b). Selective localization of messenger RNA for cytoskeletal protein MAP2 in dendrites. Nature 336, 674–677.

Garner, C.C. & Matus, A. (1988). Different forms of microtubule-associated Protein 2 are encoded by separate mRNA transcripts. J. Cell Biol. 106, 779–783.

Garner, C.C., Matus, A., Anderton, B., & Calvert, R. (1989). Microtubule-associated proteins MAP5 and MAP1x: Closely related components of the neuronal cytoskeleton with different cytoplasmic distribution in the developing brain. Molec. Brain Res. 5, 85–92.

Garner, C.C., Garner, A., Huber, G., Kozak, C., & Matus, A. (1990). Molecular cloning of microtubule-associated protein 1 (MAP1a) and microtubule-associated protein 5 (MAP1b): Identification of distinct genes and their differential expression in developing brain. J. Neurochem. 55, 146–154.

Gething, M.-J. & Sambrock, J. (1992). Protein folding in the cell. Nature 355, 33–45.

Georgieff, I.S., Liem, R.K.H., Mellado, W., Nunez, J., & Shelanski, M.L. (1991). High molecular weight tau: Preferential localization in the peripheral nervous system. J. Cell Sci. 100, 55–60.

Ginzburg, I. (1991). Neuronal polarity: Targeting of microtubule components into axons and dendrites. Trends Biochem. 16, 257–261.

Glass, J.D. & Griffin, J.W. (1991). Neurofilament redistribution in transected nerves: Evidence for bidirectional transport of neurofilaments. J. Neurosci. 11, 3146–3154.

Goedert, M. & Jakes, R. (1990). Expression of separate isoforms of human tau protein: Correlation with the tau pattern in brain and effects on tubulin polymerization. EMBO J. 9, 4225–4230.

Goedert, M., Crowther, R.A., & Garner, C.C. (1991). Molecular characterization of microtubule-associated proteins tau and MAP2. Trends in Neurosci. 14, 193–199.

Goedert, M., Spillantini, M.G., & Crowther, R.A. (1992). Cloning of a big tau microtubule-associated protein characteristic of the peripheral nervous system. Proc. Natl. Acad. Sci. USA, 89, 1983–1987.

Goldstein, M.E., Sternberger, N.H., & Sternberger, L.A. (1987). Phosphorylation protects neurofilaments against proteolysis. J. Neuroimmunol. 14, 149–160.

Goodman, S.R., Zagon, I.S., Whitfield, C.F., Casoria, L.A., Shohet, S.B., Bernstein, S.E., McLaughlin, P.J., & Laskewicz, T.L. (1984). A spectrin-like protein from mouse brain membranes: Phosphorylation of the 235,000-dalton subunit. Am. J. Physiol. 16, C61–C73.

Goodman, S.R., Krebs, K.E., Whitfield, C.F., Riederer, B.M., & Zagon, I.S. (1988). Spectrin and related molecules. CRC Critical Reviews in Biochemistry 23, 171–234.

Gordon-Weeks, P.R. (1991). Evidence for microtubule capture by filopodial actin filaments in growth cones. NeuroReport 2, 573–576.

Gordon-Weeks, P.R. & Mansfield, S.G. (1992). Assembly of microtubules in growth cones: The role of microtubule-associated proteins. In: The Nerve Growth Cone (Letourneau, P.C., Kater, S.B., Macagnon, E.R., eds.), pp. 55–65, Raven Press Ltd., New York.

Greene, L.A., Liem, R.K.H., & Shelanski, M.L. (1983). Regulation of a high molecular weight microtubule-associated protein in PC12 cells by nerve growth factor. J. Cell Biol. 96, 76–83.

Greene, L.A. (1989). A new neuronal intermediate filament protein. Trends Neurosci. 12, 228–230.

Guadano-Ferraz, A., Riederer, B.M., & Innocenti, G.M. (1990). Developmental changes in the heavy subunit of neurofilaments. Devel. Brain Res. 56, 244–256.

Hagestedt, T., Lichtenberg, B., Wille, H., Mandelkow, E.M., Mandelkow, E. (1989). Tau protein becomes long and stiff upon phosphorylation: Correlation between paracrystaline structure and degree of phosphorylation. J. Cell Biol. 109, 1643–1651.

Halpain, S. & Greengard, P. (1990). Activation of NMDA receptors induces rapid dephosphorylation of the cytoskeletal protein MAP2. Neuron 5, 237–246.

Hammarback, J.A., Obar, R.A., Hughes, S.M., & Vallee, R.B. (1991). MAP1b is encoded as a polyprotein that is processed to form a complex N-terminal microtubule-binding domain. Neuron 7, 129–139.

Hargreaves, A.J., Wandosell, F., & Avila, J. (1986). Phosphorylation of tubulin enhances its interaction with membranes. Nature 323, 827–828.

Harris, A.S. & Morrow, J.S. (1988). Proteolytic processing of human brain α spectrin (fodrin): Identification of a hypersensitive site. J. Neurosci. 8, 2640–2651.

Hartwig, J.H. & Kwiatkowski, D.J. (1991). Actin-binding proteins. Curr. Opinion Cell Biol. 3, 87–97.

Heimann, R., Shelanski, M.L., & Liem, R.K.H. (1985). Microtubule-associated proteins bind specifically to the 70-kDa neurofilament protein. J. Biol. Chem. 260, 12160–12166.

Heins, S., Song, Y.H., Wille, H., Mandelkow, E., & Mandelkow, E.M. (1991). Effect of MAP2, MAP2c and tau on kinesin-dependent microtubule-motility. J. Cell Sci. 14, 121–124.

Heriot, K., Gambetti, P., & Lasek, P. (1985). Proteins transported in slow components a and b of axonal transport are distributed differently in the transverse plane of the axon. J. Cell Biol. 100, 1167–1172.

Hesketh, J.E. & Pryme, I.F. (1991). Interaction between mRNA, ribosomes and the cytoskeleton. Biochem. J. 277, 1–10.

Hirokawa, N., Glicksman, M., & Willard, M.B. (1984). Organization of mammalian neurofilament polypeptides within the neuronal cytoskeleton. J. Cell Biol. 98, 1523–1536.

Hirokawa, N. (1991). Molecular architecture and dynamics of the neuronal cytoskeleton. In: The Neuronal Cytoskeleton (Burgoyne, R.D., ed.), pp. 5–74, Wiley-Liss, Inc., New York.

Hirokawa, N., Sato-Yoshitake, R., Kobayashi, N., Pfister, K.K., Bloom, G.S., & Brady, S.T. (1991). Kinesin associates with anterogradely transported membranous organelles in vivo. J. Cell Biol. 114, 295–302.

Hisanaga, S.-I. & Hirokawa, N. (1989). The effects of dephosphorylation on the structure of the projections of neurofilament. J. Neurosci. 9, 959–966.

Hisanaga, S. & Hirokawa, N. (1990). Dephosphorylation-induced interaction of neurofilaments with microtubules. J. Biol. Chem. 265, 21852–21858.

Hisanaga, S.I., Gonda, Y., Inagaki, M., Ikai, A., & Hirokawa, N. (1990). Effects of phosphorylation of the neurofilament L protein on filamentous structures. Cell Reg. 1, 237–248.

Hisanaga, S.I., Kusubata, M., Okumura E., & Kishimot, T. (1991). Phosphorylation of neurofilament H subunit at the tail domain by CDC2 kinase dissociates the association to microtubules. J. Biol. Chem. 266, 21798–21803.

Hoffman, P.N., Griffin, J.W., & Price, D.L. (1984). Control of axonal caliber by neurofilament transport. J. Cell Biol. 99, 705–714.

Hoffman, P.N. & Cleveland, D.W. (1988). Neurofilament and tubulin expression recapitulates the developmental program during axonal regeneration: Induction of a specific β-tubulin isotype. Proc. Natl. Acad. Sci. USA 85, 4530–4533.

Hollenbeck, P.J. (1989). The transport and assembly of the axonal cytoskeleton. J. Cell Biol. 108, 223–227.

Hoshi, M., Akiyama, T., Shinohara, Y., Miyata, Y., Ogawara, H., Nishida, E., & Sakai, H. (1988). Protein-kinase C catalyzed phosphorylation of the microtubule-binding domain of microtubule-associated protein 2 inhibits its ability to induce tubulin polymerization. Eur. J. Biochem. 174, 225–230.

Huber, G. & Matus, A. (1984). Differences in the cellular distribution of two microtubule-associated proteins, MAP1 and MAP2, in rat brain. J. Neurosci. 4, 151–160.

Izant, J.G. & McIntosh, J.R. (1980). Microtubule-associated proteins: a monoclonal antibody to MAP2 binds to differentiated neurons. Proc. Natl. Acad. Sci. USA 77, 4741–4745.

Jia, W.-G., Beaulieu, C., Huan, F.L., & Cynader, M.S. (1990). Protein kinase C immunoreactivity in kitten visual cortex is developmentally regulated and input-dependent. Devl. Brain Res. 57, 209–221.

Julien, J.-P. & Mushynski, W.E. (1982). Multiple phosphorylation sites in mammalian neurofilament polypeptides. J. Biol. Chem. 257, 10467–10470.

Kadowaki, T., Nishida, E., Kasuga, M., Akiyama, T., Takaku, F., Ishikawa, M., Sakai, H., Kathuria, S., & Fujita-Yamaguchi, Y. (1985). Phosphorylation of fodrin (nonerythroid spectrin) by the purified insulin receptor kinase. Biochem. Biophys. Res. Commun. 127, 493–500.

Kaplan, M.P., Chin, S.S.M., Fliegner, K.S., & Liem, R.K.H. (1990). Alpha-internexin, a novel neuronal intermediate filament protein, precedes the low molecular weight neurofilament protein (NF-L) in the developing rat brain. J. Neurosci. 10, 2735–2748.
Kim, H., Strong, T.V., & Anderson, S.J. (1991). Evidence for tau expression in cells of monocyte lineages and its in vitro phosphorylation by v-fms kinase. Oncogene 6, 1085–1087.
Kindler, S., Schulz, B., Goedert, M., & Garner, C.C. (1990). Molecular structure of microtubule-associated protein 2b and 2c from rat brain. J. Biol. Chem. 265, 19679–19684.
Knops, J., Kosik, K.S., Lee, G., Pardee, J.D., Cohen-Gould, L., & McCologue, L. (1991). Overexpression of tau in a nonneuronal cell induces long cellular processes. J. Cell Biol. 114, 725–733.
Kondo, H., Takahashi, H., Abe, H., Watanabe, M., & Takahashi, Y. (1991). Differential changes in expression of the neurofilament triplet protein-immunoreactivity in Purkinje cells of the cerebellum during postnatal development of rats. Arch. Histol. Cytol. 54, 437–445.
Kosik, K.S., Orecchio, L.D., Bakalis, S., & Neve, R.L. (1989). Developmentally regulated expression of specific tau sequences. Neuron 2, 1389–1397.
Krebs, K.E., Zagon, I.S., & Goodman, S.R. (1987a). Amelin and Synapsin I are 4.1 related spectrin binding proteins in brain. Brain Res. Bull. 18, 793–798.
Krebs, K.E., Zagon, I.S., & Goodman, S.R. (1987b). Amelin: a 4.1-related spectrin-binding protein found in neuronal cell bodies and dendrites. J. Neurosci. 7, 3907–3914.
Kyriakis, J.M. & Avruch, J. (1990). pp54 microtubule-associated protein 2 kinase. J. Biol. Chem. 265, 17355–17363.
Lasek, R.J. & Hoffman, P.N. (1983). The neuronal cytoskeleton, axonal transport and axonal growth. Cell Mot. 1, 1021–1049.
Lasek, R.J., Garner, J.A., & Brady, S.T. (1984). Axonal transport of the cytoplasmic matrix. J. Cell Biol. 99, 212s–221s.
Lazarides, E. & Nelson, W.J. (1983a). Erythrocyte and brain forms of spectrin in cerebellum: Distinct membrane-cytoskeletal domains in neurons. Science 220, 1295–1296.
Lazarides, E., & Nelson, W.J. (1983b). Erythrocyte form of spectrin in cerebellum: Appearance at a specific stage in the terminal differentiation of neurons. Science 222, 931–933.
Lee, G., Neve, R.L., & Kosik, K.S. (1989). The microtubule binding domain of tau protein. Neuron 2, 1615–1624.
Lee, G. (1990). Tau protein: An update on structure and function. Cell Mot. Cytoskel. 15, 199–203.
Lee, V.M.-Y., Carden, M.J., & Trojanowski, J.Q. (1986). Novel monoclonal antibodies provide evidence for the in situ existence of a nonphosphorylated form of the largest neurofilament subunit. J. Neurosci. 6, 850–858.
Leterrier, J.F., Liem, R., & Shelanski, M. (1982). Interaction between neurofilaments and microtubule-associated proteins: A possible mechanism for interorganellar bridging. J. Cell Biol. 95, 982–986.
Leto, T.L., Fortugno, D., Barton, D., Yang-Feng, T., Francke, U., Harris, A.S., Morrow, J.S., Marchsi, V.T., & Benz, E.J., Jr. (1988). Comparison of nonerythroid α-spectrin genes reveals strict homology among diverse species. Molec. Cell. Biol. 8, 1–9.
Letourneau, P.C. (1983). Differences in the organization of actin in the growth cones compared with the neurites of cultured neurons from chick embryos. J. Cell Biol. 97, 963–973.
Levine, J. & Willard, M. (1981). Axonally transported polypeptides associated with the internal periphery of many cells. J. Cell Biol. 90, 631–643.
Lewis, S.A., Wang, D., & Cowan, N.J. (1988). Microtubule-associated protein MAP2 shares a microtubule binding motif with tau protein. Science 242, 936–939.
Lindwall, G. & Cole, R.D. (1984). Phosphorylation affects the ability of tau protein to promote microtubule assembly. J. Biol. Chem. 259, 5301–5305.
Loomis, P.A., Howard, T.H., Castleberry, R.P., & Binder, L.I. (1990). Identification of nuclear tau isoforms in human neuroblastoma cells. Proc. Natl. Acad. Sci. USA 87, 8422–8426.

Luca, F.C., Bloom, G.S., & Vallee, R.B. (1986). A monoclonal antibody that cross-reacts with phosphorylated epitopes on two microtubule-associated proteins and two neurofilament polypeptides. Proc. Natl. Acad. Sci. USA 83, 1006–1010.

Ma, Y., Zimmer, W.E., Riederer, B.M., Bloom, M.L., Barker, J.E., & Goodman, S.R. (1992). The complete amino acid sequence for brain β spectrin (β fodrin): Relationship to globin sequence. Mol. Brain Res. 18, 87–99.

Marc, C., Clavel, M.-C., & Rabié, A. (1986). Non-phosphorylated and phosphorylated neurofilaments in the cerebellum of the rat: An immunocytochemical study using monoclonal antibodies. Development in normal and thyroid-deficient animals. Devl. Brain Res. 26, 249–260.

Mareck, A., Fellous, A., Francon, A., & Nunez, J. (1980). Changes in composition and activity of microtubule-associated proteins during brain development. Nature 284, 353–355.

Marugg, R.A. & Baier-Kustermann, W. (1988). Electroelution of high yield purified neurofilament constituent proteins; monoclonal antibodies directed against structurally defined NF binding-sites. In: Structure and Functions of the Cytoskeleton, Vol. 171, (Rousset, B.A.F., ed.), pp 19–24, Colloque INSERM/John Libbey/Eurotext Ltd., Paris.

Masliah, E., Iimoto, D.S., Saitoh, T., Hansen, L.A., & Terry, R.D. (1990). Increase immunoreactivity of brain spectrin in Alzheimer disease: A marker for synapse loss? Brain Res. 531, 36–44.

Matsudaira, P. (1991). Modular organization of actin crosslinking proteins. Trends Biochem. Sciences 16, 87–92.

Matus, A., Huber, G., & Bernhardt, R. (1983). Neuronal microdifferentiation. In: Cold Spring Harbor Symposia on Quantitative Biology. Cold Spring Harbor Laboratory, Vol. 158, pp. 775–782.

Matus, A. (1988a). Microtubule-associated proteins: Their potential role in determining neuronal morphology. Ann. Rev. Neurosci. 11, 29–44.

Matus, A. (1988b). Neurofilament protein phosphorylation—where, when and why. Trends Neurosci. 11, 291–292.

McMahon, A.P., Giebelhaus, D.H., Champion, J.E., Bailes, J.A., Lacey, S., Carritt, B., Henchman, S.K., & Moon, R.T. (1987). cDNA cloning, sequencing and chromosome mapping of a non-erythroid spectrin, human α-fodrin. Diff. 34, 68–78.

McQuarrie, I.G., Brady, S.T., & Lasek, R.J. (1989). Retardation in the slow axonal transport of cytoskeletal elements during maturation and aging. Neurobiol. Aging 10, 359–365.

Meichsner, M., Doll, T., Reddy, D., Weisshaar, B., & Matus, A. (1993). The low molecular weight form of microtubule-associated protein 2 is transported into both axons and dendrites. Neurosci. 54, 873–880.

Mitchison, T. & Kirschner, M. (1988). Cytoskeletal dynamics and nerve growth. Neuron 1, 761–772.

Murthy, A.S. & Flavin, M. (1983). Microtubule assembly using microtubule-associated protein MAP2 prepared in defined states of phosphorylation with protein kinase and phosphatase. Eur. J. Biochem. 137, 37–46.

Nelson, W.J. & Lazarides, E. (1984). The pattern of expression of two ankyrin isoforms demonstrate distinct steps in the assembly of the membrane skeleton in neuronal morphogenesis. Cell 39, 309–320.

Nixon, R.A, Lewis, S.E., Dahl, D., Marotta, C.A., & Drager, U.C. (1989). Early posttranslational modifications of the three neurofilament subunits in mouse retinal ganglion cells: Neuronal sites and time course in relation to subunit polymerization and axonal transport. Molec. Brain Res. 5, 93–108.

Nixon, R.A. (1991). Axonal transport of cytoskeletal proteins. In: The Neuronal Cytoskeleton (Burgoyne, R.D., ed.), pp. 283–307, Wiley-Liss, Inc., New York.

Nixon, R.A. & Sihag, R. (1991). Neurofilament phosphorylation: A new look at regulation and function. Trends in Neurosci. 14, 501–506.

Noble, M., Lewis, S.A., & Cowan, N.J. (1989). The microtubule binding domain of microtubule-associated protein MAP1B contains a repeated sequence motif unrelated to that of MAP2 and tau. J. Cell Biol. 109, 3367–3376.

Nunez, J. (1986). Differential expression of microtubule components during brain development. Dev. Neurosci. 8, 125–141.

Nunez, J. (1988). Immature and mature variants of MAP2 and tau proteins and neuronal plasticity. Trends Neurosci. 11, 477–479.

Oblinger, M.M. (1987). Biochemical composition of posttranslational processing of the mammalian high-molecular-weight neurofilament protein in vivo. J. Neurosci. 7, 2510–2521.

Oblinger, M.M., Argasinski, A., Wong, J., & Kosik, K.S. (1991). Tau gene expression in rat sensory neurons during development and regeneration. J. Neurosci. 11, 2453–2459.

Okabe, S., Hirokawa, N. (1988). Microtubule dynamics in nerve cells: analysis using microinjection of biotinylated tubulin into PC12 cells. J. Cell Biol. 107, 651–664.

Okabe, S. & Hirokawa, N. (1990). Turnover of fluorescently labelled tubulin and actin in the axon. Nature 343, 479–482.

Okabe, S. & Hirokawa, N. (1991). Actin dynamics in growth cones. J. Neurosci. 11, 1918–1929.

Olmsted, J.B. (1986). Microtubule-associated proteins. Ann. Rev. Cell Biol. 2, 421–457.

Pachter, J.S. & Liem, R.K.H. (1985). Alpha-internexin, a 66-kD intermediate filament-binding protein from mammalian central nervous tissues. J. Cell Biol. 101, 1316–1322.

Papandrikopoulou, A., Doll, T., Tucker, R.P., Garner, C.C., & Matus, A. (1989). Embryonic MAP2 lacks the cross-linking sidearm sequences and dendritic targeting signal of adult MAP2. Nature 340, 650–652.

Papasozomenos, C.S. & Binder, L.I. (1987). Phosphorylation determines two distinct species of tau in the central nervous system. Cell Mot. Cytoskel. 8, 210–226.

Portier, M.-M., deNéchaud, B., & Gros, B. (1984). Peripherin: A new member of the intermediate filament protein family. Dev. Neurosci. 6, 325–344.

Portier, M.-M. (1992). Le cytosquelette neuronal: Aspects structuraux, fonctionnels et dynamiques. Rev. Neurol. 148, 1–19.

Purkinje, J.E. (1838). Neueste Untersuchungen aus der Nerven- und Hirn-anatomie. In: Berichte über die Versammlung deutscher Naturforscher und Aerzte in Prag im September 1837 (Sternberg, K., & Von Krombolz, J.V., eds.), pp. 177–180, Prague: Hasse Publishing House.

Reddy, D.A., Brugg, B., & Matus, A. (1991). Antisense oligonucleotides directed against microtubule-associated protein 5 inhibit process formation in nerve growth factor induced PC12 cells. J. Cell Biol. 115, 341a.

Reinsch, S.S., Mitchison, T.J., & Kirschner, M.W. (1991). Microtubule polymer assembly and transport during axonal elongation. J. Cell Biol. 115, 365–380.

Riederer, B. & Matus, A. (1985). Differential expression of distinct microtubule-associated proteins during brain development. Proc. Natl. Acad. Sci. USA 82, 6006–6009.

Riederer, B., Cohen, R., & Matus, A. (1986a). MAP5: A novel microtubule-associated protein under strong developmental regulation. J. Neurocytol. 15: 763–775.

Riederer, B.M., Zagon, I.S., & Goodman, S.R. (1986b). Brain spectrin(240/235) and brain spectrin(240/235E). Two distinct spectrin subtypes with different location within mammalian neural cells. J. Cell Biol. 102, 2088–2097.

Riederer, B.M., Zagon, I.S., & Goodman, S.R. (1987). Brain spectrin(240/235) and brain spectrin(240/235E): Differential expression during mouse brain development. J. Neurosci. 7, 864–874.

Riederer, B.M., Lopresti, L.L., Krebs, K.E., Zagon, I.S., & Goodman, S.R. (1988). Brain spectrin(240/235) and brain spectrin(240/235E): Conservation of structure and location within mammalian neural tissue. Brain Res. Bull. 21, 607–616.

Riederer, B.M. (1990). Some aspects of the neuronal cytoskeleton during brain development. Eur. J. Morphol. 28, 347–378.

Riederer, B.M., Guadano-Ferraz, A., & Innocenti, G.M. (1990). Differences in distribution of microtubule-associated protein 5a and 5b during cat cerebral cortex and corpus callosum development: Dependence on phosphorylation. Devel. Brain Res. 56, 235–243.

Riederer, B.M. & Innocenti, G.M. (1991). Differential distribution of tau proteins in developing cat cerebral cortex and corpus callosum. Eur. J. Neurosci. 3, 1134–1145.

Riederer, B.M. (1992). Phosphorylation of some cytoskeletal proteins during brain development. Histochem. J. 24, 783–790.

Riederer, B.M. (1992b). Generation of a polyclonal antibody to aldehyde fixation resistant epitopes in neurofilament proteins. Brain Res. Bull. 30, 623–627.

Riederer, B.M., Monnet-Tschudi, F., Honegger, P. (1992). Development and maintenance of the neuronal cytoskeleton in aggregated cell cultures of fetal rat telencephalon, and influence of elevated potassium ion concentration. J. Neurochem. 56, 649–658.

Riederer, B.M. & Innocenti, G.M. (1992). MAP2 isoforms in developing cat cerebral cortex and corpus callosum. Eur. J. Neurosci. 4, 1376–1386.

Riederer, B.M., Porchet, R., & Marugg, R. (1995). Appearance and distribution of the three neurofilament subunits during cat cerebellum development. Manuscript submitted.

Riederer, B.M. (1995). Differential phosphorylation of MAP1b during postnatal development of the cat brain. J. Neurocytol. 24, 45–54.

Robinson, P.A. & Anderton, B.H. (1988). Neurofilament probes—A review of neurofilament distribution and biology. Rev. Biochem. 2, 1–40.

Robson, S.J. & Burgoyne, R.D. (1989). Differential localization of tyrosinated, detyrosinated, and acetylated α-tubulins in neurites and growth cones of dorsal root ganglion neurons: implications for axonal transport of tubulin. Cell Motil. Cytoskel. 12, 273–282.

Roder, H.M. & Ingram, V.M. (1991). Two novel kinases phosphorylate tau and the KSP site of heavy neurofilament subunits in high stoichiometric ratios. J. Neurosci. 11, 3325–3343.

Sano, M., Nishiyama, K., & Kitajiama, S. (1990). A nerve growth factor-dependent protein kinase that phosphorylated microtubule-associated proteins in vitro: Possible involvement of its activity in the outgrowth of neurites form PC12 cells, J. Neurochem. 55, 427–435

Sato-Yoshitake, R., Shiomura, Y., Miyasaka, H., & Hirokawa, N. (1989). Microtubule-associated protein 1B: Molecular structure, localization, and phosphorylation-dependent expression in developing neurons. Neuron 3, 229–238.

Sattilaro, W. (1986). Interaction of microtubule-associated protein 2 with actin filaments. Biochem. 25, 2003–2009.

Schoenfeld, T.A., McKerracher, L., Obar, R., Vallee, R.B. (1989). MAP1a and MAP1b are structurally related microtubule associated proteins with distinct developmental patterns in the CNS. J. Neurosci. 9, 1712–1730.

Schultze, H. (1878). Axenzylinder und Ganlienzelle: Mikroskopische Studien über die Struktur der Nervenfaser und Nervenzelle bei Wirbelthieren. Arch. Anat. Physiol. 2, 259–287.

Seeds, N.W., Gilman, A.G., Amanos, T., & Nirenberg, M. (1970). Regulation of axon formation by clonal lines of a neural tumor. Proc. Natl. Acad. Sci. USA 66, 160–167.

Seger, R., Ahn, N.G., Boulton, T.G., Yancopoulos, G.D., Panayotatos, A. N., Radziejewska, E., Ericson, L., Bratlien, R.L., Cobb, M.H., & Krebs, E.G. (1991). Microtubule-associated protein 2 kinases, ERK1 and ERK2, undergo autophosphorylation of both tyrosine and threonine residues: implications for their mechanism of activation. Proc. Natl. Acad. Sci. USA 88, 6142–6146.

Selden, S.C. & Pollard, T.D. (1983). Phosphorylation of microtubule-associated proteins regulates their interaction with actin filaments. J. Biol. Chem. 258, 7064–7071.

Shaw, G. & Weber, K. (1982). Differential expression of neurofilament triplet proteins in brain development. Nature 298, 277–279.

Shaw, G., Osborn, M., & Weber, K. (1986). Reactivity of a panel of neurofilament antibodies on phosphorylated and dephosphorylated neurofilaments. Eur. J. Cell Biol. 42, 1–9.

Shaw, G. (1991). Neurofilament proteins. In: The Neuronal Cytoskeleton (Burgoyne, R.D., ed.), pp. 185–214, Wiley-Liss, Inc., New York.

Shimo-Oka, T. & Atsumi, S. (1986). Localization of α-spectrin in chicken and monkey ventral horns by immunoelectron microscopy. J. Neurocytol. 15, 715–723.

Sihag, R.K. & Nixon, R.A. (1991). Identification of Ser-55 as a major protein kinase A phosphorylation site on the 70-kDa subunit of neurofilaments. J. Biol. Chem. 266, 18861–18867.

Singer, H. (1992). The cytoskeleton and mRNA localization. Curr. Op. Cell Biol. 4, 15–19.

Sobue, K. & Kanda, K. (1989). Alpha-actinins, calspectin (brain spectrin or fodrin), and actin participate in adhesion and movement of growth cones. Neuron 3, 311–319.

Speicher, D.W. & Marchesi, V.T. (1984). Erythrocyte spectrin is comprised of many homologous triple helical segments. Nature 311, 177–180.

Speicher, D.W. (1986). The present status of erythrocyte spectrin structure: The 106-residue repetitive structure is a basic feature of an entire class of proteins. J. Cell Biochem. 30, 245–258.

Spooner, B.S. & Holladay, C.R. (1981). Distribution of tubulin and actin in neurites and growth cones of differentiating nerve cells. Cell Motil. Cytoskel. 1, 167–178.

Steiner, B., Mandelkow, E.M., Biernat, J., Guske, N., Meyer, H.E., Schmidt, B., Mieskes, G., Soellinger, H.D., Drechsel, D., Kirschner, M.W., Goedert, M., & Mandelkow, E. (1990). Phosphorylation of microtubule-associated protein tau: Identification of the site for Ca^{2+}-calmodulin dependent kinase and the relationship with tau phosphorylation in Alzheimer tangles. EMBO J. 9, 3539–3544.

Steinert, P.M. & Roop, D.R. (1988). Molecular and cellular biology of intermediate filaments. Ann. Rev. Biochem. 57, 593–625.

Sternberger, L.A. & Sternberger, N.H. (1983). Monoclonal antibodies distinguish between phosphorylated and nonphosphorylated forms of neurofilaments in situ. Proc. Natl. Acad. Sci. USA 80, 6126–6130.

Steward, O. & Reeves, T.M. (1988). Protein-synthetic machinery beneath postsynaptic sites on CNS neurons: Association between polyribosomes and other organelles at the synaptic site. J. Neurosci. 8, 176–184.

Steward, O. & Ribak, C.E. (1986). Polyribosomes associated with synaptic specializations on axon initial segments: Localization of protein-synthetic machinery at inhibitory synapses. J. Neurosci. 6, 3079–3085.

Szaro, B.G., Whitnali, M.H., & Gainer, H. (1990). Phosphorylation-dependent epitopes on neurofilament proteins and neurofilament densities differ in axons in the corticospinal and primary sensory dorsal column tracts in the rat spinal cord. J. Comp. Neurol. 302, 220–235.

Theurkauf, W. & Vallee, R.B. (1983). Extensive cAMP-dependent and cAMP-independent phosphorylation of microtubule-associated protein 2. J. Biol. Chem. 258, 7883–7886.

Torre, E.R. & Steward, O. (1992). Demonstration of local protein synthesis within dendrites using a new cell culture system that permits the isolation of living axons and dendrites from their cell bodies. J. Neurosci. 12, 762–772.

Tucker, R.P., Binder, L.I., & Matus, A.I. (1988a). Neuronal microtubule-associated proteins in the embryonic avian spinal cord. J. Comp. Neurol. 271, 44–55.

Tucker, R.P., Binder, L.I., Viereck, C., Hemmings, B.A., & Matus, A.I. (1988b). The sequential appearance of low- and high-molecular weight forms of MAP2 in the developing cerebellum. J. Neurosci. 8, 4503–4512.

Tucker, R.P., Garner, C.C., & Matus, A. (1989). In situ localization of microtubule-associated protein mRNA in the developing and adult rat brain. Neuron 2, 1245–1256.

Tucker, R.P. (1990). The roles of microtubule-associated proteins in brain morphogenesis: A review. Brain Res. Rev. 15, 101–120.

Tytell, M., Brady, S.T., & Lasek, R.J. (1984). Axonal transport of a subclass of tau proteins: Evidence for the regional differentiation of microtubules in neurons. Proc. Natl. Acad. Sci. USA 81, 1570–1574.

Vale, R.D. (1987). Intracellular transport using microtubule-based motors. Ann. Rev. Cell Biol. 3, 347–378.

Vallee, R.B., Shpetner, H.S., & Paschal, B.M. (1989). The role of dynein in retrograde axonal transport. Trends in Neurosci. 12, 66–70.

Van der Loos, H. (1967). The history of the neuron. In: The Neuron (Hydén, H., ed.), pp.1–47, Elsevier, Amsterdam.

Viereck, C. & Matus, A. (1989). The expression of phosphorylated and non-phosphorylated forms of MAP5 in the amphibian CNS. Dev. Brain Res. 508, 257–264.

Viereck, C., Tucker, R.P., & Matus, A. (1989). The adult rat olfactory system expresses microtubule-associated proteins found in the developing brain. J. Neurosci. 9, 3547–3557.

Vitadello, M. & Denis-Donini, S. (1990). Expression of neurofilament proteins in granule cells of the cerebellum. Brain Res. 509, 47–54.

Watson, D.F., Hoffman, P.N., Fittro, K.P., & Griffin, J.W. (1989). Neurofilament and tubulin transport slows along the course of mature motor axons. Brain Res. 477, 225–232.

Weisenberg, R.C., Flynn, J., Gao, B., Awodi, S., Skee, F., Goodman, S.R., & Riederer, B.M. (1987). Microtubule gelation-contraction: Essential components and relationship to slow axonal transport. Science 238, 1119–1122.

Willard, M. & Simon, C. (1983). Modulation of neurofilament axonal transport during the development of rabbit retinal ganglion cells. 35, 551–559

Winkelmann, J.C. & Forget, B.G. (1993). Erythroid and nonerythroid spectrins. Blood 81, 3173–3185.

Wolda, S.L. & Moon, R.T. (1990). Variable expression in human cells of mRNA transcripts encoding α-fodrin isoforms. J. Cell Biol. 111, 47a.

Woodhams, P.L., Calvert, R., & Dunnett, S.B. (1989). Monoclonal antibody G10 against microtubule-associated protein 1x distinguishes between growing and regenerating axons. Neurosci. 28, 49–59.

Yaffe, M.B., Farr, G.W., Miklos, D., Horwich, A.L., Sternlicht, M.L., & Sternlicht, H. (1992). TCP1 complex is a molecular chaprone in tubulin biogenesis. Nature 358, 245–248.

Yamada, K.M., Spooner, B.S., & Wessels, N.K. (1970). Axon growth: Roles of microfilaments and microtubules. Proc. Natl. Acad. Sci. USA 66, 1202–1212.

Yamamoto, H., Fukunaga, K., Tanaka, E., & Miyamoto, E. (1983). Ca^{2+}- and calmodulin-dependent phosphorylation of microtubule-associated protein 2 and tau factor, and inhibition of microtubule assembly. J. Neurochem. 41, 1119–1125.

Zagon, I.S., Higbee, R., Riederer, B.M., & Goodman, S.R. (1986). Spectrin subtypes in mammalian brain: An immunoelectron microscopic study. J. Neurosci. 6, 2977–2986.

Zauner, W., Kratz, J., Staunton, J., Feick, P., & Wiche, G. (1992). Identification of two distinct microtubule binding domains on recombinant rat MAP1b. Eur. J. Cell Biol. 57, 66–74.

Zimmer, W.E., Ma, Y., Zagon, I.S., & Goodman, S.R. (1992). Developmental expression of brain β-spectrin isoform messenger RNAs. Brain Res. 594, 75–83.

REGULATION OF ADHERENS JUNCTION PROTEIN LEVELS:

ITS ROLE IN CELL MOTILITY AND TUMORIGENICITY

Avri Ben-Ze'ev, José Luis Rodríguez Fernández, Ursula Glück, Daniela Salomon, Margot Zöller, and Benjamin Geiger

Advances in Molecular and Cell Biology
Volume 12, pages 143–163.

ISBN: 1-55938-845-5

I. INTRODUCTION

Cell-cell and cell-extracellular matrix (ECM) contacts are central determinants of cell and tissue morphology (Hay, 1981; Takeichi, 1991). These interactions of the cell with its environment are mediated by numerous transmembrane receptors of the integrin and cadherin families of receptors (Takeichi, 1991; Hynes, 1992). In the cytoplasmic domain, many of these adhesive interactions are linked to the microfilament system by plaque proteins which are especially prominent in adherens type junctions (AJ) (Burridge et al., 1988; Geiger and Ginsberg, 1991; Tsukita et al., 1992), thus forming a structural link between the extracellular and the intracellular domain of the plasma membrane. Earlier observations have demonstrated effects on gene expression, cell growth, motility, and differentiation in response to altered cell-cell and cell-ECM interactions (Ben-Ze'ev, 1991; 1992). In the last decade, the number of components that have been shown to localize in AJ has steadily increased (Burridge et al., 1988; Geiger and Ginsberg, 1991; Tsukita et al., 1992; Sadler et al., 1992; Ben-Ze'ev, 1992). In addition to proteins which are involved in establishing the structural and thus more mechanical interaction between the outside and inside of the cell, AJ contain a considerable number of regulatory molecules including tyrosine kinases, protein kinase C, and various oncogene products and proteases, which reside in the cytoplasmic side of the plaque area (Burridge et al., 1988; Geiger and Ginsberg, 1991; Luna and Hitt, 1992). These findings suggested that the junctional areas of cell-cell and cell-ECM interaction are directly involved in signal transduction and thereby elicit long range effects on cells.

Tyrosine phosphorylation of AJ proteins has been proposed to be a major mechanism in this signal transduction, based on studies showing abundant tyrosine phosphorylation-dephosphorylation activity in AJ of both normal and transformed cells (Rohrschneider, 1980; Comoglio et al., 1984; Maher et al., 1985; Volberg et al., 1992). The recently characterized adhesion plaque tyrosine kinase $p125^{FAK}$ (Schaller et al., 1992) has raised wide interest since this protein is rapidly phosphorylated on tyrosine by both cytokine stimulation and by adhesion. It proceeds through activation of integrins in fibroblasts adhering to the ECM (Guan et al., 1991; Burridge et al., 1992) and in platelet activation by soluble factors (Lipfert et al., 1992), but it is also apparent in the transformation of fibroblasts by RSV (Guan and Shalloway, 1992), and after stimulation of neuropeptide receptors by various soluble components (Zachary and Rozengurt, 1992). This tyrosine phosphorylation, in turn, enhances the kinase activity of $p125^{FAK}$. These studies therefore suggest that signal transduction by cell-ECM interaction and by different cytokines may converge at an early stage, in adhesion plaques. They also imply that there is a common mechanism for the effects on cell shape, motility, growth and differentiation resulting from either cell adhesion and/or soluble factor action on cells.

The assembly of AJ apparently proceeds through an initial binding of the transmembrane contact receptor to its extracellular ligand (an ECM protein sequence, or a homologous cell adhesion (CAM) receptor (Geiger et al., 1990, 1992a)). However, recent studies have clearly demonstrated that following this initial step, the recruitment and binding of the plaque and cytoskeletal components to the clustered receptor constitutes an important maturation step in the formation of stable AJ (La Flamme et al., 1992; Geiger et al., 1992a,b). The binding of cytoskeletal components to the transmembrane contact receptor may be required for the long range effects on cells (Geiger et al., 1992a). This maturation and/or stabilization step in AJ formation can be controlled both by post-transcriptional mechanisms (i.e., phosphorylation-dephosphorylation), or by regulating the amount of a particular junctional protein.

In this essay we summarize studies demonstrating AJ modulation through changes in the level of expression of junctional components, and show that this modulation has a dramatic effect not only on cell structure, but also on cell motility and the tumorigenic ability of cells.

II. GROWTH STIMULATION AND AJ PROTEIN EXPRESSION

Previous studies have shown that those changes in cell adhesion to the ECM, or to neighboring cells, which are common to growth activation, differentiation, and transformation, are associated with alterations in both cytoskeletal protein assembly and expression (Ben-Ze'ev, 1984, 1986, 1991; Ungar et al., 1986; Bendori et al., 1987). For example, growth-activation of quiescent 3T3 cells by serum or purified growth factors is followed by a rapid assembly of microfilaments into stress fibers terminating in large vinculin-containing AJ (Ridley and Hall, 1992). In addition to microfilament and vinculin assembly, the expression of vinculin, which is considered a constitutive cellular protein, is also increased in quiescent 3T3 cells treated with serum factors (Ben-Ze'ev et al., 1990; Bellas et al., 1991). Vinculin transcription is undetectable in quiescent 3T3 cells, but its transcription is extensively, yet transiently, induced within 15 to 30 min after growth-factor stimulation (Ben-Ze'ev et al., 1990; Bellas et al., 1991), and is followed by an increase in vinculin RNA level (Figure 1A) and protein synthesis (Ben-Ze'ev et al., 1990; Bellas et al., 1991). This transient increase in vinculin expression is independent of new protein synthesis (Ben-Ze'ev et al., 1990).

The relevance of this change in vinculin expression to growth-stimulation of cells is supported by results showing a similar transient increase in vinculin RNA level in regenerating adult liver following partial hepatectomy (Figure 2A).

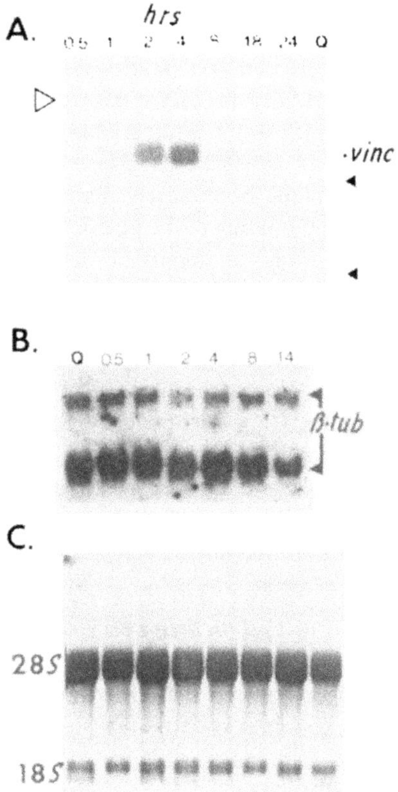

Figure 1. Transient increase in vinculin RNA in serum-stimulated quiescent cells. Quiescent (Q) 3T3 cells were stimulated with serum and after 30 min, 1, 2, 4, 8, 18, and 24 h total RNA was extracted, separated on agarose gel and transferred to nitrocellulose. The blots were stained with methylene blue (**C**) and hybridized to ^{32}P-mouse vinculin cDNA (**A**), and rat β-tubulin cDNA (**B**). The large arrowhead points to a putative precursor for vinculin mRNA. Note the transient increase in vinculin RNA levels (**A**), but not in β-tubulin RNA levels. (Reproduced from Ben-Ze'ev et al., 1990).

Vinculin organization and expression are also modulated during the differentiation of granulosa cells (Ben-Ze'ev and Amsterdam, 1987), adipocytes (Rodríguez Fernández and Ben-Ze'ev, 1989), smooth muscle cells (Belkin et al., 1988), and in migrating corneal epithelial cells (Zieske et al., 1989). Additional studies have demonstrated a transient induction in the expression of genes coding for other proteins which are involved in the formation of the ECM-microfilament complex

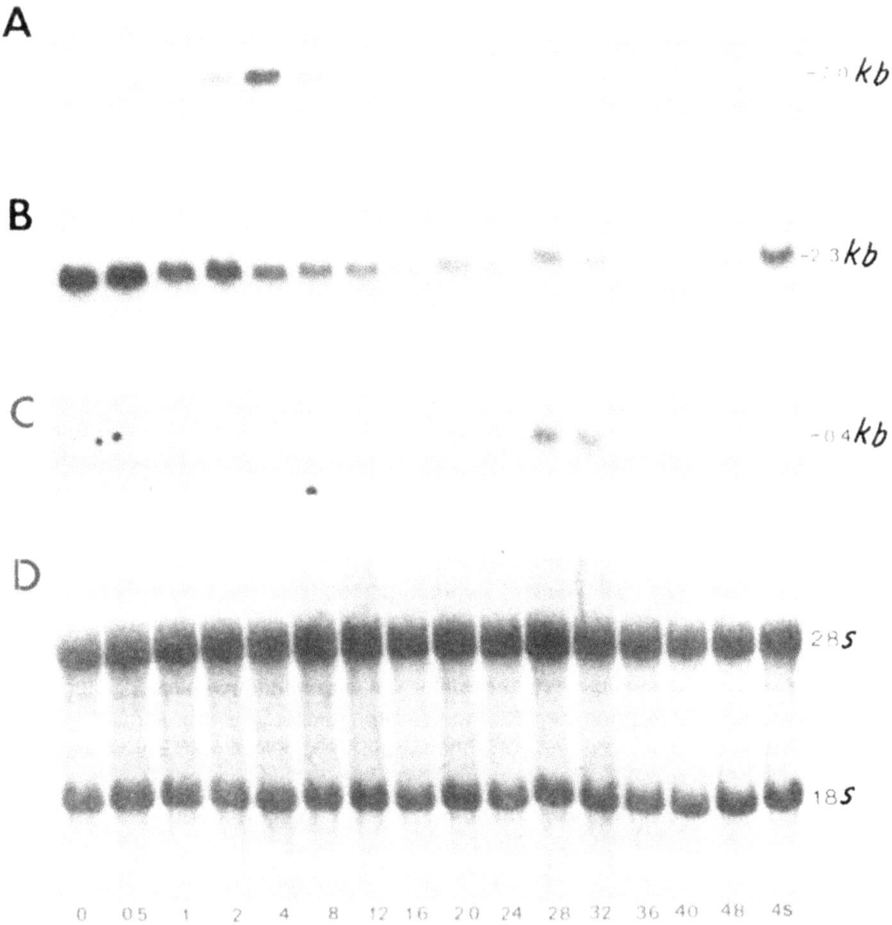

Figure 2. Induction of vinculin expression in regenerating liver. RNA was prepared from rat livers at various times (0–48 hours) after partial hepatectomy, after sham operation (4s), and from intact rats (0). Twenty micrograms of total RNA, per lane, were separated and analyzed by Northern blotting, followed by staining of the blot with Methylene blue to reveal the position of 18S and 28S rRNAs (**D**). The same blot was sequentially hybridized with ^{32}P-labeled cDNAs to vinculin (**A**), albumin (**B**), and histone H4 (**C**). The times after partial hepatectomy are labeled under each lane. The transient increase in vinculin RNA level (**A**) was not observed in tissue specific RNAs such as albumin (**B**), and was different from the increase in the S-phase marker histone RNA (**C**). (Reproduced from Glück et al., 1992).

in both growth-factor stimulated fibroblasts (Rysek et al., 1989; Bellas et al., 1991; Glück et al., 1992), in growth-stimulated hepatocytes in culture (Ben-Ze'ev et al., 1988; Mooney et al., 1992), and in regenerating liver (Glück et al., 1992). These include ECM components such as fibronectin, the transmembrane receptor β_1-integrin, and components of adhesion plaques and actin-binding proteins such as α-actinin, talin, and tropomyosin. The correlation between the regulation of AJ protein expression, in both cultured cells and *in vivo*, suggests that this modulation of cytoskeletal gene expression may have a physiologically relevant function.

III. OVEREXPRESSION AND ANTISENSE TRANSFECTION OF VINCULIN AFFECT CELL MOTILITY

To determine the role of changes in the expression of AJ proteins in cell function, 3T3 cells were transfected with a full length chicken vinculin cDNA construct and clones expressing stably different levels of the transgene were isolated (Rodríguez Fernández et al., 1992a). 3T3 cells expressing the transgene at 20% of the endogenous vinculin level displayed major changes in cell structure (Figure 3).

Actin stress fibers and vinculin containing plaques were significantly more abundant in such cells (Figure 3C, D) than in neo^r control clones (Figure 3A, B). Moreover, the motility of individual cells overexpressing vinculin measured by their ability to form phagokinetic tracks was dramatically reduced (Figure 3F compared to 3E). Furthermore, the rate at which such cells migrate and close an artificial wound introduced in a confluent monolayer is also reduced several fold (Figure 4).

To test the effect of reduced vinculin expression on cell motility, in contrast to the effect of vinculin overexpression described above, the antisense transfection method was used. An antisense expression-vector construct containing a 1-kb mouse vinculin cDNA, corresponding to the amino terminal of the protein (Ben-Ze'ev et al., 1990), was co-transfected into 3T3 cells with the neo^r gene, and G418 resistant clones expressing decreased vinculin levels were isolated (Figure 5 I, lanes 2, 5, and 12–14).

Cells with low levels of vinculin had a round phenotype (Figure 5 II, C, E, and F), poorly organized microfilaments, and small vinculin positive plaques (Figure 5 II, B). Moreover, individual cells with reduced vinculin content displayed an increase in their ability to locomote on colloidal gold, and a higher capability to form colonies in soft agar (Table 1).

These studies show that the movement of fibroblasts can be either enhanced or suppressed by decreasing or elevating, respectively, the expression of a single adhesion plaque protein vinculin. In addition, cells which display decreased vinculin have a rounded morphology, increased motility, a decrease in the assembly of

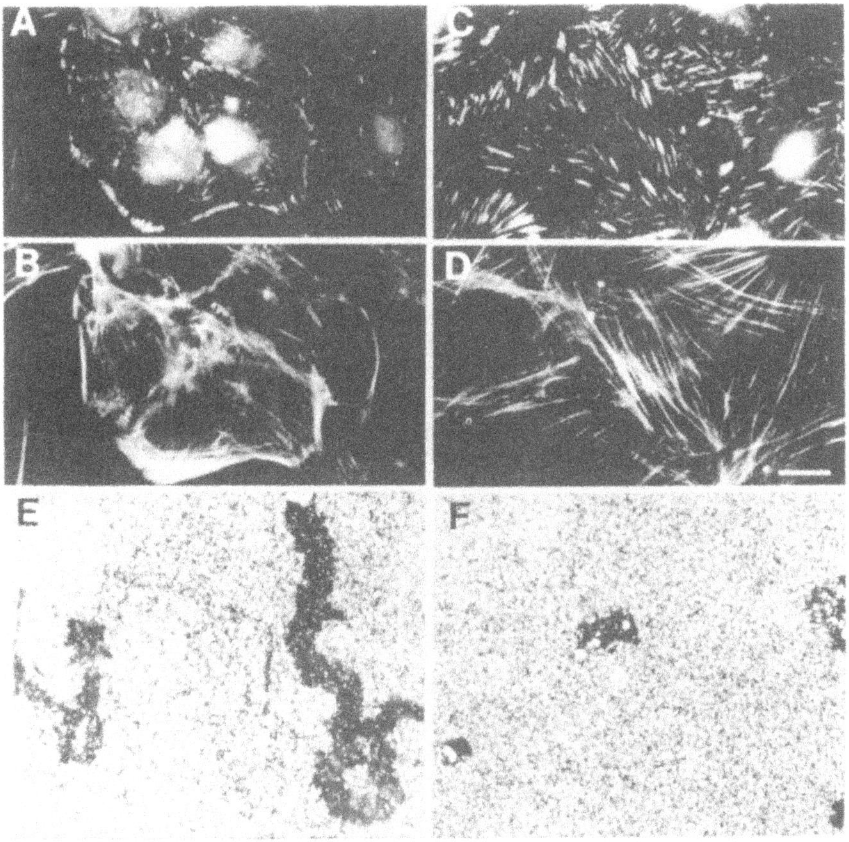

Figure 3. Increased stress fiber and vinculin assembly and decreased motility in 3T3 cells overexpressing vinculin. Double fluorescent labeling for vinculin (**A, C**) and actin (**B, D**) and dark field visualization of phagokinetic tracks (**E, F**) of 3T3 cells transfected with chicken vinculin (**C, D, F**), or with the neo^r selective marker (**A, B, E**). The level of chicken vinculin in the transfected cells is about 20% relative to the endogenous protein. The bar in **D** indicates 10 μm. (Reproduced from Geiger et al., 1992a).

actin filaments, and an increase in their ability to grow in suspension culture in agar. These properties are characteristic of transformed fibroblasts (Shin et al., 1975).

In other studies, another microfilament-associated protein gelsolin, which severs actin filaments, was overexpressed in 3T3 cells. This overexpression resulted in increased cell motility (Cunningham et al., 1991). In contrast, increasing the expression of actin binding protein (ABP-280) which enhances actin polymerization, was shown to decrease cell locomotion (Cunningham et al., 1992). These results are consistent with the view that modulations in the expression of individual proteins which are involved in the stabilization of microfilament-membrane complexes can regulate cell morphology and motility.

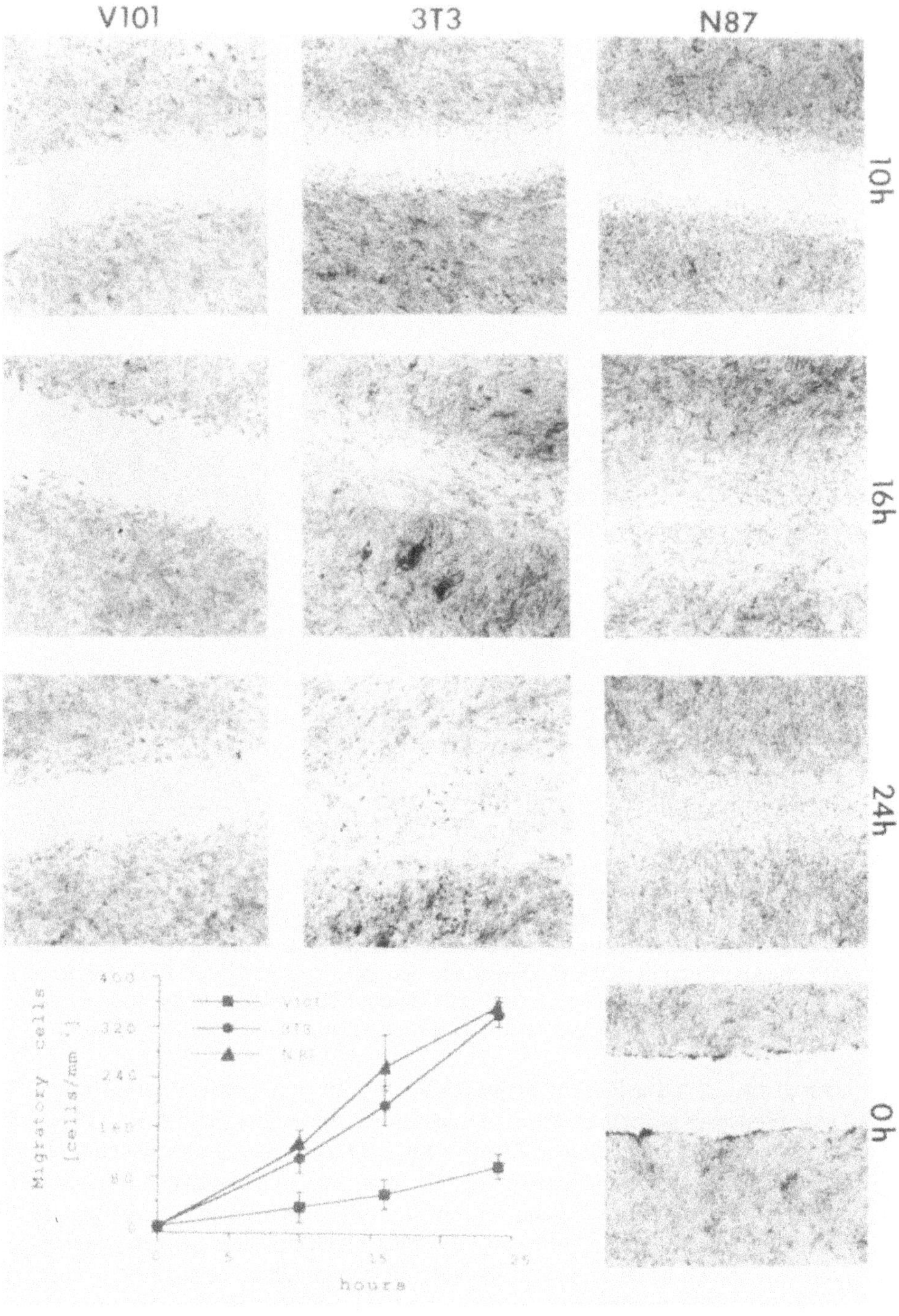
V101
3T3
N87
10h
16h
24h
0h

Figure 4. Reduced ability to migrate into a "wound" of vinculin-transfected cells. 3T3 cells, transfected cells with the neo^r gene alone (N87), and cells transfected with chicken vinculin (V101), were cultured to confluence in 24-well dishes. A wound was introduced in the monolayers with a sharp plastic scriber and the cultures were incubated in fresh medium. At various times after the wounding, the cultures were fixed and stained with Giemsa. The number of cells migrating into an area of 1 mm^2 was determined randomly in 10 different areas along the wound for each cell type at each time point. The decrease in the motile ability of the vinculin-overexpressing clone V101 is not apparent in neither untransfected 3T3 cells, nor in the neo^r N87 3T3 clone. (Reproduced from Rodríguez Fernández et al., 1992a).

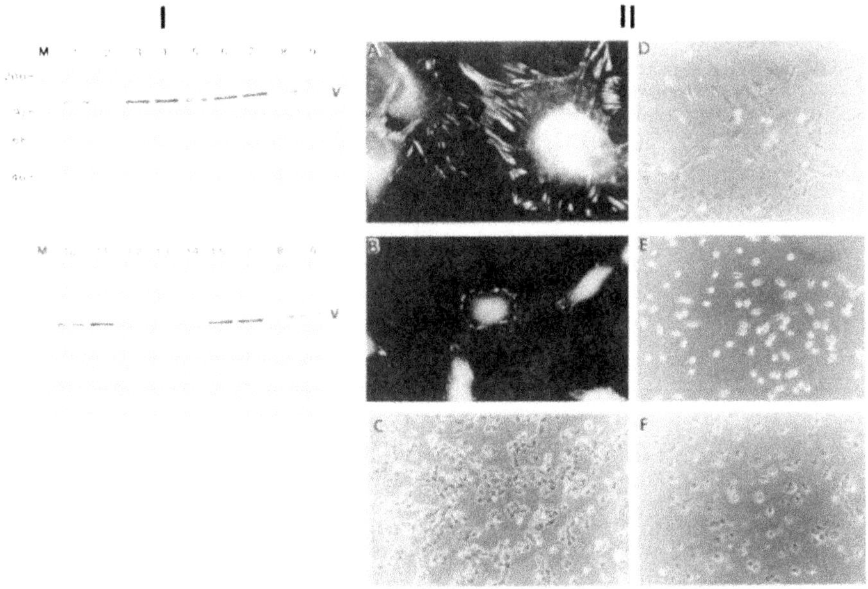

Figure 5. Antisense vinculin transfection confers a decrease in 3T3 cell size. 3T3 cells were co-transfected with a 1kb antisense mouse vinculin cDNA (Ben-Ze'ev et al., 1990) inserted in antisense orientation in an expression vector and with the neo^r gene. G418 resistant colonies were screened for reduced vinculin expression (**I**) and organization (**II**) by immunoblotting and immunofluorescence with anti vinculin antibody. The cells of clones expressing decreased vinculin levels (**I**, lanes 1, 2, and 12–14) had a round shape (**II, C, E, F**), and small peripheral vinculin-positive plaques (**II, B**), in contrast to neo^r control clones which had a flat phenotype with large vinculin plaques (**II, A, D**). Lanes 1–5 and 10–15 are independent clones transfected with antisense vinculin; lane 7, 3T3 control; lane 8, 3T3 one third the amount of protein loaded; lane 9, SV40-transformed 3T3 (SVT2) cells.

Table 1. Phenotypic and Motile Properties of Antisense-Vinculin Transfected 3T3 Clones

Cell Line	*Vinculin Levels % of 3T3*	*Morphology on Plastic*	*No. of Colonies in Agar /10 ^{4} Cells*	*Motility μm/day*
3T3	100	flat	ND	476±240
A1	69.6	flat	ND	ND
A4	106.7	flat	4	ND
A5	31.6	round	188	741±288
B5	16.6	round	575	900±344
B6	11.0	round	143	873±362
C1	102.7	flat	ND	418±185
C4	98.8	flat	20	ND
C5	43.4	flat	ND	ND
C6	10.3	round	113	926±423
D3	108.3	flat	ND	ND
D4	58.4	flat	6	ND
D5	78.9	flat	ND	ND

Note: Neo^r colonies expressing different levels of vinculin after antisense-vinculin transfection were analyzed for cell morphology on plastic, ability to form colonies in soft agar, and motility on colloidal gold-coated coverslips. Note that clones which express vinculin at 30% and lower of the control 3T3 vinculin level have a round morphology, increased ability to grow in agar, and increased motility.

IV. SUPPRESSION OF TUMORIGENICITY IN CELLS TRANSFECTED WITH VINCULIN cDNA

Transformation of cells is accompanied by changes in both cell growth and cell morphology. The structural changes are characterized by altered adhesive and motile properties, as well as by deterioration of the microfilament system (Pollack et al., 1975; Ben-Ze'ev, 1985; Raz and Ben-Ze'ev, 1987). The changes in cell morphology and cytostructure often correlate with an increase in the ability of tumorigenic cells to grow in semisolid medium in suspension, a phenomenon known as anchorage independence (Stoker et al., 1968; Shin et al., 1975). Furthermore, the loss of microfilament bundles in tumorigenic cells is associated with a decreased expression of several microfilament-associated proteins including tropomyosin (Matsumura and Yamashiro-Matsumura, 1986), gelsolin (Vandekerkhove et al., 1990), and vinculin (Raz et al., 1986).

While the changes in cell morphology and cell growth are both characteristics of tumor cells, the causal relationships between the alterations in cytostructure and malignant transformation are still unknown. To assess whether changes in cytoskeletal organization and cell adhesion are responsible for the loss of adhesion-dependent growth control and tumorigenicity, the expression of AJ proteins was modulated in tumorigenic epithelial and fibroblast cell lines, and their effect on the transformed cell phenotype was examined.

We have used SV40-transformed 3T3 cells (SVT2) which express several fold less vinculin, α-actinin and actin (Figure 6A compared to 6E), and a highly

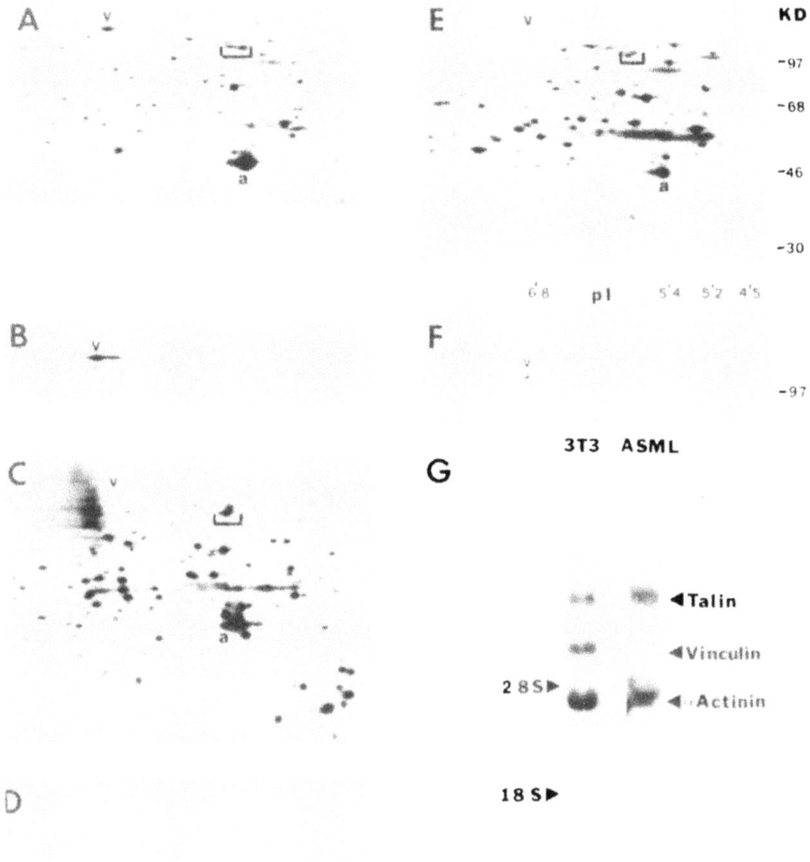

Figure 6. Levels of vinculin and α-actinin expression in 3T3, SVT2, and ASML cells. Cells were labeled for 20 hours with 100 μCi/ml of ^{35}S-methionine and equal amounts of radioactive total cell protein were separated by 2-D gel electrophoresis and transferred to nitrocellulose. The pattern of total cell proteins was visualized by exposing the blots to X-ray film (**A**, **C**, **E**). The blots were incubated with a broad-range anti vinculin antibody followed by alkaline phosphatase-bound anti mouse IgG (**B**, **D**, **F**). Total RNA was extracted from 3T3 and ASML cells and equal amounts of RNA were separated on agarose gel, blotted onto nitrocellulose and hybridized sequentially with mouse vinculin, human α-actinin, and mouse talin cDNAs (**G**). 3T3 (**A**, **B**); ASML (**C**, **D**); SVT2 (**E**, **F**); a, actin; v, vinculin. The brackets in **A**, **C**, and **E** mark the position of α-actinin. Note the decrease in vinculin, α-actinin and actin expression in SVT2 cells (**E**) compared to 3T3 (**A**), while ASML cells show a decreased vinculin (**C**), but not α-actinin or talin RNA levels compared to 3T3 cells (**G**). (Reproduced from Rodríguez Fernández et al., 1992b).

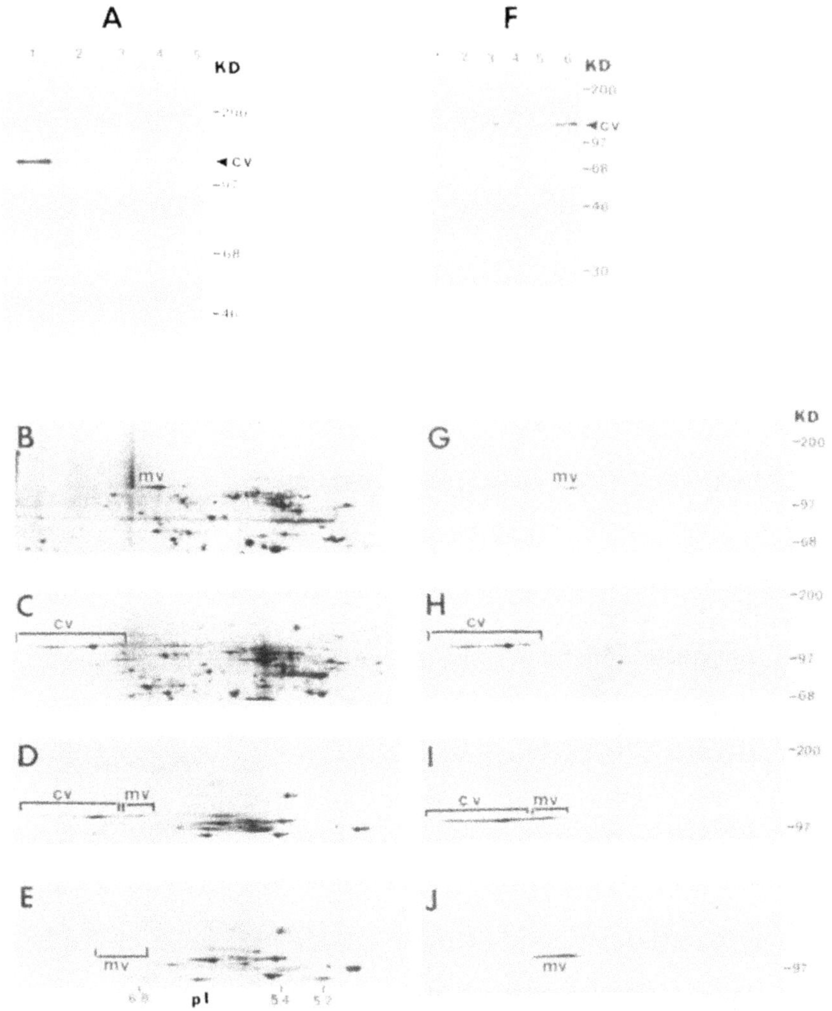

Figure 7. Levels of vinculin expressed in transfected SVT2 and ASML cells. Non-transfected 3T3 cells (**B**, **G**), chicken vinculin transfected ASML cells (**C**, **H**), chicken vinculin transfected SVT2 cells (**D**, **I**), and neo^r control SVT2 cells (**E**, **J**) were labeled for 20 hours with ^{35}S-methionine, total cell protein lysates were separated by 2-D gel electrophoresis and blotted onto nitrocellulose. Autoradiography of the blots (**B–E**). Immunodetection of vinculin on the same blots with a broad range anti vinculin antibody (**G–J**). Individual chicken vinculin transfected clones of ASML (**A**, lanes 1–5) and SVT2 (**F**, lanes 1–6) were labeled for 3 hours with 200 μCi/ml of ^{35}S-methionine, and 5×10^6 c.p.m. of total cell protein were used for immuno-precipitation with an antibody specific for chicken vinculin from each clone. Clone B 12 (**A**, lane 1, **C** and **H**); clone B 42 (**A**, lane 2); clone C 62 (**A**, lane 3); clone B 31 (**A**, lane 4); control, untransfected ASML (**A**, lane 5); control, untransfected SVT2 (**F**, lane 1); neo^r SVT2 control (**F**, lane 2, **E** and **J**); clone D 34 (**F**, lane 3); clone D 41 (**F**, lane 4); clone D 43 (**F**, lane 5); clone D 44 (**F**, lane 6, **D** and **I**). cv, chicken vinculin; mv, mouse vinculin. (Reproduced from Rodríguez Fernández et al., 1992b).

malignant ASML adenocarcinoma (Matzku et al., 1983) which expresses no detectable vinculin at either the RNA (Figure 6G) or protein level (Figure 6C and D).

An expression vector containing the full length chicken vinculin cDNA was introduced into these cells together with the neo^r gene, and G418 resistant colonies expressing different levels of the transgene were isolated (Figure 7).

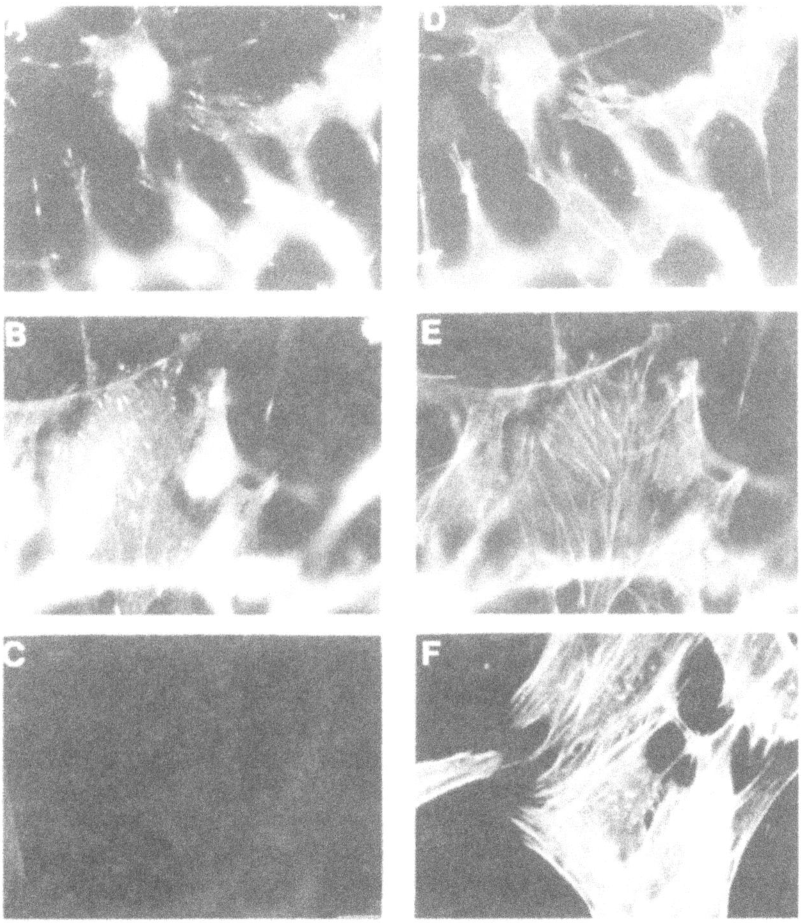

Figure 8. Organization of vinculin and actin in 3T3, SVT2, and vinculin-transfected SVT2 cells. Cells grown on coverslips were fixed with paraformaldehyde, permeabilized with Triton X-100 and immuno-stained with a broad range (**A**), or with a chicken vinculin specific (**B** and **C**) anti-vinculin antibody followed by rhodamine-labeled anti-mouse IgG. Actin filaments were visualized with FITC-labeled phalloidin (**D-F**). SVT2 cells (**A**, **D**); chicken vinculin transfected SVT2 cells, clone D 44 (**B**, **E**); 3T3 cells (**C**, **F**). (Modified from Rodríguez Fernández et al., 1992b).

Table 2. Tumorigenicity of SVT2 Clones Expressing Different Vinculin Levels

	*Vinculin Level**		*Tumor Incidence*		*Growth in Agar†*
Clone	*Chicken*	*Mouse*	*(%)*		*(%)*
3T3	0	750	0/6	(0)	< 0.01
neo 1	0	ND	6/6	(100)	11.0±0.8
neo 3	0	140	6/6	(100)	13.3±1.0
SVT2	0	135	6/6	(100)	12.3±0.5
D 34	344	170	1/6	(16)	5.7±1.0
D 43	135	ND	5/6	(83)	ND
D 41	80	ND	5/6	(83)	ND
D 44	670	160	0/6	(0)	3.6±1.5

Notes: Balb/c mice were injected with 5×10^6 SVT2 cells per animal. The table shows the tumor incidence 46 days following injection. At this time, all the animals injected with the neo^r control and non-transfected SVT2 cells died with very large tumors (>5 cm in diameter), while in animals injected with clone D 44 which expresses the highest level of vinculin no palpable tumors were formed.

*Arbitrary O.D. units of the chicken vinculin levels were obtained by densitometer scanning of the immunoprecipitates separated as shown in Figure 7F. The level of mouse vinculin in the SVT2 clones is about 4 times lower than in 3T3 cells (see Figure 6A and E).

†Cells were seeded in duplicates at 250 and 10^3 cells per 35 mm dish in soft agar and the number of colonies (percent) formed per plate was determined. ND, not done. (Reproduced from Rodríguez Fernández et al., 1992b).

The chicken vinculin expressed in the SVT2 clones was integrated into adhesion plaques, and by immunofluorescence its distribution was indistinguishable from that of the endogenous protein. SVT2 and ASML cells expressing the highest levels of the transgene (Figure 7H and I, respectively) had a flatter phenotype with more abundant adhesion plaques when compared to neo^r-transfected control SVT2 clones (Figure 8E compared to 8D). The ASML adenocarcinoma expressing high levels of chicken vinculin showed an about 2-fold increase in the cross section length of cell substrate contact sites (Rodríguez Fernández et al., 1992b).

The tumorigenic and malignant metastatic capabilities of clones expressing different levels of vinculin was determined in both syngeneic animals and in nude mice. The tumorigenic ability of SVT2 clones expressing vinculin at a level similar to, or higher than, 3T3 cells was completely suppressed in syngeneic animals (Table 2), and was dramatically reduced in nude mice (Rodríguez Fernández et al., 1992b). In addition, the ability of these cells to form colonies in soft agar was also reduced several fold (Table 2).

Syngeneic rats injected in the foot pad with ASML adenocarcinoma cells expressing the highest levels of vinculin showed a significantly extended life span (Figure 9), suggesting that the malignant metastatic ability of these cells was markedly inhibited.

The analysis of several independent vinculin-transfected ASML clones revealed effects on both tumorigenicity (development of tumors at the site of injection of

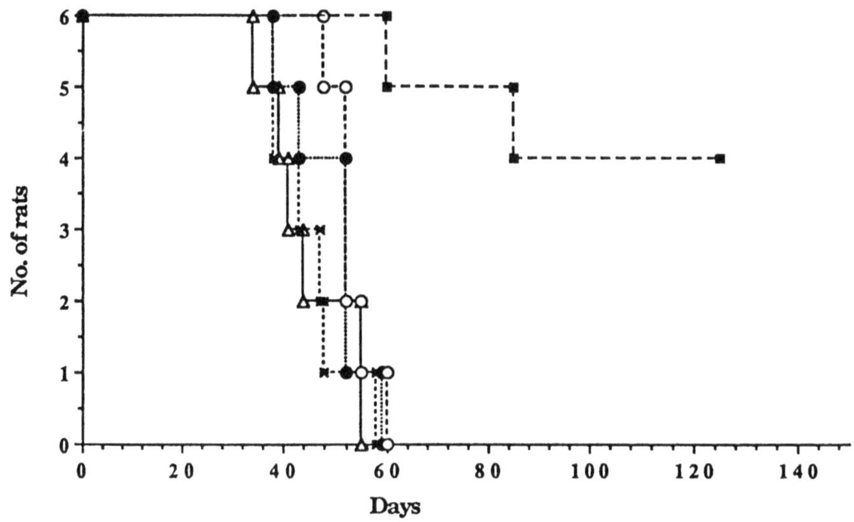

Figure 9. Rat survival after injection with ASML cells expressing different chicken-vinculin levels. Groups of 6 rats were injected in the footpad with 5×10^5 cells of ASML clones expressing different levels of chicken vinculin. The survival time of individual rats, in days, was determined. Δ----Δ, untransfected ASML cells; ×----×, clone B 12 r; ■---■, clone B 12; ○----○, clone B 31; ●----●, clone C 62. For levels of vinculin in these clones see Figure 7. Note that clone B 12 which expresses the highest level of vinculin (Figure 7) is the least malignant, while a revertant obtained from this clone, B 12 r, which has lost vinculin expression regained its highly malignant phenotype. (Reproduced from Rodríguez Fernández et al., 1992b).

cells), as well as on the number of lung metastases formed, in correlation with the level of vinculin expressed by the different clones (Rodríguez Fernández et al., 1992b). Revertants which were obtained from such clones after prolonged passage in culture, and which have lost the expression of vinculin, regained the highly malignant phenotype (Figure 9 clone B 12 r). These results strongly suggest that the restoration of vinculin content to that found in non-transformed cells was effectively suppressing the tumorigenic and malignant metastatic ability of both mesenchymal and epithelial cells.

V. SUPPRESSION OF TUMORIGENICITY IN MALIGNANT CELLS TRANSFECTED WITH ALPHA-ACTININ cDNA

The analysis by 2-D gel electrophoresis of major cytoskeletal proteins expressed in SVT2 cells revealed that in addition to decreased vinculin expression, SVT2 cells also show a reduction in the level of α-actinin (Figure 6E compared to 6A). This

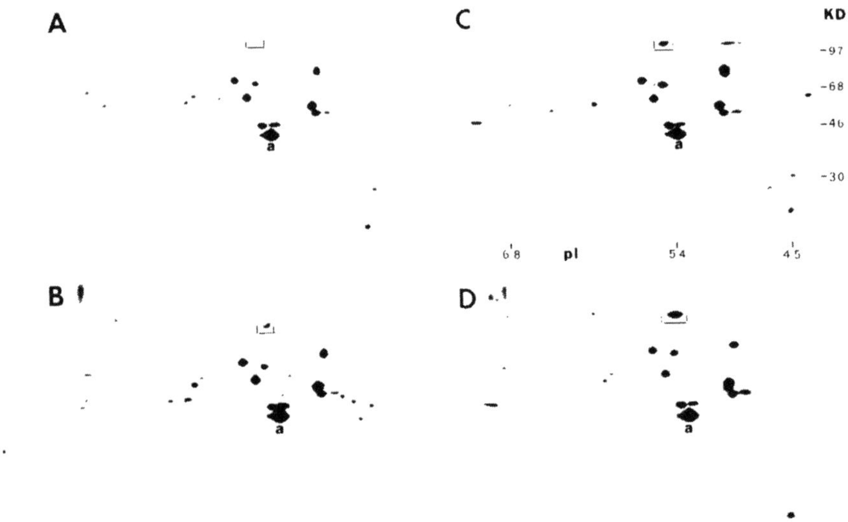

Figure 10. Levels of α-actinin expressed by different SVT2 clones transfected with human α-actinin cDNA. SVT2 cells were co-transfected with the neo^r gene and with a full length human α-actinin cDNA. Neo^r control and clones expressing high levels of α-actinin were labeled with ^{35}S-methionine and analyzed by 2-D gel electrophoresis followed by autoradiography. a, actin; the bracket marks the position of α-actinin. **A**, SVT2 neo^r control; **B–D**, α-actinin transfected clones: **B**, Sα1; **C**, Sα8; **D**, Sα29. Note the high levels of α-actinin expressed by the α-actinin transfected clones (**B–D**) as compared to a control neo^r clone (**A**). (Reproduced from Glück et al., 1993).

decrease results from a comparable decrease in α-actinin RNA content of these cells (Glück et al., 1993).

Alpha-actinin is an abundant actin-crosslinking protein which is also found in AJ (Burridge et al., 1988; Geiger and Ginsberg, 1991; Luna and Hitt, 1992). Recently, α-actinin was shown to have an ability to form a complex with β_1-integrin (Otey et al., 1990), and with the LIM domain-containing protein zyxin which may have a role in transcription regulation (Sadler et al., 1992). It was of interest, therefore, to elevate its expression in SVT2 cells and study the effect of α-actinin overexpression on the transformed phenotype.

A full length human cytoskeletal α-actinin cDNA inserted in an expression vector was co-transfected with the neo^r gene into SVT2 cells, and G418 resistant colonies were screened by Western blotting for higher levels of α-actinin. Representative clones expressing different levels of α-actinin were analyzed by 2-D gel electrophoresis (Figure 10) and Northern blotting (Glück et al., 1993). The endogenous mouse protein and the transfected human α-actinin are 99% homologous (Youssoufian et al., 1990), and therefore were indistinguishable by 2-D gel analysis (Figure 10).

Table 3. Tumorigenicity of SVT2 Clones Expressing Different α-actinin Levels

Cell Line	*α–Actinin level**	*Tumor Incidence in Balb/C Mice (%)*		*Growth in Agar (%)†*	*Tumor Size in Nude Mice (cm^3) ‡*
3T3	658.0	0/6	(0)	<0.01	0
SVT2	108.8	6/6	(100)	12.3±0.9	ND
Sneo	135.9	4/4	(100)	13.3±1.0	0.47±0.13
Sα16	250.1	4/5	(80)	11.5±0.8	ND
Sα1	497.3	2/4	(50)	ND	0.15±0.02
Sα8	693.0	1/5	(20)	5.1±0.7	ND
Sα29	1297.6	0/5	(0)	1.8±0.1	0.11±0.01

Notes: *Arbitrary O.D. units obtained by computerized densitometer scanning of 2-D gels as shown in Figures 7 and 10.

†% number of colonies formed in agar compared to the number of colonies formed on plastic per same number of cells seeded.

‡The size of tumors was determined 12 days after subcutaneous injection of 5×10^5 cells.

N.D., not done.

Note that cells overexpressing α-actinin at the highest level (clone Sα29) are completely suppressed in their ability to form tumors in syngeneic animals. (Reproduced from Glück et al., 1993).

The human α-actinin cDNA had a shorter 5′ non-coding region and served as the diagnostic marker for the expression of human α-actinin in the SVT2 mouse cells by Northern analysis (Glück et al., 1993). Cells expressing the transgene at a level close to or higher than 3T3 cells, showed a flatter phenotype than control SVT2 cells, but were less well spread on the substratum than 3T3 cells (Glück et al., 1993). This is not unexpected, as the expression of vinculin and actin in such cells remained decreased as in control SVT2 cells (Figure 10).

SVT2 clones expressing different levels of the transgene were analyzed for their tumorigenic ability in syngeneic animals and nude mice (Table 3). Clones showing increased levels of α-actinin displayed a dramatically suppressed ability to form tumors in both types of experimental animals.

SVT2 cells overexpressing α-actinin about 2 fold higher than 3T3 cells (clone Sα29, Table 3) were completely suppressed in their ability to form tumors in syngeneic animals, and had a several fold reduced potential to grow in soft agar.

VI. CONCLUSIONS AND PERSPECTIVES

The studies summarized in this essay provide evidence for a role of the submembranal plaque proteins of AJ in eliciting long-range effects on cells. These include both changes in cell structure, cell motility, and the tumorigenic ability of the cells. The mechanisms whereby changes in the expression of these junctional proteins affect cell behavior are most probably involving the cooperative feature of the interactions between these proteins in the process of AJ assembly, and the binding of microfilament bundles to these structures (Kreis et al., 1984; Geiger et al., 1990, 1992). According to this model, an increase in vinculin or α-actinin levels will lead to an increase in their level in the AJ of the transfected cells, which will then lead to the formation of more stable microfilament-membrane interactions, increased adhesion, and decreased motility. In contrast, a decrease in the level of an AJ component such as vinculin, obtained by antisense transfection, will lead to less actin-membrane assembly, a round cell shape, and will impose less restriction on motility. This could result in increased motility, and may influence the ability of cells to grow in soft agar.

The recruitment and immobilization of transmembrane adhesion receptors by AJ plaque components most probably constitute an important mechanism for regulating cell adhesion and motility by an "inside out" mechanism, as suggested by recent studies with chimeric cell adhesion receptors of both the integrin and cadherin family of receptors (La Flamme et al., 1992; Geiger et al., 1992b).

The observations showing that growth-activation, differentiation, and cell transformation are associated with modulations in both the organization and the expression of AJ proteins, may therefore point to important changes in AJ protein function during these vital cellular processes. Further studies involving gene targeting in somatic cells and in transgenic animals are needed for addressing the role of such changes in AJ levels during animal development and physiology.

ACKNOWLEDGMENTS

The studies from the authors' laboratory were supported by grants from the National Council for Research and Development—German Cancer Research Center Collaboration Fund, and from the Leo and Julia Forchheimer Center for Molecular Genetics at the Weizmann Institute of Science. A. Ben-Ze'ev is the Lunenfeld-Kunin Professor of Cell Biology and Genetics, and B. Geiger is the E. Neter Professor of Cell and Tumor Biology.

REFERENCES

Belkin, A.M., Ornatsky, O.I., Kabanov, A.E., Glukhova, M.A., & Koteliansky, V.E. (1988). Diversity of vinculin/meta-vinculin in human tissues and cultivated cells. J. Biol. Chem. 263, 14631–14635.

Bellas, R.E., Bendori, R., & Farmer, S.R. (1991). Epidermal growth factor activation of vinculin and β_1-integrin gene transcription in quiescent Swiss 3T3 cells. J. Biol. Chem. 266, 12008–12014.

Bendori, R., Salomon, D., & Geiger, B. (1987). Contact dependent regulation of vinculin expression in cultured fibroblasts: A study with vinculin specific cDNA probes. EMBO J. 6, 2897–2905.

Ben-Ze'ev, A. (1984). Differential control of cytokeratins and vimentin synthesis by cell-cell contact and cell spreading in cultured epithelial cells. J. Cell. Biol. 99, 1424–1433.

Ben-Ze'ev, A. (1985). The cytoskeleton of cancer cells. Biochim. Biophys. Acta. 780, 197–212.

Ben-Ze'ev, A. (1986). The relationship between cytoplasmic organization, gene expression and morphogenesis. Trends Biochem. Sci., 11, 478–481.

Ben-Ze'ev, A. (1991). Animal cell shape changes and gene expression. BioEssays. 13, 207–212.

Ben-Ze'ev, A. (1992). Cytoarchitecture and signal transduction. Critical Rev. Eukaryotic Gene Expression 2, 265–281.

Ben-Ze'ev, A. & Amsterdam, A. (1987). In vitro regulation of granulosa cell differentiation: Involvement of cytoskeletal protein expression. J. Biol. Chem. 262, 5366–5376.

Ben-Ze'ev, A., Reiss, R., Bendori, R., & Gorodecki, B. (1990). Transient induction of vinculin gene expression in 3T3 fibroblasts stimulated by serum-growth factors. Cell Regul. 1, 621–636.

Ben-Ze'ev, A., Robinson, G., Bucher, N.L.R., & Farmer, S.R. (1988). Cell-cell and cell-matrix interactions differentially regulate the expression of hepatic and cytoskeletal genes in primary cultures of rat hepatocytes. Proc. Natl. Acad. Sci. USA 85, 2161–2165.

Burridge, K., Fath, K., Kelly, T., Nuckolls, G., & Turner, C. (1988). Focal adhesions: Transmembrane junctions between the extracellular matrix and the cytoskeleton. Annu. Rev. Cell Biol. 4, 487–525.

Burridge, K., Turner, E.E., & Romer, L.H. (1992). Tyrosine phosphorylation of paxillin and pp 125^{FAK} accompanies cell adhesion to extracellular matrix: A role in cytoskeletal assembly. J. Cell Biol. 119, 893–903.

Comoglio, P.M., DiRenzo, M.F., Tarone, G., Giancotti, F.G., Naldini, L., & Marchisio, P.C. (1984). Detection of phosphotyrosine-containing proteins in the detergent-insoluble fraction of RSV-transformed fibroblasts by azobenzene phosphate antibodies. EMBO J. 3, 483–489.

Cunningham, C.C., Gorlin, J.B., Kwiatkowski, D.J., Hartwig, J.H., Janmey, P.A., Byers, R.H., & Stossel, T.P. (1992). Actin-binding protein requirement for cortical stability and efficient locomotion. Science 255, 325–327.

Cunningham, C.C., Stossel, T.P., & Kwiatkowski, D.J. (1991). Enhanced motility in NIH 3T3 fibroblasts that overexpress gelsolin. Science 251, 1233–1236.

Geiger, B. & Ginsberg, D. (1991). The cytoplasmic domain of adherens-type junctions. Cell Motil. Cytoskel. 20, 1–6.

Geiger, B., Ginsberg, D., Ayalon, O., Volberg, T., Rodríguez Fernández, J.L., Yarden, Y., & Ben-Ze'ev, A. (1992). Cytoplasmic control of cell-adhesion. Cold Spring Harbor Symp. Quant. Biol. (in press).

Geiger, B., Ginsberg, D., Salomon, D., & Volberg, T. (1990). The molecular basis for the assembly and modulation of adherens-type junctions. Cell Diff. Dev. 32, 343–354.

Geiger, B., Salomon, D., Takeichi, M., & Hynes, R. (1992b). A chimeric N-cadherin/β_1-integrin receptor which localizes to both cell-cell and cell-matrix adhesions. J. Cell Sci. 103, 943–952.

Glück, U., Kwiatkowski, D.J., & Ben-Ze'ev, A. (1993). Suppression of tumorigenicity in simian virus 40-transformed 3T3 cells transfected with α-actinin cDNA. Proc. Natl. Acad. Sci. USA 90, 383–387.

Glück, U., Rodríguez Fernández, J.L., Pankov, R., & Ben-Ze'ev, A. (1992). Regulation of adherens junction protein expression in growth-activated 3T3 cells and in regenerating liver. Exp. Cell Res. 202, 477–486.

Guan, J.L. & Shalloway, D. (1992). Regulation of focal adhesion-associated protein tyrosine kinase by both cellular adhesion and oncogenic transformation. Nature 358, 690–692.

Guan, J.L., Trevithick, J.E., & Hynes, R.O. (1991). Fibronectin/integrin interaction induces tyrosine phosphorylation of a 120 kDa protein. Cell Regul. 2, 951–964.

Hay, E.D. (ed.) (1981). In: Cell Biology of Extracellular Matrix, 417 pp., Plenum Publishing Corp., New York.

Hynes, R.O. (1992). Integrins: Versatility, modulation, and signaling in cell adhesion. Cell 69, 11–25.

Kreis, T.E., Avnur, Z., Schlessinger, J., & Geiger, B. (1984). Dynamic properties of cytoskeletal proteins in focal contacts. In: Molecular Biology of the Cytoskeleton (Borisy, G., Cleveland, D., and Murphy, D., eds.) pp. 45–57, Cold Spring Harbor Laboratory, Cold Spring Harbor.

La Flamme, S.E., Akiyama, S.K., & Yamada, K.M. (1992). Regulation of fibronectin receptor distribution. J. Cell Biol. 117, 437–447.

Lipfert, L., Haimovich, B., Schaller, M.D., Cobb, B.S., Parsons, J.T., & Brugge, J.S. (1992). Integrin-dependent phosphorylation and activation of the protein tyrosine kinase $pp125^{FAK}$ in platelets. J. Cell Biol. 119, 905–912.

Luna, E.J. & Hitt, A.L. (1992). Cytoskeleton-plasma membrane interactions. Science 258, 955–964.

Maher, P.A., Pasquale, E.B., Wang, J.Y.J., & Singer, S.J. (1985). Phosphotyrosine-containing proteins are concentrated in focal adhesions and intercellular junctions in normal cells. Proc. Natl. Acad. Sci. USA 82, 6576–6580.

Matsumura, F. & Yamashiro-Matsumura, S. (1986). Tropomyosin in cell transformation. Cancer Rev. 6, 21–39.

Matzku, S., Komitowski, D., Mildenberger, M. & Zöller, M. (1983). Characterization of BSp73, a spontaneous rat tumor and its *in vivo* selected variants showing different metastasizing capacities. Invasion Metastasis. 3, 109–123.

Mooney, D., Hansen, L., Vacanti, J., Langer, R., Farmer, S., & Ingber, D. (1992). Switching between growth and differentiation in hepatocytes: Control by extracellular matrix. J. Cell. Physiol. 151, 497–505.

Otey, C.A., Pavalko, F.M., & Burridge, K. (1990). An interaction between α-actinin and the β_1 integrin subunit in vitro. J. Cell Biol. 111, 721–729.

Pollack, R., Osborn, M., & Weber, K. (1975). Patterns of organization of actin and myosin in normal and transformed cells. Proc. Natl. Acad. Sci. USA 72, 994–998.

Raz, A. & Ben-Ze'ev, A. (1987). Cell contact and architecture of malignant cells and their relationship to metastasis. Cancer Met. Rev. 6, 3–21.

Raz, A., Zöller, M., & Ben-Ze'ev, A. (1986). Cell configuration and adhesive properties of metastasizing and nonmetastasizing Bsp73 rat adenocarcinoma cells. Exp. Cell Res. 162, 127–141.

Ridley, A.J. & Hall, A. (1992). The small GTP-binding protein rho regulates the assembly of focal adhesions and actin stress fibers in response to growth factors. Cell 70, 389–399.

Rodríguez Fernández, J.L. & Ben-Ze'ev, A. (1989). Regulation of fibronectin, integrin and cytoskeleton expression in differentiating adipocytes: Inhibition by extracellular matrix and polylysine. Differentiation 42, 65–74.

Rodríguez Fernández, J.L., Geiger, B., Salomon, D., and Ben-Ze'ev, A. (1992a). Overexpression of vinculin suppresses cell motility in Balb/c 3T3 cells. Cell Motil. Cytoskel. 22, 127–134.

Rodríguez Fernández, J.L., Geiger, B., Salomon, D., Sabanay, H., Zöller, M., & Ben-Ze'ev, A. (1992b). Suppression of tumorigenicity in transformed cells after transfection with vinculin cDNA. J. Cell Biol. 119, 427–438.

Rohrschneider, L.R. (1980). Adhesion plaques of Rous sarcoma virus-transformed cells contain the src gene product. Proc. Natl. Acad. Sci. USA 77, 3514–3518.

Rysek, R.-P., MacDonald-Bravo, H., Zerial, M., & Bravo, R. (1989). Coordinate induction of fibronectin, fibronectin receptor, tropomyosin, and actin genes in serum-stimulated fibroblasts. Exp. Cell Res. 180, 537–545.

Sadler, I., Crawford, A.W., Michelsen, J.W., & Beckerle, M.C. (1992). Zyxin and cCRP: Two interactive LIM domain proteins associated with the cytoskeleton. J. Cell Biol. 119, 1573–1587.

Schaller, M.D., Borgman, C.A., Cobb, B.S., Vines, R.R., Reynolds, A.B., & Parsons, J.T. (1992). $pp125^{FAK}$ a structurally unique protein tyrosine kinase associated with focal adhesions. Proc. Natl. Acad. Sci. USA 89, 5192–5196.

Shin, S.-I., Freedman, V.H., Risser, R., & Pollack, R. (1975). Tumorigenicity of virus-transformed nude mice is correlated specifically with anchorage independent growth in vitro. Proc. Natl. Acad. Sci. USA 72, 4435–4439.

Stoker, M., O'Neill, C.H., Berryman, S., & Waxman, V. (1968). Anchorage and growth regulation in normal and virus-transformed cells. Int. J. Cancer 3, 683–688.

Takeichi, M. (1991). Cadherin cell adhesion receptors as a morphogenetic regulator. Science 251, 1451–1455.

Tsukita, S., Tsukita, S., Nagafuchi, A., & Yonemura, S. (1992). Molecular linkage between cadherins and active filaments in cell-cell adherens junctions. Curr. Opin. Cell Biol. 4, 834–839.

Ungar, F., Geiger, B., & Ben-Ze'ev, A. (1986). Cell contact and shape dependent regulation of vinculin synthesis in cultured fibroblasts. Nature 319, 787–791.

Vandekerckhove, J., Bauw, G.K., Vancompernolle, G., Honore, B., & Celis, J. (1990). Comparative two dimensional gel analysis and microsequencing identifies gelsolin as one of the most prominent downregulated markers of transformed human fibroblast and epithelial cells. J. Cell Biol. 111, 95–102.

Volberg, T., Zick, Y., Dror, R., Sabanay, I., Gilon, C., Levitzki, A., & Geiger, B. (1992). The effect of tyrosine-specific protein phosphorylation on the assembly of adherens-type junctions. EMBO J. 11, 1733–1742.

Youssoufian, H., McAfee, M., & Kwiatkowski, D.J. (1990). Cloning and chromosomal localization of the human cytoskeletal α-actinin gene reveals linkage to the β-spectrin gene. Am J. Hum. Genet. 47, 62–72.

Zachary, I. & Rozengurt, E. (1992). Focal adhesion kinase ($p125^{FAK}$): A point of convergence in the action of neuropeptides, integrins, and oncogenes. Cell 71, 891–894.

Zieske, J.D., Bukusoglu, G., & Gipson, I.K. (1989). Enhancement of vinculin synthesis by migrating stratified squamous epithelium. J. Cell Biol. 109, 571–576.

THE KINESIN SUPERFAMILY:
VARIATIONS ON A THEME

Janet L. Cyr, Ann O. Sperry, and Scott T. Brady

Advances in Molecular and Cell Biology
Volume 12, pages 165–190.

ISBN: 1-55938-845-5

I. INTRODUCTION

Microtubules perform a variety of functions in higher eukaryotic cells, including the maintenance of cellular shape, cytoplasmic organization, and cell motility. As a major structural element of eukaryotic cells, microtubules provide far more than just a structural scaffold within the cell. The discovery and characterization of several new classes of molecular motor proteins including the kinesins and dyneins emphasize that cellular functions of microtubules include essential roles in numerous forms of intracellular motility such as chromosomal movements during mitosis and meiosis, transport of membrane-bounded organelles, nuclear migration and fusion, and ciliary beating. Abundant experimental evidence supports the proposal that microtubules serve as tracks against which molecular motors like kinesin and dynein exert force to produce many forms of intracellular movement.

Interest in molecular motor proteins and microtubules was initiated in the 1960s with the discovery of axonemal dynein, a large multisubunit microtubule-stimulated ATPase which is responsible for ciliary and flagellar beating (for review see Gibbons, 1981). Until recently, however, axonemal dyneins were the only microtubule-based motors which had been identified. The underlying molecular mechanisms for a variety of other microtubule-based motile events remained elusive. For many years investigators searched with little or no success for a cytoplasmic motor protein which could be involved in mitosis or intracellular organelle motility. This impasse in understanding microtubule-dependent intracellular motility was overcome in 1985 with the discovery of the kinesins, a second class of microtubule-based motor proteins (Brady, 1985; Vale et al., 1985a).

The existence of kinesin was initially recognized from the pharmacological properties of fast axonal transport (Lasek and Brady, 1985), which led to the identification of a novel microtubule activated ATPase involved in the movement of membrane-bounded organelles toward the plus ends of microtubules (Brady, 1985; Vale et al., 1985a,b). The identification of kinesin led to a rapid expansion in our understanding of microtubule-dependent motility within the cell. Biochemical studies demonstrated that kinesin is a heterotetramer consisting of two heavy chain and two light chain subunits with a native molecular weight of ~380 kDa (Bloom et al., 1988; Kuznetsov et al., 1988). The pharmacology and biochemistry of kinesin revealed a number of distinctive properties which could serve as the signature of a kinesin ATPase. When the kinesin subunits were cloned, the sequences revealed that the kinesins were indeed a unique class of ATPases (Yang et al., 1989; Cyr et al., 1991).

Molecular genetic analyses of the kinesin motor identified several distinctive motifs in the kinesin heavy chain and resulted in the discovery of a superfamily of kinesin-related polypeptides. Based on phenotypes for mutations in various kinesin-related proteins, the kinesin superfamily is now believed to be involved in a vast assortment of microtubule-based movements, including aspects of mitosis and meiosis (Endow et al.,1990; Enos and Morris, 1990; Hagan and Yanagida,

1990; McDonald and Goldstein, 1990; Meluh and Rose, 1990; Zhang et al., 1990; LeGuellac et al., 1991). In addition to utilizing the conventional kinesin motor for transport of membrane bounded organelles, eukaryotic cells appear to utilize kinesin-related proteins to perform a variety of specialized types of microtubule-based motility.

Kinesin-related polypeptides possess a high degree of homology to the specific regions of the kinesin heavy chain subunits which contain microtubule and ATP binding sites (for review see Endow, 1991). Aside from these regions of specific homology to the head of kinesin heavy chain, however, the other domains of kinesin-related polypeptides may have little homology to kinesin. Instead, as more and more kinesin-related polypeptides have been sequenced, patterns have emerged which allow them to be grouped into structural and functional subfamilies.

This explosion of knowledge about the kinesin superfamily has created the need for a description of its members and an analysis of microtubule-based motility generated by this protein superfamily. First, kinesin will be described as a prototype for these microtubule-based mechanochemical enzymes. Kinesin is the only member of the superfamily for which a substantial amount of biochemical, molecular genetic, and cell biological information exists. This information provides a basis for considering other members of the kinesin superfamily which represent variations on the kinesin theme. No effort has been made to provide a complete review of the kinesin literature, which has been reviewed recently (Brady, 1991; Cyr and Brady, 1992). The goal is to highlight functional and structural relationships between members of the kinesin superfamily.

II. KINESIN: THE THEME

Since the discovery of kinesins in 1985, a number of laboratories have worked to dissect the structure of the kinesin holoenzyme and to determine its role as an organelle translocator. It is now clear that kinesin is a microtubule-stimulated ATPase composed of two heavy-chain (~124 kDa) and two light-chain (~64 kDa) subunits (Brady, 1985; Bloom et al., 1988; Kuznetsov et al., 1988; Wagner et al., 1989). Multiple isoforms exist for both the heavy-chain and light-chain polypeptides differing in apparent molecular weight and isoelectric point (Pfister et al., 1989; Wagner et al., 1989). A complex of two heavy chains and two light chains is assembled to form an elongated heterotetrameric molecule of ~80 nm in length with a characteristic structure as visualized by rotary shadow electron microscopy (Figure 1; Amos, 1987; Hirokawa et al., 1989; Scholey et al., 1989). At one end of the holoenzyme two globular heads of ~40 nm are present which are joined by an extended shaft to a diffuse fan-shaped tail region (Amos, 1987; Hirokawa et al., 1989; Scholey et al., 1989). The globular heads and most, if not all, of the shaft result from distinct domains of the heavy-chain polypeptides, while the fan-shaped

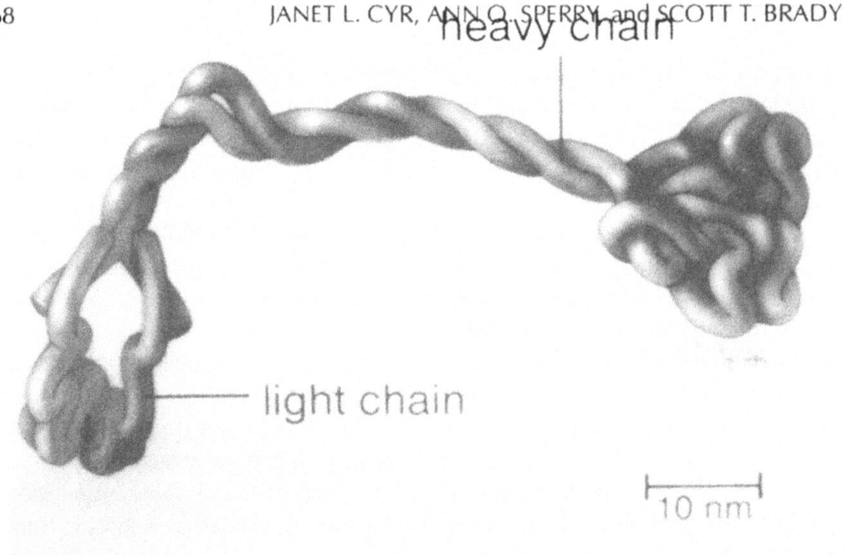

Figure 1. Molecular architecture of the kinesin heterotetramer. Rotary shadow immuno-electron microscopy studies have provided insight into the molecular architecture of the kinesin heterotetramer (Hirokàwa et al., 1989). Secondary structure predictions of the heavy and light chain primary sequences support this proposed model (Yang et al., 1989; Cyr et al., 1991). Kinesin consists of two globular heads present at one end of the molecule and connected by an alpha-helical coiled-coil stalk to a fan-shaped tail region. A kink in the stalk region is often seen which is believed to correspond to a break in the predicted alpha-helical coiled-coil region of the heavy chain. The globular heads, coiled-coil stalk, and a portion of the tail are contributed by the heavy chains of kinesin while the fan-shaped tail region contains epitopes recognized by monoclonal antibodies against the light chains. The detailed interactions between the subunits is unclear and is likely to be more complex than depicted, possibly involving an interchain coiled-coil structure (Cyr et al., 1991; Gauger and Goldstein, 1993). Figure reprinted from Cyr and Brady (1992) with permission from Humana Press, Inc.

tail region is believed to contain portions of both the heavy- and light-chain subunits (Hirokawa et al., 1989).

Consistent with a function as a microtubule-based motor molecule, early work demonstrated that biochemically purified kinesin exhibits a basal ATPase activity which is increased several-fold in the presence of microtubules (Brady, 1985; Kuznetsov and Gelfand, 1986; Cohn et al., 1987; Bloom et al., 1988; Wagner et al., 1989). One distinctive characteristic of the kinesin ATPase is the inhibitory effect of the non-hydrolyzable ATP analog 5′-adenylylimidodiphosphate (AMP-PNP) (Brady, 1985; Lasek and Brady, 1985). In the presence of near stoichiometric levels of AMP-PNP and ATP, kinesin binds stably to microtubules and exhibits no ATPase activity. The AMP-PNP stabilization of kinesin binding to microtubules is reversible in the presence of excess amounts of ATP. This property has been exploited during the initial stages of kinesin purification and represents the single most distinctive pharmacological property of this motor molecule, discriminating kinesin from the other classes of mechanochemical enzymes (Brady, 1985; Vale et al., 1985a; Bloom et al., 1988; Wagner et al., 1989).

In addition to its microtubule-stimulated ATPase activity, purified kinesin is capable of force generation *in vitro*, as visualized in microtubule-gliding assays (Vale et al., 1985a,b; Cohn et al., 1987; Murofushi et al., 1988; Saxton et al., 1988; Cohn et al., 1989). Utilizing microtubular structures of known polarity, *in vitro* microtubule gliding assays indicate that kinesin functions as a unidirectional motor responsible for movement toward the plus ends of microtubules (Vale et al., 1985a,b). This designation as a plus-ended organelle motor is consistent with a role in the movement of membrane-bounded organelles away from the perikaryon and toward the synapse as occurs in anterograde fast axonal transport. Although kinesins are widely held to be involved in organelle transport, many details about the underlying molecular mechanisms of organelle movement are only beginning to be understood.

One insight into the mechanism by which kinesin moves organelles was provided by study of the subcellular locations of kinesin within cells by immunofluorescence light microscopy (Pfister et al., 1989). Examination of several cell types with a panel of monoclonal antibodies against kinesin revealed a detergent-sensitive punctate staining pattern consistent with a stable association of kinesin with membrane-bounded organelles. This punctate staining pattern was located in close proximity to, but did not colocalize with, microtubules, thus suggesting that kinesin is stably associated with membrane-bounded organelles and transiently interacts with microtubules during force generation (Pfister et al., 1989).

Several lines of evidence provide additional support for kinesin localization to membrane-bounded organelles. Radioisotopic labeling of axonal transport in rat optic nerve followed by immunoprecipitation of axonally transported kinesin reveals that kinesin moves with organelles as a constituent of fast anterograde transport. No radiolabeled kinesin is detected moving with the microtubule-neurofilament network (Elluru et al., 1990). These data are substantiated by immunohistochemical and immuno-EM analyses of kinesin accumulation at an axonal (Dahlström et al., 1991; Hirokawa et al., 1991). Kinesin accumulates predominantly on the proximal side of a ligation at time points which are consistent with fast transport of membrane-bounded organelles. Direct evidence for a stable association of kinesin on membrane-bounded organelles has been provided by studies utilizing purified synaptic vesicles and mitochondria from bovine brain (Leopold et al., 1992). Biochemical methods and immuno-electron microscopy of synaptic vesicles and mitochondria revealed the presence of kinesin molecules on these purified organelles.

The regions of the kinesin holoenzyme responsible for microtubule binding and organelle interactions are believed to be present on different subunits of the protein

(Hirokawa et al., 1989). The heavy chains contain the ATP and microtubule binding sites (Gilbert and Sloboda, 1986; Penningroth et al., 1987; Bloom et al., 1988; Hirokawa et al., 1989; Kuznetsov et al., 1989; Yang et al., 1989) and are capable of force generation without associated light chains (Yang et al., 1990). The light chains, located at the opposite end of the kinesin molecule, are thought to be involved in organelle targeting and association (Hirokawa et al., 1989; Cyr et al., 1991), although the possibility exists that the heavy chain may also contribute to organelle binding. The function of these and other domains of the kinesin subunits will be discussed below.

A. Kinesin Heavy Chains

Purification of kinesin from various sources permitted the generation of antibodies to the kinesin heavy chains. These antibodies were used for a variety of studies including the identification of immunoreactive cDNA clones. The isolation of heavy chain cDNA clones from Drosophila in 1988 (Yang et al., 1988; Yang et al., 1989), and more recently from squid (Kosik et al., 1990), sea urchin (Wright et al., 1991), human (Navone et al., 1992), rat (Cyr, DeGrendele, and Brady, unpublished results), and *C. elegans* (Patel et al., 1991) stimulated further genetic, biochemical and structural experiments on the heavy-chain constituents. The primary sequence of the heavy chain revealed several structural and functional features consistent with previous biochemical data. First, as demonstrated by rotary shadow immunoelectron microscopy of bovine brain kinesin (Hirokawa et al., 1989) and supported by secondary structure predictions of the primary sequence (Yang et al., 1989), the heavy chains of kinesin have a three-domain structure: (1) an amino-terminal globular head domain, (2) an extended α-helical coiled-coil rod, and (3) a small globular carboxyl-terminal tail region. The three distinct structural domains of the polypeptide are thought to have distinct functional properties: microtubule and ATP binding sites; a region important for heavy-chain dimerization and perhaps heavy-chain/light-chain interactions; and a region involved in organelle binding, respectively. Although isolation of heavy-chain cDNA clones from a variety of other species revealed a high degree of identity to the originally isolated *Drosophila* sequence, the most thorough molecular genetic analyses of the kinesin heavy chains has utilized the *Drosophila* isolates (Yang et al., 1988; Yang et al., 1989; Yang et al., 1990), and therefore we will concentrate our analyses on these data as being representative.

The amino-terminus of the heavy-chain polypeptide is predicted to form a globular domain of ~50 kDa. This secondary structure prediction is consistent with rotary shadow immuno-EM analysis of purified bovine brain kinesin which demonstrated the presence of two globular heads which were immunoreactive with monoclonal antibodies to the heavy chains (Hirokawa et al., 1989). A close examination of the primary sequence of this globular head region reveals a consensus ATP binding site (GX_4GKTX_6I/V) (Yang et al., 1989). These data support cross-linking and photoaffinity labeling experiments which demonstrated that the heavy chains of the kinesin heterotetramer bind ATP (Gilbert and Sloboda, 1986; Penningroth et al., 1987; Bloom et al., 1988; Kuznetsov et al., 1989) and that the ATP binding site is localized to a ~45-kDa heavy-chain chymotryptic fragment (Kuznetsov et al., 1989).

The globular head of the heavy chain also includes the putative microtubule binding site of the kinesin holoenzyme. This subdomain, contained between residues 320 and 390, has been delineated utilizing truncated recombinant heavy-chain polypeptides in *in vitro* microtubule pelleting assays (Yang et al., 1989). The identification of a microtubule binding site within the globular head domain supports quick-freeze deep-etched electron microscopic data demonstrating that kinesin globular heads bind to taxol stabilized microtubules (Hirokawa et al., 1989).

The presence of both the microtubule and ATP binding sites within the heavy-chain globular heads suggested that this portion of the polypeptide represents the mechanochemical domain of the kinesin holoenzyme. *In vitro* microtubule gliding assays were utilized to examine this possibility and revealed that the globular head region is able to generate microtubule gliding *in vitro*, as long as an appropriate extender was present (Yang et al., 1990). Thus, this mechanochemical domain, containing the microtubule and ATP binding sites, is sufficient for force production along microtubules. The recent identification of a superfamily of molecular motors with a high degree of homology to motifs in the mechanochemical globular head of kinesin suggests that this domain has been utilized extensively throughout evolution to generate specialized microtubule motors. (The characteristics of these kinesin-related polypeptides will be discussed below.)

Distal to the heavy-chain amino-terminal globular heads, a heptad repeat pattern of hydrophobic and charged amino acids is present which extends for 393 amino acids (from Leu 437 to Gln 830 in *Drosophila*) (Yang et al., 1989) corresponding to 53 heptads. Heptad repeats are tandem repeats of seven residues which demonstrate characteristic spacing of hydrophobic and charged amino acids, and facilitate the interaction of two polypeptide helices into an α-helical coiled-coil structure (for review see Cohen and Parry, 1986; Brodsky, 1989). The existence of an extended heptad repeat pattern is *a priori* evidence of a coiled-coil α-helix. Such polypeptide supercoiling is believed to confer both stability and mechanical strength to protein complexes.

The length of the heavy-chain heptads predicts a coiled-coil rod of ~56 nm in length with a "kink" approximately 52 nm from the head domain resulting from a 24 amino acid interruption in the repeat pattern. This break occurs following the 147th amino acid (corresponding to the end of the 21st heptad) and contains a proline residue, as well as 24 amino acids which do not conform to the conventional heptad repeat (Yang et al., 1989). The remaining 32 heptads immediately follow this disruption of the repeat pattern. Although the exact amino acids in the heavy-chain rod are not conserved among different species, the presence of a heptad repeat pattern, the break in the heptads, and the proline found within this break are all strictly conserved (Kosik et al., 1990; Wright et al., 1991; Navone et al., 1992; Cyr, DeGrendele, and Brady, unpublished results). Both the length of the coiled-coil rod, and the presence of a significant break within this structure are consistent with rotary shadow electron micrographs of the kinesin molecule (Amos, 1987; Hirokawa et al., 1989; Scholey et al., 1989).

The significance of the kink in the heavy-chain extended rod is not known, but is postulated to add flexibility that may be necessary to accommodate light-chain association and/or organelle binding. The α-helical coiled-coil domain of the heavy chains was originally thought only to facilitate heavy-chain dimerization; however, identification of an α-helical coiled-coil region at the light-chain amino-terminus (Cyr et al., 1991; see light-chain characterization below) suggests that this region may also mediate heavy-chain/light-chain interactions via an interchain α-helical coiled-coil. Thus it is possible that the heptad repeats amino-terminal to the kink mediate homomeric interactions of two heavy chains, while the α-helical coiled-coil region carboxy-terminal to the kink facilitates the heteromeric interactions of a heavy chain with a light chain.

The heavy-chain α-helical coiled-coil shaft connects the amino-terminal globular heads to the small carboxyl-terminal tail region. This region contains a large number of basic residues and is predicted to contain alternating α-helical and β-sheet regions, thus is likely to be globular in nature (Yang et al., 1989). The function of this carboxyl-terminal domain is unclear although it is likely to have some interactions with the light chains based upon light chain localization to the holoenzyme tail region (Hirokawa et al., 1989). This region may also play a role in organelle binding, although this possibility remains speculative.

B. Kinesin Light Chains

Prior to sequencing and cloning of the kinesin light chains from rat brain (Cyr et al., 1991), little information was available concerning these smaller subunits. The available data included copurification of light chains with the heavy chains; the existence of multiple light-chain species differing in apparent molecular weight and isoelectric point (Wagner et al., 1989); and the immunolocalization of light-chain epitopes to the fan-shaped tail region of the kinesin holoenzyme as visualized by rotary shadow electron microscopy (Hirokawa et al., 1989). Based on these data, the light chains were thought to form part of the kinesin tail and to be involved in the interaction of kinesin with membrane-bounded organelles.

Isolation of light chain cDNA clones from rat and an analysis of the predicted primary sequences revealed the existence of three light chain isoforms differing in their primary sequence (Cyr et al., 1991). The three light-chain polypeptides, termed light chain A, B, and C, differ only in their carboxyl terminal amino acids. These polypeptide isoforms, in combination with phosphorylation and perhaps additional post-translational modifications, are believed to account for the multiple electrophoretic species of biochemically purified kinesin light chains.

In addition to revealing multiple polypeptide isoforms, primary sequence analysis of the rat light-chain cDNAs provided substantial information concerning light-chain structure. The light-chain polypeptides also have three major structural and functional domains: (1) an amino terminal α-helical coiled-coil which may mediate heavy-chain/light-chain interactions; (2) five imperfect repeats of 42 amino acids each which may play a role in organelle binding; and (3) a heterogenous carboxyl terminal tail which may function to target kinesin to specific organelle classes (Cyr et al., 1991; Cyr and Brady, 1992).

The amino-terminus of the light chain is predicted to be α-helical and contains 15 heptad repeats encompassing 106 amino acids. A single skip residue is present within the light chain heptads located after the second repeat. The importance of this skip residue is unknown and many coiled-coil-helical proteins contain occasional skip amino acids (McLachlan and Karn, 1983; Cohen and Parry, 1986). In contrast to the 24 amino acid disruption in the heptad repeat pattern of the heavy chain, a single amino acid skip is unlikely to result in a major kink in the coiled-coil. This skip residue would be predicted to alter the coiled-coil only slightly, producing subtle changes in the polypeptide–polypeptide interactions mediated by this domain (Cohen and Parry, 1986).

The involvement of the light-chain amino-terminus in polypeptide–polypeptide interactions via an α-helical coiled-coil structure is clear; the polypeptide partner to which a light chain binds is less certain. The kinesin holoenzyme is comprised of four distinct polypeptides: two heavy chain subunits and two light chain subunits. All four of the kinesin polypeptides have domains predicted to generate an α-helical coiled-coil structure (Yang et al., 1989; Cyr et al., 1991). These domains may represent a site of light-chain/heavy-chain interactions or alternatively they may mediate dimerization of two homologous subunits (two heavy chains or two light chains) which then interact with the complementary dimer via an alternative interaction site. Biochemical and electron microscopic data favor the first hypothesis, suggesting that interactions between the heavy and light chains are mediated by an interchain coiled-coil, but direct evidence for this designation is not yet available.

The light-chain amino-terminal heptad repeats are followed by an additional motif (Cyr et al., 1991). The central portion of the light chain contains a total of five imperfect repeats of 42 amino acids each. The first four repeats are in tandem, while the fifth repeat element is separated from the previous four by a 41 amino acid spacer (Cyr and Brady, 1992). This region constitutes 39% of the polypeptide. Although the function of this region is unknown, the repeats have a high degree of identity to each other (Cyr et al., 1991; 1992) and these sequences are highly conserved across species (Wedeman et al., 1993; Gauger and Goldstein, 1993) suggesting that they are important for light-chain function. This domain may represent the organelle association site of the light chains and current work is investigating this possibility.

The carboxyl-terminus of the light-chain polypeptide is the domain of sequence heterogeneity. In the rat, three light-chain variants differing only in their carboxyl-terminal amino acids are generated by the alternative splicing of a single gene (Cyr et al., 1991; Cyr and Brady, 1992). The functional significance of these light-chain polypeptide isoforms is unknown; however, the proposed role of light chains in organelle binding suggests that the small differences in these polypeptides may provide a means of targeting, and perhaps binding, of individual kinesin molecules to specific organelle classes.

Recent data indicate that the kinesin light chains are highly conserved across species (Cyr et al., 1991; Wedaman et al., 1993; Gauger and Goldstein, 1993). Similar light chains for kinesin-related proteins have not yet been identified and no homologous polypeptides are known. However, several lines of evidence suggest that the shaft and tail of the kinesin holoenzyme interact with the cargo to be moved by the kinesin motor molecule, in this case a membrane-bounded organelle. Only the kinesin motor domain is conserved throughout the kinesin superfamily, although there is considerable homology between light chains and heavy chains from a variety of species ranging from *Drosophila* to sea urchin to mammals. These data suggest that the motor domain of the kinesin heavy chain is responsible for movement along a microtubule, while regions outside the motor domain convey specificity of function to the motor.

III. KINESIN-RELATED POLYPEPTIDES: VARIATIONS ON A THEME

The isolation of kinesin heavy-chain clones from *Drosophila* and the determination of the heavy-chain primary sequence led to the unexpected discovery of several genes that encode polypeptides with a high degree of homology to the head domain of the kinesin heavy chain. Although kinesin was initially isolated and characterized using exclusively biochemical techniques, most newly identified kinesin-related proteins were identified by a combination of classical and molecular genetics. Homologies to the kinesin heavy chain were detected by comparison of sequences for proteins responsible for mutations in a variety of systems to protein databases containing the heavy-chain sequence.

In 1990 two genes were isolated by genetic complementation of quite diverse mutant phenotypes: *KAR3* by rescue of a defect in nuclear fusion during yeast mating (Meluh and Rose, 1990), and *bimC* by complementation of a mutation in spindle-pole body duplication in *Asperigillus nidulans* (Enos and Morris, 1990). Both *KAR3* and *bimC* contain a domain with striking homology to the amino-terminal motor domain of the kinesin heavy chain. Their discovery marked the emergence of the kinesin superfamily of molecular motors and was quickly followed by the isolation of other related genes from organisms as diverse as flies and frogs. The encoded polypeptide products of these genes are implicated in an assortment of microtubule-based events including the separation of spindle pole bodies in *Aspergillus nidulans* (Enos and Morris, 1990), the formation and separation of spindle pole bodies in *Schizosaccharomyces pombe* (Hagan and Yanagida, 1990), meiotic chromosome segregation in *Drosophila* (Endow et al., 1990; McDonald and Goldstein, 1990; Zhang et al., 1990), and nuclear fusion during karyogamy in *Saccharomyces cerevisiae* (Meluh and Rose, 1990).

The primary sequence similarity of the kinesin-related polypeptides to the kinesin heavy chain is confined to the mechanochemical globular head domain. Nevertheless, several of the kinesin-related polypeptides have other structural features in common with the kinesin heavy chain. For example, some of these polypeptides contain predicted α-helical coiled-coil regions which presumably facilitate dimerization. One characteristic which is not conserved in all of these polypeptides is the location of the globular head domain in the polypeptide (Figure 2). Two of the kinesin-related proteins (*ca*nd from *Drosophila* and *KAR3* from yeast) have the

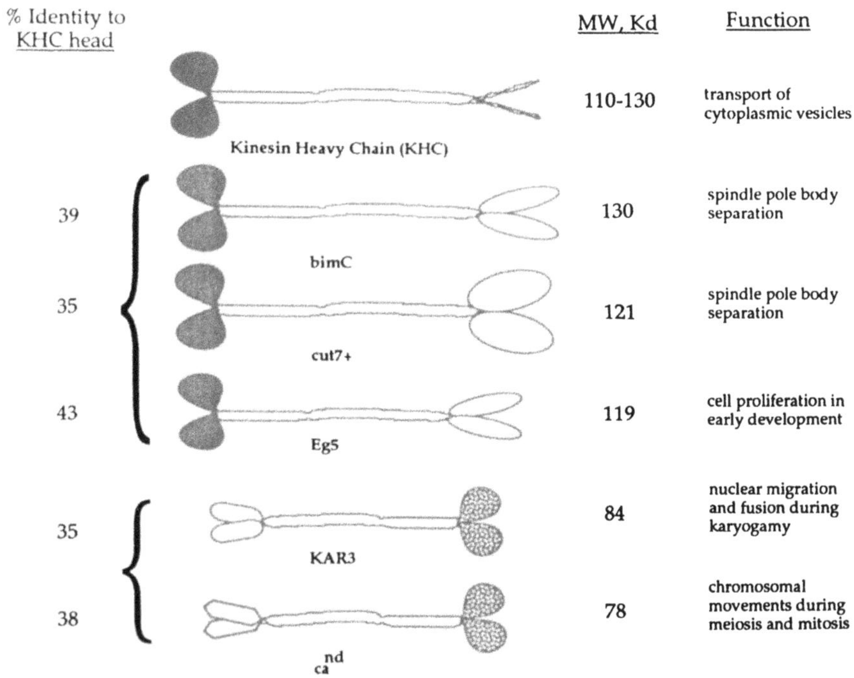

Figure 2. Subfamilies within the kinesin superfamily. Selected members of the kinesin superfamily are depicted schematically. The representation of a three domain structure of the kinesin heavy chain with a globular head and extended stalk and tail is supported by ultrastructural and molecular genetic studies as discussed in the text (Hirokawa et al., 1989; Yang et al., 1989). The drawings of the other proteins are more speculative. ~he motor domains of these proteins are located at the amino-terminus (extreme left) ~. *bimC, cut7+,* and *Eg5,* or at the carboxyl-terminus (extreme right) of *KAR3* ~he brackets indicate the proposed subfamilies within the kinesin superfa-

	Kinesin	*unc-104*	*nod*	ca^{nd}	*KAR3*	*bim C*	*cut7+*	*Eg5*	Motor Domain
Kinesin	100	41	**30**	38	35	40	36	43	amino
unc-104	41	100	30	33	31	**34**	33	35	amino
nod	**30**	30	100	24	27	34	33	31	amino
ca^{nd}	**38**	33	24	100	44	**35**	**33**	39	carboxyl
KAR3	35	31	27	44	100	**34**	31	32	carboxyl
bimC	39	34	34	**35**	34	100	59	53	amino
cut7+	**35**	33	33	**33**	31	59	100	47	amino
Eg5	**43**	35	31	39	32	53	47	100	amino

Figure 3. Pairwise comparison of kinesin superfamily members in the mechanochemical globular head domain. Predicted amino acid sequences for selected members of the kinesin superfamily were analyzed pairwise to determine the primary sequence identity. Such analyses reinforce the presence of distinct subfamilies within the kinesin superfamily which share higher sequence identity and functional properties. Alignments were performed as in Endow (1991) with slight modifications. The regions of the predicted polypetides which were analyzed are as follows: *Drosophila* kinesin heavy chain residues 1-340; *unc-104* residues 1-354; *nod* residues 1-327; *claret* residues 337-677; *KAR3* residues 375-729; *bimC* residues 72-423; *cut7+* residues 51-416; *Eg5* residues 1-358. The primary sequence identity of the *bimC, Eg5,* and *cut7+* polypeptides extends beyond the mechanochemical head region and proceeds throughout the entire length of the predicted polypeptides.

mechanochemical head domain located at the carboxyl-terminus of the polypeptide (Endow et al., 1990; McDonald and Goldstein, 1990; Meluh and Rose, 1990), in contrast to the kinesin heavy-chain polypeptide and most other kinesin-related proteins identified to date which have their motor domain at the amino-terminus.

While all kinesin-related polypeptides share the highly homologous mechanochemical domain first identified in the kinesin heavy chain, pairwise comparisons of these polypeptides reveal subgroups that show higher homology to each other than to the family as a whole (Figure 3). Kinesin-related polypeptides of a given subgroup from different species may be more closely related to one another than to kinesin or other kinesin-related proteins from the same species. The existence of subfamilies within the kinesin superfamily lends credence to the idea that kinesin subtypes exist to perform specialized functions within the cell, including chromosomal segregation, nuclear migration, and vesicular transport. Classifying members of the kinesin superfamily into structural subfamilies may illuminate their functional roles in the cell.

A. The *bimC, cut7+,* and *Eg5* Subfamily

Within the broader superfamily of kinesin-related polypeptides, *bimC* (blocked in mitosis) from *Asperigillus nidulans* (Enos and Morris, 1990), *cut7+* from the fission yeast *Schizosaccharomyces pombe* (Hagan and Yanagida, 1992), and *E* from *Xenopus laevis* (LeGuellac et al., 1991; Sawin et al., 1992) share considerable functional and structural homology. These polypeptides comprise a kinesin subfamily implicated in the early steps of mitosis. Thus, members of this subfamily can be grouped both by primary sequence homology and function.

The *Asperigillus nidulans* mutation *bimC4* is a temperature-sensitive allele which results in aberrant spindle pole body separation, spindle morphology, and nuclear division (Enos and Morris, 1990). At nonpermissive temperatures, *bimC4* mutants accumulate unseparated spindle pole bodies at one end of a unipolar spindle. The spindle pole bodies continue to duplicate and generate microtubules, but the malformed spindle fails to elongate and separate sister chromatids. The *bimC4* phenotype appears to result from an inability of spindle microtubules to slide past one another during spindle formation and an inhibition of spindle elongation.

A very similar phenotype is seen with $cut7^{ts}$. A temperature-sensitive mutation at the *cut7+* loci produces a phenotype where the mitotic spindle microtubules do not appear able to slide past one another to separate the spindle poles to form a bipolar spindle (Hagan and Yanagida, 1990). At nonpermissive temperatures, a malformed, v-shaped structure replaces the normal mitotic spindle. This structure is believed to result from the inability of microtubules originating at the spindle pole bodies to interdigitate and slide apart, thus inhibiting spindle elongation.

Unlike most kinesin-related proteins, including the *bimC* and *cut7+* proteins, the polypeptide encoded by the *Xenopus Eg5* gene was not discovered as a mutation defective in cell division. *Eg5* was identified by differential screening of a cDNA library comprised of polyA+ RNA from unfertilized *Xenopus* eggs, which was designed to isolate sequences that undergo adenylation or deadenylation following fertilization (LeGuellac et al., 1991). As a result, the functions of *Eg5* were not clearly defined upon its identification, but measurement of *Eg5* message levels during early embryonic development suggested that it also played a role in mitosis. The *Eg5* gene is expressed exclusively during a period of rapid cell proliferation prior to the midblastula transition. These patterns of expression and the high degree of primary sequence identity between *Eg5* and both *bimC* and *cut7+* suggested that its function was also in the early stages of mitosis.

Recent studies have confirmed these initial hypotheses and have demonstrated the involvement of the *Eg5* gene product in spindle development. Utilizing *in vitro* spindle assembly assays, the *Eg5* polypeptide has been shown to be involved in the organization of the mitotic spindle (Sawin et al., 1992). These studies also determined that the *Eg5* polypeptide is a plus-ended microtubule motor as demonstrated in *in vitro* microtubule gliding assays (Sawin et al., 1992).

The similarities between *cut7+*, *bimC*, and *Eg5* extend beyond functional homologies. All three members of this subfamily share a common overall structural architecture based on algorithms for structure prediction. Like the kinesin heavy chain, *cut7+*, *bimC*, and *Eg5* contain their mechanochemical globular head domains at the amino-terminus of the encoded polypeptide. The globular head is followed by a region enriched in helix-forming residues consistent with a helical rod.

Although original sequence analyses did not reveal the presence of heptad repeats in this region, subsequent examination of these sequences do suggest that a limited heptad repeat pattern may exist. As with the kinesin heavy-chain heptads, this domain may facilitate dimerization, but this prediction awaits confirmation. The carboxyl-termini of all three polypeptides contain a small domain of alternative conformation, perhaps globular in nature.

Finally, as might be expected from the structural and functional similarities, members of this subfamily exhibit a high degree of primary sequence identity. As indicated in Figure 3, a pairwise comparison of *bimC*, *cut7+*, and *Eg5* reveals 47–59% identity within the globular head domain. This high degree of sequence identity extends beyond the putative mechanochemical head and continues to regions within the putative rod region and carboxyl-terminal tail.

Two additional members of this subfamily have recently been described: the *CIN8* and *KIP1* genes from *S. cerevisiae* (Hoyt et al., 1992; Roof et al., 1992; Saunders and Hoyt, 1992). The protein products of these genes share a similar structural organization with other members of the superfamily. Both have an amino-terminal force generating head domain, an internal stalk domain containing heptad repeats characteristic of an α-helical coiled-coil structure, and a carboxy-terminal globular domain. The protein products of these genes show high primary sequence homology to the force generating head domain of the kinesin heavy chain. *CIN8* and *KIP1* polypeptides are placed in the *bimC/cut7+/Eg5* subfamily based on the higher degree of sequence homology to members of this subfamily throughout the entire molecule than to other kinesin-related proteins. The motor domains of the *KIP1* and *CIN8* gene products are 65% and 64% identical to that of the *bimC* protein, respectively. This degree of homology is significantly higher than that observed between the KIP1 polypeptide or CIN8 protein and kinesin or other kinesin-related proteins (see Figure 3). Unique among the kinesin-related proteins, *CIN8* gene product has a 90 amino acid insert in a region of low sequence conservation within the head domain. The role of this region is unknown but apparently not essential for function: yeast containing a deletion of this region display no detectable phenotype. Like other members of this subfamily, the *CIN8* and *KIP1* proteins function in the assembly of the mitotic spindle. Genetic experiments (discussed in more detail below) suggest that both proteins perform essential, possibly redundant functions in the early stages of mitosis.

The precise mechanism of action for the *bimC/cut7+* subfamily of kinesin will require biochemical analyses. However, predictions may be made on the basis of the *bimC4* and *cut7*^{ts} phenotypes (Hagan and Yanagida, 1992). One possibility is that this class of motor proteins has a motor domain at one end of the molecule and a second, nucleotide insensitive microtubule binding site(s) at the other end. Such a microtubule-based motor would be expected to generate force against one microtubule polymer while it translocates its cargo (another microtubule) during the power stroke (see Figure 4). This model cannot be tested from sequence data alone. Although these putative motors contain a mechanochemical domain at one

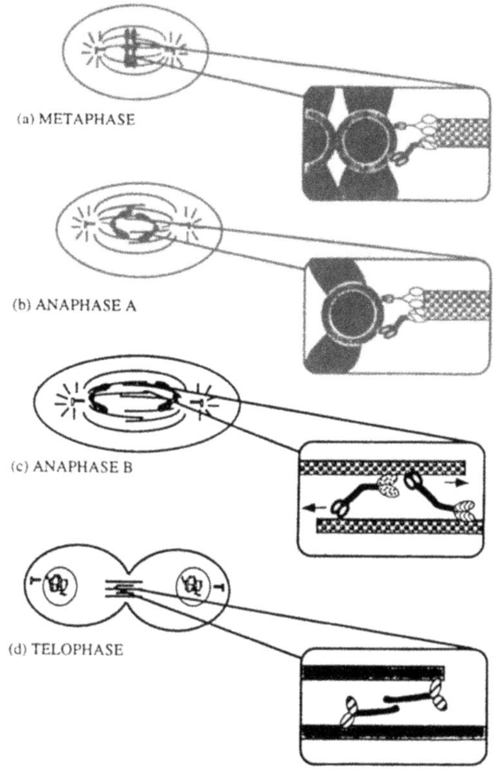

Figure 4. Motor molecules involved in mitosis. The force-generation during mitosis appears to be produced by a number of distinct microtubule-based and actin-based motor molecules. The identification of several kinesin-related motors involved in mitosis indicates that a multitude of microtubule-based motors are involved in spindle formation and chromosome separation. In addition to the kinesin-related motors, cytoplasmic dynein has also been localized to the mitotic spindle and is thought to function during mitosis. The formation of the cleavage furrow and the final separation of the two sister cells is mediated by the actin-based motor, myosin.

Possible functions of motor molecules in the various steps of mitosis are shown. This diagram is not meant to be all-inclusive, but only to highlight the involvement of several molecular motors during mitosis. The essential movements of mitosis can be divided into five steps: (1) the duplication and separation of the poles to establish the spindle (not shown); (2) chromosome oscillations at the metaphase plate [panel (**a**)] believed to be mediated by cytoplasmic dynein and kinesin-related motors; (3) the poleward movement of chromosomes during anaphase A [panel (**b**)] by cytoplasmic dynein and kinesin-related motors; (4) elongation of the spindle during anaphase B thought to be mediated by kinesin-related microtubule-based motors [panel (**c**)]; and (5) the formation of the cleavage furrow during telophase by the actin-based motor, myosin [panel (**d**)]. (continued) end, the amino acid sequence requirements for microtubule binding are very degenerate, thus the presence of a second microtubule binding site at the carboxyl-termini awaits biochemical tests. Alternatively, these polypeptides may not contain a nucleotide insensitive microtubule binding site in their carboxyl-terminus, but the protein may include additional polypeptides (i.e., light chains) which confer this binding capacity.

B. The *KAR3* and *ca^nd* Subfamily

A second subgroup of the kinesin superfamily includes the *ca^nd* (also called *ncd*) polypeptide from *Drosophila melanogaster* (Endow et al., 1990; McDonald and Goldstein, 1990) and the *Saccharomyces cerevisiae KAR3* protein (Meluh and Rose, 1990). These motor proteins are implicated in meiotic chromosome distribution and nuclear fusion following karyogamy, respectively. The *Drosophila ca^nd* (*cla*ret *nond*isjunctional) was cloned by chromosomal walking using transposon induced mutations in chromosomal movement (Endow et al., 1990). *S. cerevisiae KAR3* (karyogamy) gene was isolated by complementation of cells defective in nuclear fusion during yeast mating (Meluh and Rose, 1990).

The *ca^nd* and *KAR3* polypeptides each have a small globular domain at the amino-termini with no significant homology within this region to each other or to any other polypeptide (Endow et al., 1990; McDonald et al., 1990; Meluh and Rose, 1990). The central portion of the predicted *KAR3* polypeptide is α-helical in nature and contains a series of heptad repeats (Meluh and Rose, 1990). Conventional heptad repeats are not present within the *ca^nd* polypeptide although a heptapeptide leucine repeat is present which could mediate dimerization (Endow et al., 1990); thus both of these polypeptides are predicted to exist as dimers *in vivo*. The distinctive feature of these two members of the kinesin superfamily is the location of the conserved kinesin heavy chain-related mechanochemical domain at the carboxyl termini of the encoded polypeptides. This region of the *ca^nd* and *KAR3* polypeptides are 38% and 35% identical, respectively, to the kinesin heavy-chain amino-terminal motor domain and 44% identical to each other (Figure 3).

Although both the *ca^nd* and *KAR3* polypeptides are microtubule-based motors involved in nuclear events, the most striking characteristic of these two polypeptides is their functional architecture. The polypeptides encoded by *KAR3* and *ca^nd* are superfamily members which have a reversed domain arrangement, i.e., the

---

Figure 4. (continued) Specifically, dynein is represented as the typical oligomeric flower-pot structure while the kinesin-related proteins are shown as dimers containing three domains: the mechanochemical head domains are represented by decorated ovals, the stalk regions are elongated and filled-in, and the tail domains are unfilled ovals. Myosin is depicted as two oval heads attached to an elongated stalk.

homologous kinesin mechanochemical domain is located at the carboxyl-terminus instead of at the amino-terminus (Figures 2 and 3). For both *ca^nd^* and *Kar3* polypeptides, biochemical studies have shown that the differences in domain organization have unexpected functional consequences on the direction of force generation. *In vitro* microtubule gliding assays utilizing recombinant ca^{nd} polypeptide and demonstrated that these kinesin- and *Kar3* polypeptide-related motors support microtubule gliding with a polarity opposite to that of the conventional kinesin (McDonald et al., 1990; Walker et al., 1990; Endow et al., 1994). The significance of the correlations between the location of the mechanochemical head domain (amino- or carboxyl-terminus) and directionality of force generation in kinesin-related proteins is not known. Future biochemical and biophysical analyses will be needed to determine the molecular basis of directionality in generation of force.

C. Orphan Members of the Family: *nod, unc-104,* and *SMY1*

Several additional members of the kinesin superfamily have been identified with considerable homology to the kinesin heavy-chain and to other kinesin-related polypeptides (Figure 3), but these polypeptides do not readily lend themselves to classification into a particular structural or functional subfamily. Three kinesin superfamily orphans are *nod* from *Drosophila melanogaster* (Zhang et al., 1990), *Caenorhabditis elegans unc-104* (Otsuka et al., 1991), and *SYM1* from yeast (Lillie and Brown, 1992). These may be unique members of the superfamily or, more likely, the first identified members of additional subfamilies.

The *Drosophila nod* (no distributive disjunction) gene is necessary for proper distribution of meiotic chromosomes. The gene encoding this kinesin-related polypeptide was cloned by chromosomal walking and physical mapping of transposon induced *nod* mutants (Zhang et al., 1990). Mutations at the *nod* locus result in aberrant positioning of nonexchange chromosomes (chromosome 4) along the metaphase plate. This abnormality is present at early prometaphase, a time earlier than that seen in ca^{nd} mutants. Unlike the phenotype of ca^{nd} mutants, *nod* mutations primarily affect the distribution of nonexchanged chromosomes (Knowles and Hawley, 1991). The phenotypes of organisms heterozygotic for both mutations suggests that these two proteins interact with one another in the chromosome segregation machinery.

Mutations at the *unc-104* (uncoordinated behavior) locus in *C. elegans* manifest as paralyzed, slow growing organisms. These mutants have defects in chemical synapse formation and appear to accumulate vesicles in the perikaryon of neurons (Otsuka et al., 1991). Analyses of the *unc-104* phenotype were consistent with the hypothesis that *unc-104* was the functional equivalent of conventional kinesin heavy chains. However, the degree of identity across species was less than expected based upon the considerable sequence identity between the *Drosophila* and squid kinesin heavy chains (Otsuka et al., 1991). The isolation of the gene *unc-116* from

C. elegans clarified this apparent inconsistency. The *unc-116* polypeptide has considerable sequence identity to the previously isolated kinesin heavy chain sequences from a variety of species and is believed to be the *C. elegans* homologue to the kinesin heavy chain subunit (Patel et al., 1991). The mild paralysis exhibited by *unc-116* mutants suggests a neuronal function and supports this assignment. Thus, *unc-104* is thought to be a kinesin-related polypeptide whose function is not completely clear. This designation is more consistent with the level of sequence identity of *unc-104* to the conventional kinesin heavy chains and other kinesin-related polypeptides.

Both the *unc-104* and *nod* polypeptides display little resemblance to the kinesin heavy chain in regions other than the amino-terminal mechanochemical head domain (Zhang et al., 1990; Otsuka et al., 1991). *unc-104* is predicted to be a relatively large polypeptide with a molecular mass of 175 kDa (Otsuka et al., 1991). Following the amino-terminal globular domain a stretch of hydrophilic residues is present which are α-helical in nature. Analyses of this region do not reveal the presence of heptad repeats or a periodicity of hydrophobic residues; the possibility that this region mediates dimerization is uncertain. The carboxyl-terminus of the *unc-104* polypeptide contains two basic domains. One curious feature of *unc-104* is the presence of a RGDS (Arg-Gly-Asp-Ser) sequence within one of these basic domains. This sequence is familiar in extracellular proteins where it has been implicated in receptor-ligand interactions (Pierschbacher and Ruoslahti, 1984; Yamada and Kennedy, 1985). The function of this sequence within the *unc-104* polypeptide, which is presumably cytoplasmic in location, is uncertain.

Unlike the other members of the kinesin superfamily, the *nod* polypeptide does not display a distinctive three-domain structure (Zhang et al., 1990). Besides the amino terminal globular mechanochemical head domain, no obvious structural domains have been identified within this polypeptide. Primary sequence analyses reveal neither an extended α-helical region nor the presence of heptad repeats; so it is unlikely that this polypeptide dimerizes *in vivo*. This divergence in the overall structural architecture of the *nod* polypeptide in combination with the lower percentages of identity with other superfamily members (Figure 3) suggests that *nod* may represent the first member of a new subfamily within the kinesin superfamily.

Like *nod* and *unc-104*, the yeast kinesin-related polypeptide *SMY1* does not easily fit into an established subfamily, either by functional or by sequence similarity to other kinesin-related polypeptides. The *SMY1* gene was identified by its ability to partially substitute for a mutant myosin gene in yeast cells (Lillie and Brown, 1992). The primary sequence of the *SMY1* polypeptide clearly places it within the kinesin superfamily and it displays the three-domain structure typical of this family of proteins. However, the *SMY1* protein has the most sequence divergence from the kinesin heavy chain in the mechanochemical head domain. In addition to its sequence divergence, the *SMY1* polypeptide may also represent the most functionally divergent member of the kinesin superfamily. The ability of *SMY1* to partially complement a myosin mutation (see below) suggests that this member of the kinesin superfamily may not serve as a microtubule-based motor, but rather function in actin-based motility. Further studies are needed to determine if the *SMY1* polypeptide utilizes actin as a substrate or if it is a microtubule motor whose function overlaps that of myosin in yeast.

D. Additional Family Members: Function Unknown

The above mentioned members of the kinesin superfamily were isolated predominately by classical genetic techniques; several other members have recently been identified or their existence established by molecular biological techniques. Additional members of the kinesin-related family of proteins have been isolated utilizing known kinesin heavy-chain head sequences to amplify related sequences (McDonald and Goldstein, 1990; Stewart et al., 1990) with polymerase chain reaction protocols. Utilizing degenerate oligonucleotide primers to highly conserved regions of the globular head, multiple distinct sequences have been amplified from *Drosophila* genomic DNA (Stewart et al., 1990) or from cDNA made from various rat tissues (Sperry et al., 1992). The function of these family members are not yet known, although based on available sequences some of them appear to belong to one of the previously identified subfamilies.

In addition to the family members identified by the polymerase chain reaction, Endow and Hatsumi (1991) report that an examination of *Drosophila* polytene chromosomes with probes to the kinesin motor head domain detects greater than 30 different sites of hybridization distributed over all of the amplified chromosomes. While a more conservative estimate of kinesin-related genes has been made by Stewart et al. (1990), there appear to be at least 11 distinct genes in the *Drosophila* genome. Similarly, multiple distinct gene products have been reported in yeast, *A. nidulans*, and mammals (Enos et al., 1991; Roof et al., 1991; Sperry et al., 1992), although the number of kinesin-related genes in these systems remains to be determined. The available data indicate that the kinesin superfamily may be quite large and diverse. Recent studies have shown a similar complexity in the myosin family of molecular motors (Spudich, 1989; Korn and Hammer, 1990; Brady, 1991; Mercer et al., 1991), while insufficient information is available about the dyneins. Regardless, it appears that there will be a remarkable diversity in the number and characteristics of motor proteins available for use in any given type of motility. The identification of the remaining kinesin superfamily members and the roles they may play in intracellular microtubule-based motility will be a continuing challenge in the near future.

IV. FURTHER VARIATIONS: GENETIC INTERACTIONS AND FUNCTIONAL REDUNDANCY

The large number of kinesins and kinesin-related proteins raises the question of whether each motor performs a specific and exclusive task within the cell or whether their functions overlap. These two possibilities are not mutually exclusive and evidence from genetic studies in several species suggest that examples of both may exist. The possibility of functional overlap is illustrated well by the genetic interactions of kinesin-related proteins that are part of the mitotic machinery; i.e., ca^{nd} and *nod* from *Drosophila*; and *KAR3*, *KIP1*, and *CIN8* from *S. cerivisiae*. Confusing the issue even further are genetic experiments that demonstrate that the *SMY1* protein from yeast, another kinesin-related protein, suppresses a myosin defect (Lillie and Brown, 1992). These experiments have a number of implications for interactions between motor proteins of one class and between motors of different classes.

The *nod* and ca^{nd} proteins participate in different aspects of meiotic chromosome segregation in female flies. Mutation in the *nod* gene affects only the distribution of nonexchange chromosomes while mutation in the ca^{nd} gene affects the segregation of all chromosome pairs (Endow et al., 1990; Zhang et al., 1990). Both mutations are fully recessive; however, double mutants show higher frequencies of chromosome disjunction (Knowles and Hawley, 1991). Such unlinked noncomplementation usually indicates that two gene products physically interact. In this case, however, cytological analysis of double mutants support the proposal that these two proteins act independently but sequentially during meiosis (Knowles and Hawley, 1991). These data indicate that the *nod* protein may produce a force driving chromosomes toward the metaphase plate while the ca^{nd} polypeptide performs a general function in organizing the meiotic spindle. Therefore, noncomplementation of these loci is due to the sequential nature of meiosis: the formation of an aberrant spindle in *ncd* mutant flies enhances the effect of an added *nod* mutation and increases chromosome nondisjunction.

A similar situation exists between the kinesin-related proteins encoded by *KIP1*, *CIN8*, and *KAR3* in yeast. The complex interactions between these proteins in the chromosome segregation apparatus is indicated by the phenotypes of mutant yeast cells containing various combinations of mutations in these three genes. While complete deletion or missense mutations in any one of these genes has a mild or invisible phenotype, double mutants are more severe (Hoyt et al., 1992; Roof et al., 1992; Saunders and Hoyt, 1992). Mutant cells containing both a *CIN8* temperature sensitive allele and deletion of the *KIP1* gene arrest growth at the beginning of mitosis with duplicated but unseparated spindle-pole bodies. Such a collapse of the spindle is very reminiscent of *bimC* or $cut7^{ts}$ single mutants. Interestingly, mutations in the *KAR3* gene partially suppress the *CIN8 KIP1* double mutant (Saunders and Hoyt, 1992). This suggests that the *KAR3* protein may produce force to counteract that which is generated at the metaphase plate by the *CIN8* and *KIP1* proteins. This is consistent with the suggestion that the *KAR3* protein produces movements with the opposite polarity of most members of the kinesin superfamily.

The most unusual genetic interaction was found between the yeast kinesin-related protein *SMY1* and myosin. The *SMY1* protein was identified genetically by selecting for multicopy suppressors of a mutation in the yeast *MYO2* gene which encodes a member of the myosin superfamily (Lillie and Brown, 1992). The *MYO-2* protein has been implicated in polarized secretion along actin filaments (Johnston et al., 1991). Yeast containing the *MYO2-66* temperature-sensitive allele was transformed with a yeast genomic library on a high-copy-number plasmid. Those transformants which were able to grow at the nonpermissive temperature were selected and the suppressing genes isolated. One such gene was found to be the *SMY1* (*s*uppressor of *my*osin) gene. Unexpectedly, sequence analysis demonstrated this gene to be a member of the kinesin superfamily, although a very divergent one. Thus, a putative microtubule-based motor can functionally substitute, at least partially, for an actin-based motor. One possible explanation of this apparent contradiction is that microtubules form a minor pathway for the secretion of vesicles normally carried by the *MYO2* protein.

V. SUMMARY

The decade since the discovery and initial characterization of kinesin have witnessed an explosion in the number of force-generating enzymes in the kinesin superfamily. The newly identified kinesin-related proteins have reiterated the molecular pattern established by the kinesin heavy chain. Like kinesin, these proteins contain a distinctive three-part structure; however, the arrangement of these parts is not invariant, as seen in the *KAR3* and ca^{nd} polypeptides. All members of the superfamily share a highly conserved head domain which is proposed to provide the motive force necessary for the various movements supported by these proteins. Another feature seen in most members is a coiled-coil structure adopted by the internal stalk domain. This region has been shown to mediate interchain association in the kinesin heavy chain and likely plays a similar role in the kinesin-related proteins. The diversity of members of this superfamily is displayed by the dissimilar sequences of their tail domains. This region contains no significant homology between any member of the kinesin superfamily. This domain is thought to confer the specificity necessary for attachment of individual family members to the desired cargo.

The diversity of the tail domains of members of the kinesin superfamily reflects the variety of intracellular movements in which these proteins participate. Kinesin is responsible for fast axonal transport in neuronal cells and vesicular movement in other types of cells. Several family members function in the numerous movements required to segregate chromosomes correctly during mitosis and meiosis. Possible roles include establishing and maintaining the spindle, moving the chromosomes poleward during anaphase **a**, and elongating the spindle during anaphase **b** (see Figure 4). The great number of kinesin-related proteins found in a relatively short time and the detection of multiple genes in a single species suggests both that more kinesin-related proteins are yet to be found and that some may function in ways not yet envisaged.

The discovery of a large superfamily of kinesin-related motors has provided possible mechanisms for various types of microtubule-based motility events. In addition, however, their identification has also complicated our understanding of intracellular motility. The properties of kinesin-related proteins have blurred the lines separating motor molecules of different classes. For example, the finding that the *Drosophila* cd^{nd} protein is a minus-end directed motor contradicted the established dogma concerning the directionality of different classes of motor molecules: dynein produces minus-end directed movement and the kinesins perform movement toward microtubule plus-ends. The polymers upon which kinesins and myosins produce movement were also separate and well defined. Kinesin interacts exclusively with microtubules while myosin utilizes actin as a substrate. The identification of the yeast kinesin-related protein *SMY1* as a suppressor of a myosin defect has confounded this distinction as well.

While unifying our understanding of intracellular motility, many questions remain concerning the newly identified kinesin-related proteins. Although these proteins contain very high homology to the kinesin motor domain, most have not yet been conclusively shown to possess motor activity in a motility assay. The light chain constituents of the kinesin heterotetramer have been characterized at both the biochemical and molecular genetic levels. Little, however, is known about potential light chains associated with kinesin-related proteins. Clearly, many unanswered questions concerning the kinesin-related proteins await the biochemical characterization of the native polypeptides. Until recently, the organisms from which these kinesin-related proteins were identified did not easily lend themselves to the isolation of large quantities of purified protein. The ability to express recombinant polypeptides has alleviated some of these problems, but still will not address questions concerning the polypeptide composition and enzymatic properties of the native motors. The recent identification of kinesin-related molecules in mammals should facilitate the biochemical isolation of these motor proteins and initiate a thorough examination of their biophysical properties. The resulting information will provide a clearer understanding of the kinesin superfamily of motor molecules and their roles in the movement of intracellular structures along microtubules.

NOTE ADDED IN PROOF

Although the literature cited in this review was current upon submittal, during the publication delays the kinesin superfamily has expanded considerably. Unfortunately, space and time constraints prevent us from addressing the most recent data presented in the kinesin literature. Therefore, we would like to direct the reader to the following recent reviews which include a discussion of the present status of the kinesin superfamily of motor proteins.

Goldstein, L.S.B. (1993). With apologies to Sheherazade: tails of 1001 kinesin motors. Ann. Rev. Genet. 27, 319–351.

Sawin, K.E. & Endow, S.A. (1993). Meiosis, mitosis and microtubule motors. Bioessays 15(6), 399–407.

Skoufias, D.A. & Scholey, J.M. (1993). Cytoplasmic microtubule-based motor proteins. Curr. Op. Cell Bio. 5, 95–104.

Hoyt, M.A. (1994). Cellular roles of kinesin and related proteins. Curr. Op. Cell Bio. 6, 63–68.

Bloom, G.S. & Endow, S.A. (1994). Motor proteins 1: kinesins. Protein Profile. 1, 1059–1116.

ACKNOWLEDGMENTS

This work was supported by National Institute of Health Grants NS23320 and NS23868, The Council of Tobacco Research, and the Robert Welch Foundation Grant I-1077.

REFERENCES

Amos, L.A. (1987). Kinesin from pig brain studied by electron microscopy. J. Cell Sci. 87, 105–111.

Bloom, G.S., Wagner, M.C., Pfister, K.K., & Brady, S.T. (1988). Native structure and physical properties of bovine brain kinesin and identification of the ATP-binding subunit polypeptide. Biochemistry 27, 3409–3416.

Brady, S.T. (1985). A novel brain ATPase with properties expected for the fast axonal transport motor. Nature (London) 317, 73–75.

Brady, S.T. (1991). Molecular motors in the nervous system. Neuron 7, 521–533.

Brodsky, B. (1989). In: Protein Folding: Deciphering the Second Half of the Genetic Code (Gierasch, L.M. & King, J., eds.), pp. 55–62, American Assoc. for the Advancement of Science, Washington, D.C.

Cohen, C. & Parry, A.D. (1986). Alpha-helical coiled coils—a widespread motif in proteins. Trends in Biochem. Sci. 11, 245–248.

Cohn, S.A., Ingold, A.L., & Scholey, J.M. (1987). Correlation between the ATPase and microtubule translocating activities of sea urchin egg kinesin. Nature (London) 328, 160–163.

Cohn, S.A., Ingold, A.L., & Scholey, J.M. (1989). Quantitative analysis of sea urchin egg kinesin-driven microtubule motility. J. Biol. Chem. 264, 4290–4297.

Cyr, J.L. & Brady, S.T. (1992). Molecular motors in axonal transport: Cellular and molecular biology of kinesin. Molecular Neurobiology 6, 137–155.

Cyr, J.L., Pfister, K.K., Bloom, G.S., Slaughter, C., & Brady, S.T. (1991). Molecular genetics of kinesin light chains: Generation of isoforms by alternative splicing. Proc. Natl. Acad. Sci. USA 88, 10114–10118.

Dahlström, A., Pfister, K.K., & Brady, S.T. (1991). The axonal transport motor kinesin is bound to anterogradely transported organelles: Quantitative studies of fast anterograde and retrograde axonal transport in the rat. Acta Physiol. Skand. 141, 469–476.

Elluru, R.G., Bloom, G.S., & Brady, S.T. (1990). Axonal transport of kinesin in the rat optic nerve/tract. J. Cell Biol. 111, 417a.

Endow, S.A. (1991). The emerging kinesin family of microtubule motor proteins. Trends Biochem. Sci. 16, 221–225.

Endow, S.A. & Hatsumi, M. (1991). A multimember kinesin gene family in *Drosophila*. Proc. Natl. Acad. Sci. USA 88, 4424–4427.

Endow, S.A., Henikoff, S., & Soler-Niedziela, L. (1990). Mediation of meiotic and early mitotic chromosome segregation in *Drosophila* by a protein related to kinesin. Nature (London) 345, 81–83.

Endow, S.A., Kang, S.J., Satterwhite, L.L., Rose, M.D., Skeen, V.P., & Salmon, E.D. (1993). The yeast Kar3 protein is a minus-end microtubule motor protein that stabilizes microtubules preferentially at the minus ends. EMBO J. 13, 2708–2713.

Enos, A. P. & Morris, N.R. (1990). Mutation of a gene that encodes a kinesin-like protein block nuclear division in A. nidulans. Cell 60, 1019–1027.

Enos, A.P., O'Connell, M., & Morris, N.R. (1991). Kinesin-like proteins of *Aspergillus nidulans*. Cold Spring Harbor Symp. Quant. Biol. 56, 643–648.

Gauger, A.K. & Goldstein, L.S. (1993). The Drosophila kinesin light chain. Primary structure and interaction with kinesin heavy chain. J. Biol. Chem. 268, 13657–13666.

Gibbons, I.R. (1981). Cilia and flagella of eukaryotes. J. Cell Biol. 91, 1075–1245.

Gilbert, S.P. & Sloboda, R.D. (1986). Identification of a MAP-2 like ATP-binding protein associated with axoplasmic vesicles that translocate on isolated microtubules. J. Cell Biol. 103, 947–956.

Hagan, I. & Yanagida, M. (1990). Novel potential mitotic motor protein encoded by the fission yeast cut7- gene. Nature (London) 347, 563–566.

Hagan, I. & Yanagida, M. (1992). Kinesin-related cut7 protein associates with mitotic and meiotic spindles in fission yeast. Nature (London) 356, 74–76.

Hirokawa, N., Kobayashi, N., Sato-Yoshitake, R., Pfister, K.K., Bloom, G.S., & Brady, S.T. (1991). Kinesin associates with anterogradely transported membranous organelles in vivo. J. Cell Biol. 114, 295–302.

Hirokawa, N., Pfister, K.K., Yorifuji, H., Wagner, M.C., Brady, S.T., & Bloom, G. S. (1989). Submolecular domains of bovine brain kinesin identified by electron microscopy and monoclonal antibody decoration. Cell 56, 867–878.

Hoyt, M.A., He, L., Loo, K.K., & Saunders, W.S. (1992). Two saccharomyces cerevisiae kinesin-related gene products required for mitotic spindle assembly. J. Cell Biol. 118, 109–120.

Johnston, G.C., Prendergast, J.A., & Singer, R.A. (1991). The saccharomyces cerevisiae MYO2 gene encodes an essential myosin for vectoral transport of vesicles. J. Cell Biol. 113, 539–551.

Knowles, B.A. & Hawley, R.S. (1991). Genetic analysis of microtubule motor proteins in Drosophila: A mutation at the ncd locus is a dominant enhancer. Proc. Natl. Acad. Sci. USA 88, 7165–7169.

Korn, E.D. & Hammer, J.A. (1990). Myosin I. Curr. Op. Cell Biol. 2, 57–61.

Kosik, K.S., Orecchio, L.D., Schnapp, B., Inouye H., & Neve, R.L. (1990). The primary structure and analysis of the squid kinesin heavy chain. J. Biol. Chem. 265, 3278–3283.

Kuznetsov, S.A. & Gelfand, V.I. (1986). Bovine brain kinesin is a microtubule-activated ATPase. Proc. Natl. Acad. Sci. USA 83, 8530–8534.

Kuznetsov, S.A., Vaisberg, E.A., Rothwell, S.W., Murphy, D.B., & Gelfand, V.I. (1989). Isolation of a 45 kda fragment from kinesin heavy chain with enhanced ATPase and microtubule-binding activities. J. Biol. Chem. 264, 589–595.

Kuznetsov, S.A., Vaisberg, E.A., Shanina, N.A., Magretova, N.A., Chernyak, N.M., & Gelfand, V.I. (1988). The quartenary structure of bovine brain kinesin. EMBO J. 7, 353–356.

Lasek, R.J. & Brady, S.T. (1985). AMP-PNP facilitates attachment of transported vesicles to microtubules in axoplasm. Nature (London) 316, 645–647.

LeGuellac, R., Paris, J., Couturier, A., Roghi, C., & Phillippe, M. (1991). Cloning by differential screening of a *Xenopus* cDNA that encodes for a kinesin related protein. Molec. Cell. Biol. 11, 3395–3398.

Leopold, P.L., McDowall, A., Pfister, K.K., Bloom, G.S., & Brady, S.T. (1992). Association of kinesin with characterized membrane-bounded organelles. Cell Motil. Cytoskel. 23, 19–33.

Lillie, S.H. & Brown, S.S. (1992). Suppression of a myosin defect by a kinesin-related gene. Nature (London) 356, 358–360.

McDonald, H.B. & Goldstein, L.S.B. (1990). Identification and characterization of a gene encoding a kinesin-like gene in Drosophila. Cell 61, 991–1000.

McDonald, H.B., Steward, R.J., & Goldstein, L.S.B. (1990). The kinesin-like NCD protein is a minus-end directed microtubule motor. Cell 63, 1159–1165.

McLachlan, A.D. & Karn, J. (1983). Periodic features in the amino acid sequence of nematode myosin rod. J. Mol. Biol. 164, 605–626.

Meluh, P.B. & Rose, M.D. (1990). *KAR3*, a kinesin-related gene required for yeast nuclear fusion. Cell 60, 1029–1041.

Mercer, J.A., Seperack, P.K., Strobel, M.C., Copeland, N.G., & Jenkins, N.A. (1991). Novel myosin heavy chain encoded by murine dilute coat colour locus. Nature 349, 709–713.

Murofushi, H., Ikai, A., Okuhara, K., Kotani, S., Aizawa, H., Kumakura, K., & Sakai, H. (1988). Purification and characterization of kinesin from bovine adrenal medulla. J. Biol. Chem. 263, 12744–12750.

Navone, F., Niclas, J., Hom-Booher, N., Sparks, L., Bernstein, H.D., McCaffrey, G., & Vale, R.D. (1992). Cloning and expression of a human kinesin heavy chain gene: Interaction of the COOH-terminal domain with cytoplasmic microtubules in transfected CV-1 cells. J. Cell Biol. 117, 1263–1275.

Otsuka, A.J., Jeyaprakash, A., Annoveros-Garcia, J., Tang L., Fisk, G., Hartshorne, T., Franco, R., & Born, T. (1991). The *C. elegans* unc-104 gene encodes a putative kinesin heavy chain-like protein. Neuron 6, 113–122.

Patel, N., de Feo, G., & Mancullas, J.R. (1991). Unc-116 encodes *C. elegans* kinesin. Society for Neuroscience Abstracts 17, 58.

Penningroth, S.M., Rose, P.M., & Peterson, D.D. (1987). Evidence that the 116 kda component of kinesin binds and hydrolyzes ATP. FEBS Lett. 222, 204–210.

Pfister, K.K., Wagner, M.W., Stenoien, D., Bloom, G.S., & Brady, S.T. (1989). Monoclonal antibodies to kinesin heavy and light chains stain vesicle-like structures, but not microtubules, in cultured cells. J. Cell Biol. 108, 1453–1463.

Pierschbacher, M.D. & Ruoslahti, E. (1984) The cell attachment activity of fibronectin can be duplicated by small synthetic fragments of the molecule. Nature (London) 309, 30–33.

Roof, D.M., Meluh, P.B., & Rose, M.D. (1991). Multiple kinesin-related proteins in yeast mitosis. Cold Spring Harb. Symp. Quant. Biol. 56, 693–703.

Roof, D.M., Meluh, P.B., & Rose, M.D. (1992). Kinesin-related proteins required for assembly of the mitotic spindle. J. Cell Biol. 118, 95–108.

Saunders, W.S. & Hoyt, M.A. (1992). Kinesin related proteins required for structural integrity of the mitotic spindle. Cell 70, 451–458.

Sawin, K.E., LeGuellec, K., Philippe, M., & Mitchison, T.J. (1992). Mitotic spindle organization by a plus-end-directed microtubule motor. Nature (London) 359, 540–543.

Saxton, W.M., Porter, M.E., Cohn, S.A., Scholey, J.M., Raff, E.C., & McIntosh, J.R. (1988). *Drosophila* kinesin: Characterization of microtubule motility and ATPase activity. Proc. Natl. Acad. Sci. USA 85, 1109–1113.

Scholey, J.M., Heuser, J., Yang, J.T., & Goldstein, L.S.B. (1989). Identification of globular mechanochemical heads of kinesin. Nature (London) 338, 355–357.

Sperry, A.O., Degrendele, H.C., Cyr, J.L., & Brady, S.T. (1992). A multimember kinesin gene family in rat. Mol. Biol. Cell 3 (suppl), 258a.

Spudich, J.A. (1989). In pursuit of myosin function. Cell Reg. 1, 1–11.

Stewart, R.J., Pesavento, P.A., Woerpel, D.N., & Goldstein, L.S.B. (1990). Identification and partial characterization of six members of the kinesin superfamily in Drosophila. Proc. Natl. Acad. Sci. USA 88, 8470–8474.

Vale, R.D., Reese, T.S., & Sheetz, M.P. (1985a). Identification of a novel force-generating protein, kinesin, involved in microtubule-based motility. Cell 42, 39–50.

Vale, R.D., Schnapp, B.J., Mitchison, T., Steuer, E., Reese, T.S., & Sheetz, M.S. (1985b). Different axoplasmic proteins generate movement in opposite directions along microtubules in vitro. Cell 43, 623–632.

Wagner, M.C., Pfister, K.K., Bloom, G.S., & Brady, S.T. (1989). Copurification of kinesin polypeptides with microtubule-stimulated Mg-ATPase activity and kinetic analysis of enzymatic processes. Cell Motil. Cytoskel. 12, 195–215.

Walker, R.A., Salmon, E.D., & Endow, S.A. (1990). The Drosophila claret segregation protein is a minus-end directed motor molecule. Nature (London) 347, 780–782.

Wedaman, K.P., Knight, A.E., Kendrick-Jones, J., & Scholey, J.M. (1993). Sequences of sea urchin kinesin light chain isoforms. J. Mol. Biol. 231, 155–158.

Wright, B.D., Henson, J.H., Wedamon, K.P., Willy, P.J., Morand, J.N., & Scholey, J.M. (1991). Subcellular localization and sequence of sea urchin kinesin heavy chain: Evidence for its association with membranes in the mitotic apparatus and interphase cytoplasm. J. Cell Biol. 113, 817–833.

Yamada, K.M. & Kennedy, D.W. (1985). Amino acid sequence specificities of an adhesive recognition signal. J. Cell Biochem. 28, 99–104.

Yang, J.T., Laymon, R.A., & Goldstein, L.S.B. (1989). A three domain structure of kinesin heavy chain revealed by DNA sequence and microtubule binding analyses. Cell 56, 879–889.

Yang, J.T., Saxton, W.M., Stewart, R.J., Raff, E.C., & Goldstein, L.S.B. (1990). Evidence that the head of kinesin is sufficient for force generation and motility *in vitro*. Science 249, 42–47.

Yang, J.T., Saxton, W.M., & Goldstein, L.S.B. (1988). Isolation and characterization of the gene encoding the heavy chain of *Drosophila* kinesin. Proc. Natl. Acad. Sci. USA 85, 1864–1868.

Zhang, P., Knowles, B.A., Goldstein, L.S.B., & Hawley, R.S. (1990). A kinesin-like protein required for distributive chromosome segregation in *Drosophila*. Cell 62, 1053–1062.

GENETIC DISSECTION OF *DROSOPHILA* CYTOSKELETAL FUNCTIONS

Eric Fyrberg

Advances in Molecular and Cell Biology
Volume 12, pages 191–205.

ISBN: 1-55938-845-5

I. INTRODUCTION

An important goal of contemporary cell biology is to further elaborate cytoskeletal functions. The value of this endeavor to science and to society is of course subjective, but at least three apparently sound rationales have underpinned the research for some time. The first is that enhancing our understanding of the cytoskeleton, in the general sense, will clarify cellular and developmental processes, for example the transport and assembly of organelles, acquisition of cell polarity, receipt and transduction of various extracellular signals, karyokinesis, cytokinesis, and alterations of cell morphology. The second is that knowledge gained will facilitate understanding how cytoskeletal defects engender human health problems, including neuropathies and myopathies; aneuploidy; hereditary elliptocytosis and pyropoikilocytosis; and neoplasia and metastasis. It is hoped that understanding will lead to rational therapies, and also to the ability to screen affected individuals using the polymerase chain reaction (PCR) or monoclonal antibodies. Finally, detailed understanding of cytoarchitecture may lead to an improved ability to alter cell morphology and function in a purposeful manner, so as to permit expression of high levels of specific proteins, grow in a defined culture media, adhere to various substrates, or fuse with particular cell types.

The crux of further understanding cytoskeletal functions is to better define the roles of individual proteins. This endeavor poses problems for a number of approaches. Traditional biochemical methods, wherein proteins are analyzed outside of their normal environs, are limited because cytoskeletal functions are strongly context-dependent. Cellular approaches, for example immunolocalization of cytoskeletal components within live or fixed cells, typically only correlate those proteins to a process, and do not necessarily provide causal insights. Genetic techniques offer the ability to eliminate or modify particular proteins and observe the phenotypic consequences, and hence to provide mechanistic details. If animals having aberrant cytoskeletal proteins can be subjected to rigorous biochemical and cell biological analyses, prospects for novel insights are enhanced.

The fruit fly, *Drosophila melanogaster*, is arguably the best understood and genetically most tractable metazoan. It is an excellent subject for detailed multidisciplinary studies of cytoskeletal structure and dynamics, especially in light of prior cytological studies of both actin- and tubulin-based cytoskeletal arrays (Karr and Alberts, 1986; Warn and Warn, 1986). *Drosophila* cytoskeletal functions are presently being dissected using classical and reverse genetic approaches, although the endeavor suffers from the fact that cytoskeletal protein abnormalities sometimes confer nondescript or cell-lethal phenotypes. Notwithstanding this difficulty, defects in chromosome segregation, cell division, cell shape alterations, blastoderm formation, cytoplasmic transport, and vesicle transport and recycling have been recently shown to be due to perturbations of particular cytoskeletal and cytoskeleton-based motor proteins. In several instances these correlations produced significant new insights. In the following paragraphs I will review some of the major progress during the prior two years. Consideration of these results demonstrates that molecular genetics has significantly enhanced biochemical and cell biological approaches and, consequently, the field is gravitating toward more integrated investigations.

Although most cell biologists are somewhat familiar with fruit flies, it may be necessary to have access to reviews of the relevant biology. For a recent description of the *Drosophila* cytoskeleton, refer to Fyrberg and Goldstein (1990). For more general aspects of *Drosophila* biology, consult Roberts (1986), Ashburner (1989), and Lawrence (1992).

II. KINESIN-LIKE PROTEINS AND MYOSINS: ROLES IN MITOSIS AND CYTOKINESIS

For some time cytoskeletal elements have been suspected to mediate two of the most fundamental events in growing cells, chromosome separation during mitosis, and the pinching of one cell to form two during cytokinesis. In neither case do we understand the mechanistic aspects of cytoskeletal involvement in detail, particularly the several new microtubule- and microfilament-based motors that have been discovered since the mid-1980s.

Much recent karyokinesis and cytokinesis research has been focused on defining the roles of kinesin-like proteins, cytoplasmic dyneins, and myosins. In order to investigate this issue comprehensively it is useful to establish the number and chromosomal locations of genes that encode these polypeptides, and to define the roles of each. In the case of the kinesin superfamily of proteins, Endow and Hatsumi (1991) and Stewart et al. (1991) have used the polymerase chain reaction in conjunction with degenerate oligonucleotide primers to amplify kinesin heavy chain-like genes (KLPs) from *Drosophila*. These authors find that the kinesin superfamily is large, minimally consisting of 11 members. Nine of these—KLP1-6, Ncd, Nod, and conventional kinesin heavy chain (KHC)—prove to be expressed within many different cell and tissue types. This is a highly significant finding, as it shows that, within a given cell, microtubule-based motility is divided amongst many motors. *In situ* hybridization of these genes to polytene chromosomes has established their locations, opening the way for genetic analyses of their cellular roles.

Indeed, molecular genetic investigations of individual *Drosophila* kinesin-like proteins are proceeding rapidly, and have already demonstrated their involvement in meiosis and mitosis. The *non-claret disjunctional*, (*ncd*) locus encodes a "flipped" kinesin-related protein, while the *no distributive segregation* (*nod*), locus encodes a conventional kinesin relative. Both are required for aspects of chromosome segregation (Endow et al., 1990; McDonald et al., 1990; Zhang et al., 1990). *ncd* mutants have frequent meiotic nondisjunction and associated chromosome loss only in females. Also, the eggs of such mothers lose mitotic chromosomes, typically maternally derived, during the cleavage divisions. *nod* mutations engender nondisjunction and loss of nonrecombinant chromosomes during meiosis I and subsequent mitotic divisions. Neither mutation is lethal when homozygous, corroborating the co-expression results described above, and further attesting that kinesin superfamily member functions overlap to such a degree that eliminating the functions of any one results in a very limited phenotype. Finally, a mutant allele of the KLP-2 gene has recently been characterized by Heck et al. (personal communication) and shown to confer spindle pole body separation defects. In sum, this research has identified components of the superfamily involved in movement of chromosomes and other organelles. Continued analyses of mutant derivatives, and combinations thereof, are expected to reveal in detail the roles of each motor within intact cells.

Genetic methods are also defining novel proteins required for fidelity of mitotic chromosome segregation. Karess and Glover (1989) previously defined a gene, *rough deal*, required for mitotic chromosome segregation, and Williams et al. (1992) have recently characterized a second such gene, *lethal (1) zw10*. Mutant alleles of both genes disrupt chromosome segregation in many *Drosophila* cell types, resulting in high levels of aneuploidy in larval brains. Williams et al. (1992) report that during metaphase the Zw10 antigen is part of an unidentified filamentous structure that appears to be associated with kinetochore microtubules, while at anaphase onset the protein rapidly redistributes to kinetochores of separating chromosomes. Williams et al. (1992) sequenced Zw10 cDNAs and found that the gene encodes a protein that has no significant homology to any extant polypeptide sequence, save for the occurrence of several phosphorylation sites that may be responsible for the alteration of localization during mitosis. Interestingly, the redistribution of Zw10 antigen does not occur in *rough deal* mutants, implying that both loci may be involved in the same process (R. Karess and M. Goldberg, personal communication).

Cytoplasmic dynein is also of central importance to chromosome segregation. This motor protein was reported by two groups to be associated with mitotic spindles and kinetochores (Pfarr et al., 1990; Steuer et al., 1990) during anaphase separation. Based upon this observation and the fact that cytoplasmic dynein is a minus end-directed motor, it may play a principal role in chromosome separation during anaphase. A family of *Drosophila* sequences that encodes dynein-related polypeptides has been isolated by T. Hays et al. (personal communication) using PCR amplification of genes encoding proteins related to cytoplasmic dynein of sea urchin, but as yet no details as to the functions of individual cytoplasmic dynein isoforms have been reported. It will be very interesting to compare phenotypes attributable to kinesin-like proteins versus cytoplasmic dynein defects, and also to analyze various double mutant combinations.

Considerable progress in defining functions of non-muscle myosins has been reported recently by R. Karess, D. Kiehart, and their collaborators. Kiehart's group previously described the characterization of a *Drosophila* cytoplasmic myosin heavy-chain gene (Kiehart et al., 1989; Ketchum et al., 1990), and has more recently characterized two cytoplasmic myosin light-chain genes of the fly (Karess et al., 1991, and personal communications), while Karess and collaborators focused on isolating *Drosophila* mutants having abnormal mitosis or cytokinesis (for a description of how this is done refer to Ripoll et al., 1987; Glover, 1989; Karess and Glover, 1989). Together, the two groups were able to correlate a myosin regulatory light-chain gene mutation with defective cytokinesis.

A variety of evidence suggests that the contractile ring of dividing cells is comprised of actin and myosin, and that progress of the cleavage furrow is driven by actin-myosin interactions analogous to those occurring during muscle contraction (Schroeder, 1990). This model is supported, *in vivo*, by reverse genetic experiments in which the myosin heavy chain was eliminated from axenic *Dictyostelium* cells, that consequently are unable to divide (DeLozanne and Spudich, 1987; Knecht and Loomis, 1987). However, the mechanisms by which the assembly and functioning of actomyosin networks within cleavage furrows are regulated are for the most part conjectural, and a comprehensive set of mutants having defective cytokinesis would be very useful for further elucidating these processes.

Karess et al. (1991) recently described one such defect. They isolated a number of *Drosophila* mutants wherein cell divisions within the highly proliferative imaginal disc and brain tissues were abnormal, and established that one, *spaghetti-squash*, was defective in cytokinesis. By aceto-orcein staining they showed that many mutant larval brain cells are polyploid, typically ranging from diploid to octaploid, but have roughly the same percentage of anaphase figures as do normal brain cells, implying that cytokinesis, rather than mitosis, is aberrant. The morphology of mutant brain cells also suggested that cytokinesis is defective, as cleavage furrows are frequently absent well after chromosomes have migrated to opposite poles. Normal cytokinesis does occur in some instances, however, indicating that the mutant allele produces at least some functional product.

The map position of the mutant was found to coincide with the known location of the gene that encodes the non-muscle regulatory myosin light chain (the 5D subdivision). Molecular characterization of the allele revealed that a transposon had inserted within the second exon, reducing steady state mRNA levels to roughly one tenth of normal. Karess et al. (1991) thus provide the first direct demonstration that regulatory myosin light chain is essential for cytokinesis, the involvement of cytoplasmic myosin heavy chain having been shown previously, as stated above, by Knecht and Loomis (1987) and DeLozanne and Spudich (1987). Since the regulatory myosin light chain is known to influence both the assembly and contractility of myosin in a phosphorylation-dependent manner (refer to the review by Tan et al., 1992), we do not formally know how the mutation disrupts cytokinesis, and this issue presumably will be the subject of future investigations of the *spaghetti-squash* mutant.

III. NON-MUSCLE MYOSIN: ROLE IN ALTERING CELL SHAPE DURING DORSAL CLOSURE

Within the larger context of cell and developmental biology, cytokinesis may be thought of as one type of alteration in cell shape. Hence, we might expect, based upon the results of Karess et al. (1991) described above, that myosin would be involved in several types of cell shape changes that have been documented during gastrulation. Indeed, a good deal of circumstantial evidence suggests that myosin plays a significant role in altering cell shape during morphogenetic movements, but the evidence, in the main, is derived from pharmacological and immunolocalization studies, and only recently have reverse genetic studies in *Dictyostelium* and *Drosophila* begun to contribute.

One difficulty inherent to genetic investigations of cytoskeletal involvement in early development is that the complement of mRNAs and proteins placed in animal eggs by the mother is sufficient to direct many key events, even in embryos unable to synthesize a cytoskeletal component because of a zygotic mutation. Hence, it can be difficult to perturb early cytoskeletal involvement in embryogenesis using genetic methods. Maternal mutations (mutations within genes of the mother that block her ability to synthesize a particular egg protein, and consequently perturb embryo development) may be analyzed to circumvent this difficulty, but they are useful only if they primarily affect the egg cytoplasm and largely spare the mother's somatic cells. An alternative approach which avoids this problem is to analyze cell shape changes that occur after the maternal supply of cytoskeletal proteins begins to be depleted. For example, midway through embryogenesis, and well after the completion of gastrulation, epidermal cells located along both sides of the embryo undergo a profound shape change that cause them to extend toward the dorsal region and spread over the underlying squamous epithelium, a process called dorsal closure. Recently, Young et al. (1992) have shown that cytoplasmic myosin heavy chain plays a principal role in this transformation, thus providing some of the first direct evidence that the cytoskeleton implements cell shape alterations.

The crux of the Young et al.'s research entailed correlating defects in myosin heavy chain with aberrant embryonic development. Some time ago Nusselein-Volhard and collaborators demonstrated that the recessive embryonic lethal mutation *zipper* conferred defects in dorsal closure. All *zipper* alleles result in embryos that have dorsal apertures that presumably result from the inability of lateral epidermis to undergo the shape alterations that spread the cells over the underlying epithelium. *Zipper* maps to a region (60EF) of chromosome 2 where the cytoplasmic myosin heavy-chain gene is known to reside (Kiehart et al., 1989), but in no case was it clear whether *zipper* alleles perturb this myosin isoform. Kiehart et al. induced a new series of mutations in this chromosome region and showed that one is due to the introduction of a nonsense codon that eliminates the carboxy-terminal 77 residues of myosin. None of the new alleles complements the extant *zipper* mutants, confirming that myosin is indeed the affected gene.

By staining the affected lateral epidermal cells with anti-spectrin, it is clear that, in *zipper* embryos, they cannot undergo the cell shape alteration. Furthermore, by double staining the affected embryos with anti-myosin and anti-actin antibodies some details of myosin's role have become clear. It appears from the pattern of staining that myosin is concentrated at the leading edges of the cells involved in closure, and that spreading of the cell sheets is implemented by a molecular pursestring mechanism. Unlike cytokinesis, however, the pursestring is formed by the cooperative activities of myosin concentrated in the leading edges of many juxtaposed cells. Actin is not concentrated in this manner, and its distribution is not affected in *zipper* embryos, strongly suggesting that the observed defects in myosin concentration and/or distribution are primary, not secondary. It will be interesting to further explore the roles of myosin in other gastrulation movements, if specific mutations can be employed, since other work from Kiehart's laboratory (see Lutz and Kiehart, 1991) suggests the involvement of myosin heavy chain in several earlier cytoskeletal transitions within *Drosophila* embryos.

IV. THE CYTOSKELETON AND FORMATION OF THE CELLULAR BLASTODERM

Ever since the pioneering studies of Warn (1980, 1983) and Karr and Alberts (1986) revealed the complexity and dynamic alterations of the *Drosophila* embryonic cytoskeleton, researchers have speculated about its possible roles in developmental events, including movement of nuclei to the egg surface, maintenance of nuclear spacing, invagination of membranes to form the cellular blastoderm, interactions with the centrosome so as to maintain proper chromosome segregation, and alterations of cell shapes that are essential for proper gastrulation. As discussed above, many embryonic events are controlled by the maternal genome. Following fertilization, *Drosophila* nuclei replicate without cytokinesis deep within the egg, a process termed superficial cleavage, and only later do they migrate outwards into the cortical cytoplasm, or periplasm. Subsequently, membranes descend around each nucleus to form the cellular blastoderm. The cellularization process is accompanied by restructuring the cortical actin cytoskeleton to form a network of hexagons, each surrounding a nucleus, suggesting that actin and myosin may either template or drive the process of membrane invagination, or cellularization. A simple model for the mechanism by which cellular membranes form is that the cytoskeletal network contracts in a net-like manner, thus pulling membranes down between the nuclei (Schejter et al., 1992). This hypothetical mechanism is once again derived from the contractile ring hypothesis, but in this case a network of rings, rather than a single ring, drives the process.

Cellularization during blastoderm formation is largely controlled by the zygotic, rather than maternal, genome, hence it can be studied conveniently using classical genetic approaches. E. Wieschaus and collaborators are systematically investigating the genetic instructions that specify the transitions of cytoskeletal morphology that accompany cellularization. Although the work is in its infancy, several remarkable results have emerged. One is that only a small number of gene products directs the rather profound cytoskeletal transitions that accompany cellularization. Wieschaus and Sweeton (1988) and Merrill et al. (1988) studied the development of embryos having particular genomic regions missing. They found only three loci, *nullo*, *serendipity-a*, and *bottleneck*, that affected the restructuring of the actin cytoskeleton. None of the three can be mutated to prevent the transition. Rather, defective alleles lead to partial disruption of the network that results in many incorrectly shaped cells within the blastoderm. These phentoypes suggest that the proteins encoded by the three genes are needed to stabilize the network in the face of contractile forces, rather than to initiate or maintain contraction.

All three of the genes defined by this study have been cloned and sequenced. They all encode relatively small proteins, ranging in size from 24 to 56 kDa, having no similarities to any sequences presently deposited in databases. Both Nullo and Bottleneck proteins are highly charged with basic residues and *nullo* contains a potential N-terminal myristoylation site. Further biochemical work will be needed to establish how these proteins implement alterations in cytoskeletal morphology.

Very recently the Wieschaus group, in collaboration with that of K. Miller, has begun to investigate the role of maternally supplied proteins in implementing transitions of the actin-based cytoskeleton. Postner et al. (1992) have analyzed cytoskeletal defects observed in embryos from homozygous *sponge* mothers, and found that microfilament caps fail to form over interphase blastoderm nuclei, and that mitotic spindles are abnormally positioned and frequently fused. Misoriented spindles lead to aneuploidy, defects in cellularization, and embryonic lethality. This work is probably the strongest evidence that microfilament network morphology is intimately related to mitotic spindle position and separation, and also that *in vivo* individual proteins have a profound effect in implementing cytoskeletal form.

V. OOPLASMIC TRANSPORT: ROLES OF ACTIN FILAMENTS

Transport of cytoplasm is a principal feature of oogenesis in *Drosophila*. Within each egg chamber of its polytrophic ovary cytoplasm flows from fifteen polyploid nurse cells to the oocyte via a series of cytoplasmic bridges. The process has been shown to depend upon actin filament integrity via drugs that depolymerize actin arrest transport (Guzeit, 1986). A rather different transport process occurs in telotrophic insect ovaries, where microtubule-based motors have been shown recently to drive vesicular flow (Anastasi et al., 1990). The mechanism by which microfilaments implement transport within *Drosophila* ovaries is unknown. One means by which to gain insight into the mechanism and its regulation is to characterize female-sterile mutants in which ooplasmic flow does not occur. Cooley et al. (1988) found that such mutants are rare, but succeeded in identifying several by examining rhodamine-phalloidin labeled actin filaments in egg chambers of female sterile mutants. Remarkably, one of these, *chickadee*, has no apparent egg chamber microfilament networks. More remarkable still is the fact that the affected gene encodes *Drosophila* profilin (Cooley et al., 1992).

Cooley et al. (1992) began their molecular genetic analysis by recovering a P-element insertion-induced allele of *chickadee*. A portion of the affected gene was recovered by plasmid rescue, and subsequently used to screen a *Drosophila* library. The gene was discovered to encode 1.0 and 1.2 kb transcripts, both of which specify a protein that is 40% identical to profilins from *Acanthamoeba*, *Saccharomyces cerevisiae*, or *Physarum*. All characterized *chickadee* alleles, however, perturb only the 1.0 kb transcript, which apparently accumulates only within nurse cells. Profilin is an actin and lipid binding protein that was until recently believed to function primarily as an actin monomer sequestering factor, thus maintaining a pool of unpolymerized actin. Results of Cooley et al. (1992) are inconsistent with the notion, suggesting instead that profilin must facilitate the formation of actin filaments within *Drosophila* egg chambers. Indeed, their results fit nicely with proposals of Goldschmidt-Clermont and Jamney (1991) that the effect of profilin on nucleotide exchange actually stimulates actin polymerization.

What does the research tell us about the role of actin filaments in oogenesis? In nurse cells there are two sets of actin microfilaments that could be used to generate force or serve as tracks for producing directed transport; firstly, there is a subcortical filamentous actin network present in nurse cells throughout oogenesis, and secondly, just before final nurse cell regression begins a second cytoplasmic network of actin containing filaments forms within each nurse cell. This latter network has been proposed to squeeze nurse cell cytoplasm into the oocyte. Seemingly in support of this notion, stage 11 *chickadee* nurse cells, which lack the latter network, fail to deliver all of their cytoplasm to the oocyte, although earlier stages of oogenesis appear normal. Examination of the distribution of nurse cell nuclei, however, suggests a radically different explanation for the transport failure. The crucial observation is that nurse cell nuclei in mutant egg chambers become rearranged sometime after stage 10. Ultimately, they lose their position and actually block the ring canals. Thus, the cytoplasmic actin network that fails to form in *chickadee* mutants seems to be playing a structural role, that is to anchor nurse cell nuclei so that they cannot be swept into the ring canals.

In sum, investigation of *chickadee* mutants strongly suggests that the role of the cytoplasmic nurse cell microfilaments is structural, and that it is the subcortical actin networks that generate contractile force to squeeze ooplasm into the oocyte. This model is supported by pharmacological and histological observations. First, it should be noted that cytochalasin, which prevents transport, has a dramatic effect on subcortical actin but only marginally perturbs the cytoplasmic network (Guzeit, 1986). Second, cytoplasmic myosin was found by W. Theurkauf (in Cooley et al., 1992) to be associated with subcortical actin but not with the cytoplasmic actin, consistent with the notion that it is the subcortical network that generates tension to drive ooplasmic transport.

VI. AXONAL TRANSPORT: ROLE OF KINESIN HEAVY CHAIN

Saxton et al. (1991) have isolated and partially characterized *Drosophila* mutants that either fail to synthesize, or have abnormalities in the "classical" kinesin heavy chain (KHC) encoded by the 52F subdivision gene described by Yang et al. (1988). Mutant larvae display poor mobility and tactile responses, particularly within their posterior segments. These observations are consistent with the suspected role of the protein in axonal transport (Vale et al., 1985), because axons connecting the larval central nervous system to muscles of the posterior segments are much longer than those that innervate anterior segments, and must require more transporting activity. Adults homozygous for a temperature-sensitive *khc* allele likewise have poor motility and sensory responses, again suggesting a principal role in axonal transport.

Surprisingly, ultrastructural analyses of affected larval motor neurons by Saxton and McDonald (personal communication) have not yet revealed the anticipated deficits of synaptic vesicles or mitochondria within their distal termini. Also noteworthy is the fact that kinesin is not essential in all *Drosophila* cell types, despite the fact that Saxton et al. (1991) report that it is ubiquitously distributed. Furthermore, Sheehan and Saxton (personal communication) used radiation-induced mitotic crossing-over to construct large patches of eye tissue that are homozygous for a severe defective kinesin allele and contrary to expectations the cells remained viable, and displayed no readily apparent morphological defects.

In summary, it is clear that in *Drosophila* the "classical" kinesin is present in all cell types, that it is essential for viability of the organism, and that it almost certainly plays a principal role in fast axonal transport. It is not, however, essential for generalized cellular functions, for example, mitosis. These results suggest that kinesin functions are largely redundant. Hence, the most demanding transport events within long axons would be noticeably impaired without kinesin, but analogous proteins could meet the demands of all other tissues and even mask most of the overt axonal defects, i.e., the anticipated paucity of synaptic vesicles and mitochondria. Such redundancy is proving to be commonplace for cytoskeletal proteins. Goldstein et al. (1992) has recently reported that mutations that eliminate a major microtubule associated protein, MAP205, have no apparent phenotypic consequences, and Roulier et al. (1992) provided convincing evidence that alpha-actinin function is redundant in non-muscle cells. Furthermore, recent results suggest that vesicle movement is implemented through the collaboration of both microtubule and microfilament-based networks (Lillie and Brown, 1992; Kuznetsov et al., 1992), suggesting that a defect in either transport system would impede, but not abolish, vesicular movements within axons.

VII. VESICULAR TRANSPORT AND RECYCLING: ROLES FOR DYNACTIN AND DYNAMIN

The *Drosophila Glued* locus, so named because mutant compound eye facets appear to be glued together in disorderly arrays (Meyerowitz and Kankel, 1978), was shown by Swaroop et al. (1986, 1987) to encode a 148 kilodalton protein that contains extensive alpha-helical domains and is expressed ubiquitously. Recent biochemical and molecular genetic investigations of soluble factors that activate dynein have revealed that the *Glued* protein sequence is highly similar to that of dynactin (dynein activator), a 150 kd protein that is associated with cytoplasmic dynein (Holzbaur et al., 1991; Gill et al., 1991). Since cytoplasmic dynein is implicated in retrograde organelle transport and possibly in mitosis (Steuer et al., 1990; Pfarr et al., 1990), disruptions in an associated protein might be expected to confer cell-autonomous lethality. The Gl^1 allele is dominant, and associated defects can be attenuated by supernumerary wild type gene copies, consistent with the notion that the encoded protein is part of a protein complex. All *Glued* alleles confer pleiotropic developmental defects, supporting the suggestion that the encoded protein is an essential component of a variety of cell types. Gl^1 and null *Glued* alleles confer recessive early cell-autonomous lethality (Harte and Kankel, 1982), more directly demonstrating that the corresponding protein carries out a vital function within all cells. These genetic results are in agreement with both *in situ* localization of transcripts (Swaroop et al., 1986) and RNA blot-hybridization experiments (Gill et al., 1991), that demonstrate the ubiquitous distribution of dynactin transcripts.

Further consideration of *Glued* mutants provides some hints as to the functions of the dynein complex. The Gl^1 mutation confers dominant eye and optic lobe defects in heterozygotes, and recessive embryo lethality. The disproportionate effect on architecture of axons associated with the visual system probably is indicative of a principal or specialized role in axonal transport, while the cell-autonomous lethality must be due to a more general function, perhaps membrane transport or the movement of mitotic chromosomes. Indeed, biochemical studies utilizing dynactin of bovine and avian brain show that it promotes cytoplasmic dynein-mediated motility and have also revealed that a small portion of the protein is associated with kinetochores. It would be especially informative, at this juncture, to prepare antibodies against the *Glued* protein and use them to further investigate both embryonic lethality and development of the eye and optic lobe.

Another presumed motor protein that has been shown recently to be associated with microtubules is dynamin, a 100-kDa GTP driven mechanicochemical enzyme (Schpetner and Vallee, 1988). Dynamin is related to mammalian Mx proteins and the yeast vps1/spo15 protein (Staeheli et al., 1986; Rothman et al., 1990; Obar et al., 1990). The former protein may be involved in movement of virus-containing vesicles, while the latter is known to participate in vacuole transport and separation of spindle pole bodies during meiosis. All three proteins could conceivably be involved in microtubule-based transport processes, and all are proposed to be cognate homologues on the basis of their sequence similarity.

Support that dynamin is indeed involved in vesicular transport has come from studies of the *Drosophila shibire* mutant. *shibire* mutants have a marked temperature-sensitive paralysis thought to be due to a reversible block of endocytosis that prevents membrane cycling and thus rapidly depletes synaptic vesicles. The same mutation also confers abnormalities on nonneural tissues, including blockage of endocytosis in garland cells and oocytes (Kosaka and Ikeda, 1983; Kessell et al., 1989). Two groups have reported recently the cloning of the *shibire* locus and characterization of several mutant alleles (van der Bliek and Meyerowitz, 1991, Chen et al., 1991). Remarkably, the locus encodes a protein that is 69% identical to dynamin (81% similarity if one takes into account conservative amino acid substitutions). Further attesting to the functional homology is the fact that all three N-terminal GTP-binding sequence motifs are perfectly conserved between the *shibire* protein and dynamin. Considering both the protein structural homology and the *shibire* phenotype it seems very likely that dynamin-related proteins are motors that facilitate vesicular transport during endocytosis.

VIII. SUMMARY AND CONCLUSIONS

From the prior discussion it is clear that *Drosophila* genetics is contributing significantly to our understanding of contractile, cytoskeletal, and motor proteins. As we predicted two years ago (Fyrberg and Goldstein, 1990) progress has increased due to heightened awareness of cytoskeletal issues within the genetics community. In coming years it may be anticipated that molecular genetics will continue to define new proteins that are part of, or regulate the form of, actin- and tubulin-based cytoskeletons. Concurrently, genetic studies will continue to refine our understanding of known cytoskeletal proteins. As the roles of particular protein domains are defined, hypotheses may be tested by perturbing particular regions by site-directed mutagenesis of the corresponding gene, and it can be ascertained whether the mutated protein will rescue null allele-associated phenotypes. For example, R. Karess et al. are presently defining which phosphorylated serine residues of myosin regulatory light chain are.essential for regulated cytokinesis (personal communication). Work ongoing in my own laboratory is designed to evaluate the functions of particular domains of actin. It is an especially opportune time to dissect the functioning of both myosins and kinesins in order to establish in more detail how these motors work. Finally, it is expected that *Drosophila* genetics will shortly define the role of the cytoskeleton in mRNA stabilization and localization (see, for example, Raff et al., 1990; Pokrywka and Stephenson, 1991). The collective genetic work presently in progress, or about to be initiated, should fill many gaps in our knowledge, and enrich our understanding of cell biology.

ACKNOWLEDGMENTS

Work in the author's laboratory has been generously supported by grants from the National Institutes of Health and The Muscular Dystrophy Association of America.

REFERENCES

Anastasi, A., Hunt, C., & Stebbings, H. (1990). Isolation of microtubule motors form an insect ovarian system: Characterization using a novel motility substratum. J. Cell Sci. 96, 63–69.

Ashburner, M. (1989). *Drosophila*, a laboratory handbook, 1331 pp., Cold Spring Harbor, New York.

Chen, M.S., Obar, R.A.; Schroeder, C.C., Austin, T.W., Poodry, C.A., Wadsworth, S.C., & Vallee, R.B. (1991). Multiple forms of dynamin are encoded by *shibire*, a *Drosophila* gene involved in endocytosis. Nature (London) 351, 583–586.

Cooley, L., Kelley, R., & Spradling, A. (1988). Insertional mutagenesis of the *Drosophila* genome with single P-elements. Science 239, 1121–1128.

Cooley, L., Verheyen, E., & Ayers, K. (1992). *Chickadee* encodes a profilin required for intercellular cytoplasm transport during *Drosophila* oogenesis. Cell 69, 173–184.

DeLozanne, A. & Spudich, J. (1987). Disruption of the *Dictyostelium* myosin heavy chain by homologous recombination. Science 236, 1086–1091.

Endow, S.A. & Hatsumi, M. (1991). A multimember kinesin family in *Drosophila*. Proc. Natl. Acad. Sci. USA 88, 4424–4427.

Endow, S.A., Henikoff, S., & Soler-Niedziela, L. (1990). Mediation of meiotic and early mitotic chromosome segregation in *Drosophila* by a protein related to kinesin. Nature 345, 81–83.

Fyrberg, E.A. & Goldstein, L.S.B. (1990). The *Drosophila* cytoskeleton. Ann. Rev. Cell Biol. 6, 559–596.

Gill, S.R., Schroer, T.A., Szilak, I., Steuer, E.R., Sheetz, M.P., & Cleveland, D.W. (1991). Dynactin, a conserved, ubiquitously expressed component of an activator of vesicle motility mediated by cytoplasmic dynein. J. Cell Biol. 115, 1639–1650.

Glover, D. (1989). Mitosis in *Drosophila*. J. Cell Sci. 92, 137–146.

Goldschmidt-Clermont, P.J., & Jamney, P.A. (1991). Profilin, a weak CAP for actin and RAS. Cell 66, 419–421.

Harte, P.J. & Kankel, D.R. (1982). Genetic analysis of mutations at the *glued* locus and interacting loci in *Drosophila melanogaster*. Genetics 101, 477–501.

Holzbaur, E.L., Hammarback, J.A., Paschal, B.M., Kravit, N.G., Pfster, K.K., & Vallee, R.B. (1991). Homology of the 150K cytoplasmic dynein-associated polypeptide with the *Drosophila* gene *Glued*. Nature (London) 351, 579–583.

Karess, R. & Glover, D. (1989). *Rough-deal*: A gene required for proper mitotic segregation in *Drosophila*. J. Cell Biol. 109, 2951–2961.

Karess, R.E., Chang, Z-j., Edwards, K.A., Kulkarni, S., Aguilera, I., & Kiehart, D.P. (1991). The regulatory light chain of nonmuscle myosin is encoded by *spaghetti-squash*, a gene required for cytokinesis in *Drosophila*. Cell 65, 1177–1189.

Karr, T.L. & Alberts, B.M. (1986). Organization of the cytoskeleton in early *Drosophila* embryos. J. Cell Biol. 102, 1494–1509.

Kessel, I., Holst, B.D., & Roth, T.F. (1989). Membranous intermediates in endocytosis are labile, as shown in a temperature-sensitive mutant. Proc. Natl. Acad. Sci. USA 86, 4968–4972.

Ketchum, A.S., Stewart, C.T., Stewart, M., & Kiehart, D.P. (1990). Complete sequence of the *Drosophila* nonmuscle myosin heavy chain transcript: Conserved sequences in the myosin tail and differential splicing in the 5′ untranslated sequence. Proc. Natl. Acad. Sci. USA 87, 6316–6320.

Kiehart, D.P., Lutz, M.S., Chan, D., Ketchum, A.S., Laymon, R.A., Nguyen, B., & Goldstein, L.S. B. (1989). Identification of the gene for fly non-muscle myosin heavy chains are encoded by a gene family. EMBO J. 8, 913–922.

Knecht, D. & Loomis, W. (1987). Antisense RNA inactivation of myosin heavy chain expression in *Dictyostelium* discoideum. Science 236, 1081–1086.

Kosaka, T. & Ikeda, K. (1983). Reversible blockage of membrane retrieval and endocytosis in the garland cell of the temperature-sensitive mutant of *Drosophila melanogaster, shibire ts1*. J. Cell Biol. 97, 499–507.

Kuznetsov, S.A., Langford, G.M., & Weiss, D.G. (1992). Actin-dependent organelle movement in squid axoplasm. Nature (London) 356, 722–724.

Lawrence, P. (1992). The making of a fly. London: Blackwell Scientific Publications.

Lillie, S.H. & Brown, S.S. (1992). Suppression of a myosin defect by a kinesin-related gene. Nature 356, 358–361.

Lutz, D.A. & Kiehart, D.P. (1991). Disruption of cortical nuclear and cytoplasmic organization in *Drosophila* embryos by microinjection of antimyosin. J. Cell Biol. 105, 127a.

McDonald, H.B., Stewart, R.J., & Goldstein, L.S.B. (1990). The kinesin-like *ncd* protein of *Drosophila* is a minus end-directed microtubule motor. Cell 63, 1159–1165.

Merrill, P.T., Sweeton, D., & Wieschaus, E. (1988). Requirements for autosomal gene activity during precellular stages of *Drosophila melanogaster*. Devel. 104, 495–509.

Meyerowitz, E.M. & Kankel, D.R. (1978). A genetic analysis of visual system development in *Drosophila melanogaster*. Dev. Biol. 62, 112–142.

Obar, R.A., Collins, C.A., Hammarback, J.A., Schpetner, H.S., & Vallee, R.B. (1990). Molecular cloning of the microtubule-associated mechanochemical enzyme dynamin reveals homology with a new family of GTP-binding proteins. Nature (London) 347, 256–261.

Pfarr, C.M., Coue, M., Grissom, P.M., Hays, T.S., Porter, M.E., & McIntosh, J.R. (1990). Cytoplasmic dynein is localized to kinetochores during mitosis. Nature 345, 263–265.

Pokrywka, N.J., & Stephenson, E.C. (1991). Microtubules mediate the localization of bicoid RNA during *Drosophila* oogenesis. Development 113, 55–66.

Postner, M., Miller, K., & Wieschaus, E. (1992). Maternal effect mutations of the *sponge* locus. J. Cell Biol. 119, 1205–1218.

Raff, J.W., Whitefield, W.G.F., & Glover, D.M. (1990). Two distinct mechanisms localize cyclin B transcripts in synctial *Drosophila* embryos. Development 110, 1249–1261.

Ripoll, P., Casal, J., & Gonzalez, C. (1987). Toward a genetic dissection of mitosis in *Drosophila*. BioEssays 7, 204–210.

Roberts, D.B. (1986). *Drosophila*, a practical approach. Oxford, IRL Press, 295 pp.

Rothman, J.L., Raymond, C.K., Gilbert, T., O'Hara, P. J., & Stevens, T.H. (1990). A putative GTP binding protein homologous to interferon-inducible Mx proteins performs an essential function in yeast protein sorting. Cell 61, 1063–1074.

Roulier, E.M., Fyrberg, C., & Fyrberg, E. (1992). Perturbations of *Drosophila* alpha-actinin cause muscle paralysis, weakness, and atrophy but do not confer obvious nonmuscle phenotypes. J. Cell Biol. 116, 911–922.

Saxton, W.M., Hicks, J., Goldstein, L.S.B., & Raff, E.C. (1991). Kinesin heavy chain is essential for viability and neuromuscular functions in *Drosophila* but mutants show no defects in mitosis. Cell 64, 1093–1022.

Schejter, E.D., Rose, L.S., Postner, M.A., & Wieschaus, E. (1992). The role of the zygotic genome in the restructuring of the actin cytoskeleton at the cycle 14 transition during *Drosophila* embryogenesis. Cold Spring Harbor Symp. Quant. Biol. (in press).

Schpetner, H.S. & Vallee, R.B. (1989). Identification of dynamin, a novel mechanochemical enzyme that mediates interactions between microtubules. Cell 59, 421–432.

Schroeder, T.E. (1990). The contractile ring and furrowing in dividing cells. Ann. N.Y. Acad. Sci. 582, 78.

Staeheli, P., Haller, O., Boll, W., Lindenmann, J., & Weissman, C. (1986). Mx protein: Constitutive expression in 3T3 cells transformed with cloned Mx DNA confers selective resistance to influenza virus. Cell 44, 147–158.

Steuer, E.R., Wordeman, L., Schroer, T.A., & Sheetz, M.P. (1990). Localization of cytoplasmic dynein to mitotic spindles and kintochores. Nature 345, 266–268.

Stewart, R.J., Pesavento, P.A., Woerpel, D.N., & Goldstein, L.S.B. (1991). Identification and partial characterization of six members of the kinesin superfamily in *Drosophila*. Proc. Natl. Acad. Sci. USA 88, 8470–8474.

Swaroop, A., Sun, J-W., Paco-Larson, M.L., & Garen, A. (1986). Molecular organization and expression of the genetic locus *Glued* in *Drosophila melanogaster*. Mol. Cell Biol. 6, 833–841.

Swaroop, A., Swaroop, M., & Garen, A. (1987). Sequence analysis of the complete cDNA and encoded polypeptide for the *Glued* gene of *Drosophila melanogaster*. Proc. Natl. Acad. Sci. USA 84, 6501–6505.

Tan, J.L., Ravid, S., & Spudich, J.A. (1992). Control of nonmuscle myosins by phosphorylation. Ann. Rev. Biochem. 61, 721–759.

Vale, R.D., Reese, T.S., & Sheetz, M.P. (1985). Identification of a novel force-generating protein, kinesin, involved in microtubule-based motility. Cell 42, 30–50.

van der Bliek, A.M., & Meyerowitz, E.M. (1991). Dynamin-like protein encoded by the *Drosophila* shibire gene associated with vesicular traffic. Nature (London) 351, 411–414.

Warn, R.M., Bullard, B., & Magrath, R. (1980). Changes in the distribution of cortical myosin during the cellularization of the *Drosophila* embryo. J. Exp. Embryol. Morph. 57, 167–176.

Warn, R.M. & Magrath, R. (1983). F-actin distribution during the cellularization of the *Drosophila* embryo visualized with FL-phalloidin. Exp. Cell Res. 143, 103–144.

Warn, R.M. & Warn, A. (1986). Microtubule arrays present during the syncitial and cellular blastoderm stages of the early *Drosophila* embryo. Exp. Cell Res. 163, 201–210.

Wieschaus, E. & Sweeton, D. (1988). Requirements for X-linked zygotic gene activity during cellularization of early *Drosophila* embryos. Devel. 104, 483.

Williams, B.C., Karr, T.L., Montgomery, J.M., & Goldberg, M.L. (1992). The *Drosophila L (1) zw10* gene product, required for accurate mitotic chromosome segregation, is redistributed at anaphase onset. J. Cell Biol. 118, 759–774.

Yang, J.T., Saxton, W.M., & Goldstein, L.S.B. (1988). Isolation and characterization of the gene encoding the heavy chain of *Drosophila* kinesin. Proc. Natl. Acad. Sci. USA 85, 1864–1868.

Young, P.E., Pesacreta, T.C., and Kiehart, D.P. (1991). Dynamic changes in the distribution of cytoplasmic myosin during *Drosophila* embryogenesis. Development 111, 1–14.

Young, P. E., Richman, A.M., Ketchum, A.S., & Kiehart, D.P. (1992). Morphogenesis in *Drosophila* requires nonmuscle myosin heavy chain function. Genes & Development 7, 29–41.

Zhang, P., Knowles, B.A., Goldstien, L.S.B., & Hawley, R.S. (1990). A kinesin-like protein required for distributive chromosome segregation in *Drosophila*. Cell 62, 1053–1062.

KERATIN DISEASES

E. Birgitte Lane

Advances in Molecular and Cell Biology
Volume 12, pages 207–227.

ISBN: 1-55938-845-5

I. INTRODUCTION

Intermediate filaments have come out of the woods. Within the last two years, we have finally been able to "see" the consequences of cytoskeleton of intermediate filaments functioning, or not functioning, in the dramatic effects of their unscheduled breakdown and structural failure. But why did it take us so long to accept what they do? Perhaps because the obvious answer—that they have a structural role—was not seen as important or interesting.

The roles of the other two filamentous systems of the cytoskeleton, actin and tubulin, are well defined. These two systems have evolved a large complement of associated proteins which moderate their macromolecular assemblies to reorganize cell shape and internal structure. Actin and tubulin are clearly essential for life itself, and their genes appear almost unchanged in every organism examined. Their kinetics are complex, interesting and fairly well understood. Early days of cytoskeleton research centered around the biopolymers of actin and tubulin, and promised a precision that held the same attractions as molecular biology today—the lure of a biological question for which there is a finite and absolute answer. Intermediate filaments, on the other hand, appeared to be messy and confusing. They have nevertheless always been good candidates for a true cytoskeleton. Despite many years' speculations on a direct structural role for intermediate filaments, it has not been possible to demonstrate this theory in an experimental context. The very name, *intermediate filaments*, probably reflects the ambiguity with which they were viewed at the time, and the subsequent (and ongoing) emergence of a gene family of daunting complexity doubtless played a part in limiting the interest to a minority of scientists for many years.

A much clearer prediction of filament structure and function evolved with molecular biology, one which has now been dramatically vindicated by the recent discoveries of genetic defects in keratins which cause skin fragility. We can now no longer deny the function of intermediate filaments: they *are* "the cytoskeleton." They define the plasticity, rigidity and resilience of a cell, and thus determine the range of functional habitat in which a given cell type can be most effective. As much as, and perhaps more than any end-point differentiation product, it is now believed that the expression of specific intermediate filaments can be predicted to determine tissue and structure.

In this personal account, the recent advances in understanding the keratin-dependant skin fragility syndromes will be assessed in the context of information that has been accumulated over the last decade on intermediate filament proteins.

II. DEFINING THE INTERMEDIATE FILAMENT MULTIGENE FAMILY

Our understanding of the cytoskeleton grew substantially out of the use of indirect immunofluorescence, i.e., the tracking of antibodies bound to a subcellular component or system with a second antibody carrying a fluorescent tag. This allowed structures to be viewed clearly against a complex but invisible background. Using immunofluorescence to look at intermediate filaments in cells, it was clear that the structures they formed were very different from (and more heterogeneous than) microfilaments and microtubules. In the late 1970s, immunologically distinct variants were recognized in muscle (desmin), fibroblasts and endothelial cells (vimentin), neurons (neurofilaments), and glial cells (GFAP), and finally epithelial cells (keratins). In the early 1980s, the perceived complexity of the whole family of intermediate filament proteins rose rapidly with extensive analysis of epithelial tissues by gel electrophoresis. Protein and cDNA sequencing then clarified the relationships between the different members of this growing family, and paved the way for manipulative genetic engineering experiments (see below)

It is now recognized that intermediate filaments fall into a number of subgroups, types I–V or VI, based on sequence comparisons. These groupings are quite different from the early immunological ones, which mostly identified isoforms within one type (type III). Keratins consist of two groups, type I and type II, as had been earlier suggested from studies on sheep wool keratins (Parry et al., 1977). Following this nomenclature, the type III group contain the best known intermediate filament proteins: vimentin (in mesenchymal cells and others), desmin (in all kinds of muscle), GFAP (glial fibrillary acidic protein, in astroglial cells), and peripherin (in peripheral neurons). The type IV group are the neurofilaments: NF-L, NF-M, and NF-H are the low-, middle-, and high-molecular weight neurofilament proteins. Recently α-internexin has been added to the list, either in type IV or III. Sequence analysis revealed that the proteins which make up the nuclear lamina lining the nuclear membrane are also intermediate filament proteins. The constitutive and ubiquitous B-type lamins (lamin B_1 and lamin B_2), together with the A-type lamins (lamin A and lamin C) expressed as cells differentiate, now constitute the type V group. A type VI group was suggested for a very large intermediate filament protein named nestin, found in neuroepithelial stem cells (Lendahl et al., 1990). Some recently discovered but rather divergent intermediate filament-like proteins in lens cells may yet constitute another somewhat distantly-related group.

Of all these groups, it is thought that the lamins are probably the oldest, or the closest to an ancestral intermediate filament-like protein, and may indicate the functional baseline from which intermediate filaments evolved. They have certain protein sequence characteristics and gene organization features which set them apart from other cytoplasmic intermediate filaments in vertebrates, but which link them to intermediate filaments sequenced from invertebrates (Weber et al., 1989). One can imagine that reinforcement of a nuclear envelope could underlie a very early evolutionary selection for a cytoskeleton. The evolution of a nuclear envelope in an ancestral "progenote" would serve to compartmentalize DNA and so permit the development of a timing control on DNA replication and division, which in turn would enable intermitotic cell specialization (cellular differentiation) for different functions.

The most diverse groups of all intermediate filaments are the keratins. About three-quarters of all intermediate filament genes identified in human tissues encode keratins of the type I and type II groups. Defined and specific type I-type II pairs of keratins are co-expressed in different epithelial cells, to give an overall keratin expression profile in an epithelium that is specific for its state of differentiation. The significance of two groups of keratins is that these intermediate filaments are obligatory heteropolymers, and they have to form type I-type II heterodimers at the first stage of assembly. Single, undimerized keratins of either type are rapidly degraded in the cytoplasm (Domenjoud et al., 1988; Kulesh et al., 1989). This feature has been used to study mechanisms of filament assembly in cultured cells (Lu and Lane, 1990), and it could also be a simple mechanism for post-translational quantitative regulation of keratins *in vivo*. Although there is partial dependence on copolymerization in neurofilaments, among the intermediate filaments only keratins are exclusively heteropolymeric. Copolymerization may have evolved as a way of strengthening the filament networks. It may also be a way of ensuring stability of the system by buffering it against sporadic defects arising from mutations. Our experiments have suggested that an intact keratin can to an extent carry a defective partner into polymerization (Lu and Lane, 1990), and it now seems likely that mutant keratins may not be uncommon (see below).

III. TISSUE SPECIFICITY OF KERATINS

By the early 1980s, the monoclonal antibody revolution started by Köhler and Milstein was beginning to open the floodgates for a new category of tissue-type markers, as monoclonal antibodies to human keratins became available (Debus et al., 1982; Gown and Vogel, 1982; Lane, 1982; Tseng et al., 1982; Woodcock-Mitchell et al., 1982). Immunohistochemistry became a valuable source of interesting information on tissue-specific expression of intermediate filament proteins, although this information was sometimes difficult to decode. Availability of monoclonal antibodies greatly contributed to the documentation of intermediate filament expression patterns, allowing a finer *in situ* analysis than that obtained from biochemical analysis of homogenized tissue. However, this avenue of investigation was not readily accessible to experimental manipulation.

Besides their obvious value in immunohistochemistry for diagnostic pathology, these antibodies, when used with immunoblotting techniques, allowed rapid identification of proteins of interest. Underlying rules emerged which began to make sense of what was clearly an exquisite degree of tissue specificity. The systematic biochemical identification of around thirty human keratins, and their numerical cataloging (Moll et al., 1982), could now be translated into a smaller number of phenotypic keratin pairs, reflecting coexpression of paired specific keratins as the hallmarks of distinct epithelial differentiation pathways (Sun et al., 1984). Integrating data from these two approaches together, immunohistochemistry and immunoblotting, led to a much better appreciation of the complexity of keratin filament expression patterns in different organs and cell types.

The picture which emerged was as follows: Two pairs of the keratins appear to be primary, or essential, keratins. First, K8 (type II) and K18 (type I) are the primary keratins of embryonic and simple epithelial cells. This epithelial phenotype usually has a secretory function, and such cells can also, but not necessarily, express at least K7 (type II) and/or K19 (type I). Second, type II K5 and the type I K14 are the primary keratins of epithelial cells that can build up into multiple layers, such as stratified squamous epithelia, and also the basal cells in two-layered glandular epithelia. This epithelial phenotype can probably be equated with an undifferentiated keratinocyte (Purkis et al., 1990), and is not generally secretory. When such cells differentiate, they can synthesize any of a wide range of secondary or differentiation-specific keratin pairs depending on the body site: epidermal cells make K1 and K10, mucosal cells make K4 and K13, corneal cells make K3 and K12, and keratinocytes make K6 and K16 as part of a stress response. Although there appear to be no restrictions preventing any type I keratin tested from dimerizing with any type II keratin *in vitro* (Hatzfeld and Franke, 1985), type I/type II coexpression patterns observed *in vivo* are fairly specific, suggesting that there is functional selectivity somewhere within the amino acid sequence differences between one keratin and another within the same group or type. However, although this co-expression of pairs rule is true most of the time, and has been a most useful paradigm in understanding keratin functional regulation, exceptions are seen, such as early on in some differentiation pathways and sometimes during carcinogenesis. Type II keratins are often, if not always, expressed slightly ahead of their co-expressed type I partner, as epithelial cells begin to differentiate (Roop et al., 1987). This is probably a normal regulatory sequence of events in epithelial cells, and may reflect a differential stringency in the gene regulation of type I versus type II keratins (Stasiak et al., 1989).

IV. THE STRUCTURE OF THE CYTOPLASMIC ROPE

Amino acid sequencing of intermediate filament proteins suggested a common structural theme (Geisler and Weber, 1982) which has withstood the test of time, and has been reinforced as DNA sequences for keratins and other intermediate filaments were elucidated (Hanukoglu and Fuchs, 1982; Hanukoglu and Fuchs, 1983). Most intermediate filament genes have now been sequenced and located in the genome, and all the data are consistent. A consensus diagram of intermediate

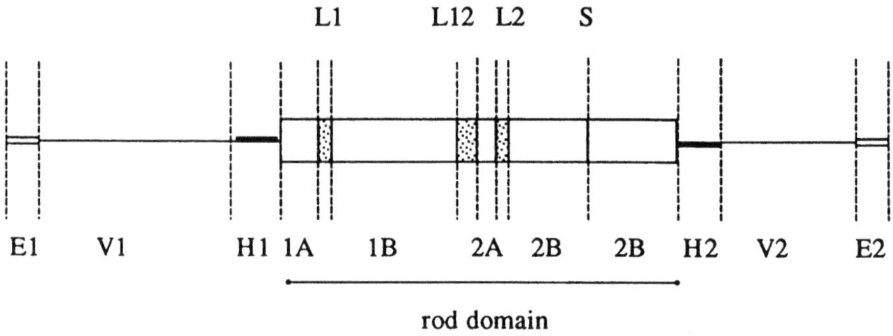

Figure 1. Diagram to show the consensus structure of intermediate filaments, indicating the current nomenclature for recognized subdomains. This drawing is based on type II keratins, and indicates (above) the subdomains interrupting the α-helix of the rod domain, and below, the nomenclature for the other subdomains.

filament polypeptide structure as represented by a type II keratin is shown in Figure 1, with current nomenclature for the domains and subdomains taken from Steinert and colleagues (Steinert and Parry, 1985).

A long central rod domain is common to all intermediate filaments, as an α-helical region of around 310 amino acid residues, which is interrupted at three or four highly conserved points by nonhelical linker stretches (L1, L12, L2) and an apparent helix polarity reversal (the skip residue, or stutter; S in Figure 1). At either end of the rod is a nonhelical domain, the head (amino-terminal end) and the tail (carboxy-terminal end). The sequences of the rod domains of isoforms within each type are quite conserved, and the classification into types I–V is primarily based on the rod domain sequence. Head and tail domains are much more variable, and have therefore been predicted to carry the functional fine-tuning between different keratin pairs. The rod domain is the structural subunit of the 10 nm wide intermediate filaments that can be seen by electron microscopy. Its sequence in any intermediate filament protein always shows a heptad periodicity of hydrophobic residues located at positions 1 and 4 (*a* and *d*) of every 7 (*a*–*g*) residues. Such periodicity is diagnostic of a protein that forms a coiled-coil with a similar protein (Cohen and Parry, 1986; Crick, 1953) and is a common structural theme among cytoskeleton proteins. The hydrophobic residues then lie in a strip along one side of the helical cylinder, rendering a single polypeptide chain unstable in an aqueous environment until two similar chains are twisted together to sequester the hydrophobic face along the internal seam between the two. This heptad repeat pattern of hydrophobic or bulky residues is an important indicator when making structural predictions about filament proteins, and can be identified in a primary amino acid sequence, using a cardboard cut-out frame.

Keratins thus form a coiled-coil type I-type II heterodimer (Hatzfeld and Weber, 1990; Steinert, 1990), and type III intermediate filaments such as desmin and vimentin form homopolymers. The protection against proteolysis that comes from rod domain interactions has been shown by the stability of expressed rod-domain-only protein fragments, which will not, however, polymerize to filaments (Lu and Lane, 1990). Some keratins may resist proteolysis for some time by aggregation, in particul, K1 and K10, which have long sticky end domains (Kartasova et al., 1993).

In addition to this heptad repeat feature, charged patches also occur along and around the external surfaces (Geisler et al., 1982) and these favor higher order interactions both between dimers to form tetramers, and between parallel and antiparallel strands. Thus the intermediate filament can be visualized as gradually assembling as a sort of submicroscopical cytoplasmic rope. Computer predictions and X-ray crystallography have suggested that, in the 10-nm filament, strands may be staggered in their lateral registration. With such variation, the details of precise molecular interactions are very hard to predict. Stewart and colleagues have presented another way of looking at registration, by analyzing rod domain packing from paracrystals of recombinant type III GFAP-cII fusion protein (Stewart et al., 1989). In these paracrystals, the GFAP rod domains were packed in alternating antiparallel stands, with helix 1A, 1B and 2A (Geisler et al., 1992) overlapping

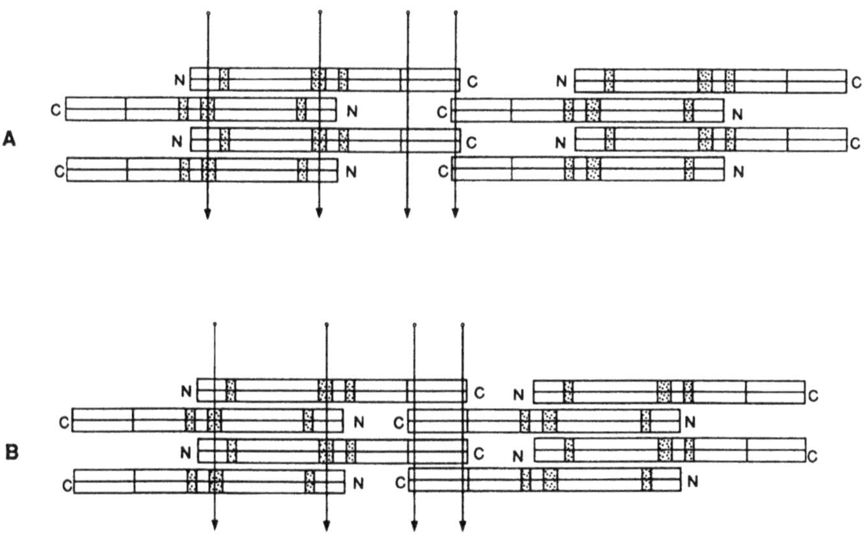

Figure 2. Diagram of one current theory about how the α-helices may interact in two dimensions in the assembled filament. Data taken from Stewart et al. (1989) and from Geisler et al. (1992). Vertical lines indicate positions of mutations which are or may be in clusters. (**A**): Packing suggested by Stewart et al. (1989) from studies of type III paracrystals; (**B**) tighter packing would bring another mutation pair in close proximity, and could be an alternative, extrapolating from lamin studies (Heitlinger et al., 1992).

head-to-tail (i.e., with 2A, 1B, and 1A, respectively) and a short overlap at the carboxy-terminal end of the rod. One of the current views on rod domain packing in filament assembly is summarized in Figure 2. Earlier studies had suggested that the most stable assembly intermediate of these filamentous proteins was probably an antiparallel tetramer (Quinlan et al., 1984); Stewart's model suggests that the tetramers overlap their carboxy-terminal ends to elongate the polymer. The main attractions of this model are that the regions predicted to overlap from the paracrystal model are the most highly conserved in amino acid sequence, and the non-overlapping bridge in helix 2 is the most variable. In particular, the small but highly significant overlap at the carboxy-terminal end involves the most conserved of all intermediate filament sequence motifs, sometimes called the *helix termination peptide*, which constitutes the boundary consensus sequence that terminates the rod domain, i.e., EIATY$^R/_K$ (X)LLEGEE. If the paracrystal model is a realistic reflection of high-affinity interactions in GFAP, then it is likely to be widely applicable among intermediate filament proteins since these regions are the most highly conserved ones throughout.

None of the helix packing models describe conclusively the arrangement of heads and tails, and how they are packed, and yet we need more information on the end domains if we are to understand how the tissue specificity works. Tissue-specific sequencing characteristics of the head and tail domains suggest that these regions define the extent and strength of the filament network, and may be important for structure modulation. There is also an increasing body of evidence suggesting an interaction of the terminal domains with helix boundary sequences which have not yet been assimilated into popular molecular models (Birkenberger and Ip, 1990; Djabali et al., 1991; Kouklis et al., 1991; Herrmann et al., 1992).

V. ATTEMPTS TO "KNOCK OUT" FILAMENTS IN CELLS

The hallmark of intermediate filament proteins has always been their stability and inertia, their long half-life. In attempts to demonstrate the function of this family of proteins, attempts to ablate them have been various and sometimes ingenious. Combinations of intensely cytotoxic drugs sometimes achieved some disruption, but in conjunction with other effects. Microinjection of monoclonal antibodies to certain types of epitopes (most still not identified) produced mild to dramatic disruption of the intermediate filament cytoskeleton but the cells would usually carry on metabolizing, migrating, and dividing (Gawlitta et al., 1981; Klymkowsky, 1981; Klymkowsky et al., 1983; Tölle et al., 1985; Tölle et al., 1986; Boller et al., 1987). Occasionally a cell line would be reported which apparently had no intermediate filaments at all (e.g., Paulin-Levasseur et al., 1989). In parallel with such published work, more distinct intermediate filaments were being discovered and sequenced, and by 1993 the total safely (excluding the marginal, unconfirmed, or unclassified candidates) reached 28 keratins, 4 type III proteins, 4 neurofilament proteins, and 4 lamins. Could this heterogeneity really reflect a need for subtly different isoforms in each tissue type, or was it just a reflection of evolutionary redundancy? Did this redundancy reflect a system evolved for *cis*-regulation, or multigene transcriptional control in blocks?

VI. PRISING APART FILAMENT ASSEMBLY

The most successful wedge to prise apart the process of intermediate filament assembly has been the type of mutant protein which has a dominant negative influence on the existing filament structure. Transfection of keratins or neurofilaments with large chunks deleted from either end have given useful information when the deletions produced dominant disruptor, or *antimorph*, phenotypes. Dominant disruptors arise whenever deletions removed the (type-conserved) boundary peptides of the consensus α-helical rod domain, but not if the deletions only removed parts of the nonhelical head and tail domains. Serial deletions from either the extreme carboxy-terminus (Albers and Fuchs, 1987) or the amino-terminus (Albers and Fuchs, 1989) of human keratin 14 expressed in epithelial cells showed that successively truncated keratins continued to be incorporated into the preexisting filament network, until the engineered deletions were so large as to encroach upon either end of the rod domain. At this point truncated proteins began to function as dominant disruptor mutants, causing the collapse of the endogenous keratin network. Similar results were obtained for neurofilament proteins NF-L and NF-M (Gill et al., 1990; Wong and Cleveland, 1990). This suggested that the boundary sequences of the helical rod domain are essential for subunit interactions for filament assembly, and this is supported by experiments in which peptides mimicking these sequences have been added to filaments or filament assembly mixtures, causing assembly to fail (Kouklis et al., 1991; Herrmann et al., 1992). Antibodies binding to terminal or near-terminal end-domain sequences have also been used *in vitro* to disrupt filament assemblies (Ip, 1988; Birkenberger and Ip, 1990; Kouklis et al., 1991), either by competing for the intermediate filament protein in binding to an essential interaction site or by steric hindrance of packing. Clearly, interference by binding of even a small proportion of competitor molecules which recognize major assembly-docking or high affinity sites, but fail to bridge the gap between sites at the two ends of the α-helical rod domain (see Figure 2), will disrupt filament integrity.

End domain removal experiments are particularly informative when performed in the absence of a preexisting carrier filament network. Microinjected intermediate filament protein will align on preexisting filaments with which they are already competent (Miller et al., 1991), and the helical rod domains on their own can become incorporated into a preexisting acceptor network of the same filament protein sort (Lu and Lane, 1990; Lu et al., 1993), even though they are incapable of forming a new filament network of their own. When end-domain truncation experiments were done in the absence of a preexisting filament network, paired keratins with truncations limited to the tail domains (rod domains intact) were unable to make a normal stable filament network by themselves (Lu and Lane, 1990). The detection of this effect depends on the proteins being expressed at fairly low copy numbers, as it is when the DNA is introduced by retroviral infection. Conventional transfection of tailless keratins appears to lead to higher expression levels, at which this instability is masked (Bader et al., 1991).

A number of studies have now addressed the question of domain function in filament assembly, mostly by expressing intermediate filament genes with pieces missing, or by swapping domains (McCormick et al., 1991) in various tissue culture cells. The effects of removing the non-helical end domains suggested that head domains were essential for polymerization of the homopolymer filaments (Traub and Vorgias, 1983; Kaufmann et al., 1985; Gill et al., 1990) and of the heteropolymer types (Lu and Lane, 1990; Wong and Cleveland, 1990). Tail domain removal had a more subtle effect, reducing the stability or efficiency with which the three-dimensional network was formed.

Overall, manipulation of intermediate filaments in tissue culture cells has produced many dramatic changes in filament organization, but little effect on the cells. Although this suggested to some authors that intermediate filaments had no vital function, this seems unlikely in view of their tissue specificity and the timing of their expression during embryogenesis. Major changes in intermediate filament expression patterns accompany major cell translocation and morphogenetic events during embryogenesis. An alternative interpretation is that their function lay in three-dimensional differentiated tissue, and that they were not required by cells growing rapidly, in the absence of differentiation, on a rigid plastic substrate.

VII. TRANSGENIC ANIMALS: ANOTHER DIMENSION

The ability to introduce foreign genes into the germline of mice has opened an awesome range of possibilities. For intermediate filaments, most of the obvious experiments have by now been attempted, but the results have often raised more questions than they have answered. The experiments fall into three categories: (1) excessive overexpression of the normal gene; (2) induced expression of an abnormal gene; and (3) complete ablation of expression of a gene.

Overexpression of a hair keratin in mice led to brittle hairs and periodically bald animals (Powell and Rogers, 1990). The excess (unpaired) keratin accumulated as non-filamentous aggregates in hair cells. This correlates with the ready aggregation of suprabasal keratins, even unpaired ones, when they are overexpressed (Kartasova et al., 1993). Overexpression of neurofilament genes in mice, both NF-H and NF-L, has recently been reported to produce a phenotype reminiscent of human neurodegenerative diseases (Côté et al., 1993; Xu et al., 1993). It is difficult to track integrity of single neurones, but one prediction must be that disorganization of filaments,

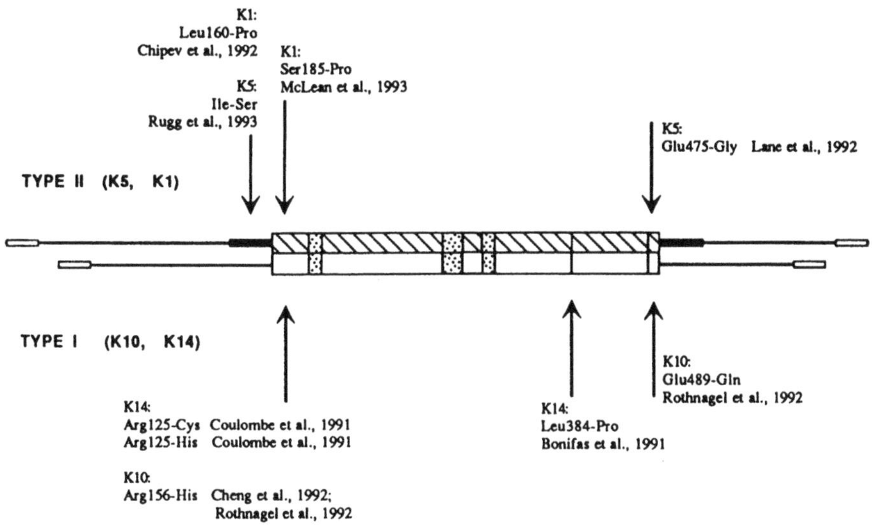

Figure 3. Summary of the keratin mutations described in type II (above) and type I (below) keratins from studies of EBS and BCIE patients with skin fragility syndromes.

especially aggregates, might weaken the cells or their axons and lead to cell fragmentation and broken neuronal connections.

Expressed abnormal transgenes which give a detectable phenotype, without deletion of the normal gene, are by definition dominant. End deletion experiments with keratins K14 (Vassar et al., 1991) and K10 (Fuchs et al., 1992), when transferred to mice, again gave dominant disruptor phenotypes as found in cultured cells. Although producing an extreme and lethal phenotype, this was recognizably similar to human skin blistering diseases. This was one of the experimental routes to the discovery of causal keratin mutations in blistering diseases (Coulombe et al., 1991).

The least ambiguous types of transgenic experiment ought to be the total ablation of a primary keratin. This was recently achieved for the keratin 8 gene (Baribault and Oshima, 1991; Baribault et al., 1993). Keratins 8 and 18 form the first embryonic keratins, expressed in the oocyte and preimplantation blastocyst. They persist as the sole keratins into a few adult cell types (hepatocytes, parts of kidney tubules), and are expressed in many other small epithelial cell types alongside additional later-expressed keratins. Keratins 8 and 18 are probably the oldest keratin pair in evolutionary terms (Waseem et al., 1990) and their genes are both located on the same chromosome (chromosome 12). Other keratin pairs are spatially separated, with all type II genes on chromosome 12 and type I genes on chromosome 17. Because of the well-established requirement for both type I and type II keratin to form filaments, deleting keratin 8 would leave the early mammalian embryo with no intermediate filaments, until approaching implantation when vimentin (Jackson et al., 1981) and keratin 19 (Kimber and Lane, unpublished observations) are synthesized. Thus it could reasonably be expected that deletion of this gene would be lethal, notwithstanding the fact that K8 negative F9 cells could still form embryoid bodies in tissue culture (Baribault and Oshima, 1991). The point at which K8 deletion would prove fatal would be very important however, since this should provide a vital clue to the essential function, if any, of these ancient and conserved keratins. Consequently, the possible outcome of these experiments was a matter for much speculation at recent conferences. It transpired the K8 *null/null* genotype was indeed progressively lethal, with most embryos (but not all) finally dying around mid-gestation. The explanation for the partial effect is still not clear, but it is clear that many embryos survive well past implantation.

VIII. MICE CAN'T TALK: ADVANTAGES OF CLINICAL STUDIES

Although these results are fascinating, they are extreme and artificial, and difficult to relate to a naturally evolved biological system. We need to know how intermediate filaments function naturally, what are the limitations on them, and what are the evolutionary pressures that have shaped these sequences to what they are today. Spontaneous viable mutations are very valuable study material to attempt to answer such questions. When these mutations are in people, we can also learn a lot more about the effects of these mutations than we can from mice; variations in the population show us the fringes of the possible; humans live for a long time and modify their behavior in ways that we can (mostly) understand. Thus, much excitement was generated when the first spontaneous intermediate filament mutations were finally discovered in a group of hereditary skin blistering conditions. The fact that these diseases were discovered in humans means that we have a huge information resource available not only from clinical records but from the patients themselves.

Skin fragility syndromes is a term to cover two groups (with several clinical subgroups) of hereditary blistering diseases, which turn out to be closely related by their origin in keratin sequence defects. Point mutations in keratins are the causative genetic defects in a group of hereditary skin fragility syndromes, as discovered first in *epidermolysis bullosa simplex* (EBS) and now subsequently in *bullous congenital ichthyosiform erythroderma* [BCIE, or BIE; this condition has also sometimes just been referred to under the more general term of *epidermolytic hyperkeratosis* (EH, or EHK)]. In these blistering diseases the skin is fragile and blisters on mild mechanical trauma, such as rough handling or friction from garment seams. The two conditions are usually inherited as autosomal dominant traits. In both conditions, the epidermal keratinocytes are rendered fragile by mutations in one of their keratin genes which compromise the strength or assembly properties of the cytoskeleton. This reduces the resilience of the cell, so that it cannot resist the stress as well as wild type keratinocytes.

The distinction between the two subgroups BCIE and EBS lies in the identity of the keratins affected. EBS is caused by mutations in K5 or K14 (Bonifas et al., 1991; Coulombe et al., 1991; Ishida-Yamamoto et al., 1991; Cheng et al., 1992; Lane et al., 1992), the primary keratins synthesized in basal cells of stratified squamous epithelia and appendageal basal cells. BCIE is caused by mutations in K1 and K10 (Bonifas et al., 1992; Cheng et al., 1992; Chipev et al., 1992; Ishida-Yamamoto et al., 1992; Rothnagel et al., 1992b), the secondary differentiation-specific keratins expressed in the suprabasal cell layers of cornifying stratified squamous epithelia, i.e., mainly in the epidermis.

The two conditions look different because of the consequences of rupturing cells in the basal layer (the proliferative compartment of the epidermis) as opposed to cells in the suprabasal layers (which are postmitotic, already committed to differentiation). EBS appears as fluid-filled blisters, often in clusters (Dowling-Meara form (McGrath et al., 1992)) from anywhere on the body to only on hands and feet (Weber-Cockayne form); otherwise the skin looks normal. BCIE on the other hand gives rise to areas of very thick scaley skin, which appears not to be sloughed off as normal. There are a wide range of variations in the precise form which either of these diseases can take; some variants are well-recognized and clinically categorized. It is likely that these variations all represent points on a spectrum of varying severity, but it is not yet clear whether any of the phenotypes ever reflects involvement of additional genes other than keratins.

IX. EPIDERMOLYSIS BULLOSA SIMPLEX

The first discoveries of spontaneous keratin mutations were made by three groups working independently. The groups each used different approaches and obtained three different results, but all three came together in a simple single-minded demonstration of the true cytoskeletal role of keratin intermediate filaments in epithelial cells. Observing the results of expressed mutations in people lends another dimension to the whole study of keratins; we can ask the affected people what their skin feels like, and how strong it is. We can also start to study a huge range of structural subtleties due to a number of sequence variations which are present in the population.

A. Mutant Keratin 14

One of the first two groups to publish an identified keratin mutation was Coulombe et al. (1991), who found a mutation in the α-helical subdomain 1A of keratin 14 (Figure 3), in three apparently unrelated patients with the Dowling-

Meara form of EBS. Although they were not able to obtain data from other family members to prove conclusively that these mutations were causative, subsequent events have lent a high degree of probability to the accuracy of this interpretation. The group did however reproduce a similar phenotype by transfection of a keratin with the same mutation introduced artificially in culture, and they did manage to culture some cells from the patients. Subsequently this group has found one more EBS mutation in this position (Cheng et al., 1992).

B. Linkage Analysis

Simultaneously another group described a mutation found by linkage analysis (Bonifas et al., 1991) in a family with EBS (reported as the Koebner variant). The mutation was in helix 2B of K14 and gave a Leucine to Proline change, in a region fairly close to the skip residue. Fortuitously this mutation created a new restriction site by which the group was able to rapidly screen a large number of controls and relatives, and thus prove unequivocally that the mutation was the causative one and not an irrelevant polymorphism.

C. Protein Alterations

Following extensive EM analysis of the abnormal K5/K14 filaments in Dowling-Meara EBS patients (Ishida-Yamamoto et al., 1991), a mutation in the helix termination peptide of keratin 5, the highly conserved sequence that marks the consensus end of the helical rod domain of all intermediate filaments, was discussed (Lane et al., 1992). We originally identified this as an abnormal antibody reactivity with one of two K5 allotypes within a (coincidentally) polymorphic family; the antibody in question, aIF (Pruss et al., 1981), is known to recognize the conserved helix termination peptide. K5 is the coexpressed keratin partner of K14, and between them they constitute the primary keratins of complex/stratifying epithelia and basal cells. They are the major keratins in the basal keratinocytes of skin and appendageal epithelia, and their synthesis diminishes as cells move upwards into the epibasal layers (on their way to terminal differentiation to form the dead impervious stratum corneum); once out of the basal layer, synthesis of keratins K1 and K10 takes over.

Faults in either of the two copolymerizing keratins, K5 or K14, could cause a pathological weakness in the skin cells and, similarly, a mutation at either end of the rod domain could be equally disruptive. At the present time, with only three mutation sites identified, it is not clear whether there is a requirement for one or the other end of the rod domain in type I or type II keratins, and whether or not there are "hotspots" of mutations associated with these diseases.

X. BULLOUS CONGENITAL ICHTHYOSIFORM ERYTHRODERMA

The identification of clinical effects of a point mutation in just one of the keratins of an epithelial cell type—i.e., pronounced fragility and inability to resist normal wear and tear—it was clear that other skin diseases, particularly blistering defects, might also be due to keratin defects. The other two forms of epidermolysis bullosa (junctional EB and dystrophic EB) are thought to be different diseases, since the blisters form at extracellular cleavage planes in the skin, and research is well on the way to identifying the genes responsible as type VII collagen (for junctional EB) and nicein/kallinin/epiligrin (for recessive dystrophic EB). However there is another histologically related condition, *bullous congenital ichthyosiform erythroderma*, in which the epidermal keratinocytes rupture higher up in the stratum spinosum and this appears to be due to a defect caused by point mutations in keratins K1 and K10, the secondary, differentiation-specific keratins that are specifically sythensized in suprabasal cells.

Selectivity for either end in type I or II does not seem to apply, as mutations have now been found in both type I and type II genes at both ends of the rod domain. Located at the beginning of the rod domain, Rothnagel et al. (1992b) found 2 cases of the same point mutation as Coulombe et al. (1991) had reported in EBS patients—in two apparently unrelated BCIE cases—in the rod domain 1A segment of keratin 1. Cheng et al. (1992) found a third example of the same mutation in EBS. It is not known whether these three patients could be related. We have recently found a different mutation in this domain in K1, a type II keratin, in BCIE (McLean et al., 1993).

Chipev et al. (1992) also found a mutation in the head domain of keratin 1. The region concerned was a short pre-helical stretch of conserved amino acids found in all type II keratins, a domain named H1 (homologous-1) by Steinert and Parry (1985; see Figure 1). This mutation was a Leucine to Proline, again potentially very disruptive of secondary structure and, interestingly, it also potentially creates an additional phosphorylation site, in a region which has been shown to contain phosphoamino acids in other intermediate filament proteins. Another mutation in this region of keratin 5 has recently been detected in a EBS-Weber Cockayne family, and this too creates a potential new phosphorylation site as a predicted substrate for protein kinase C (Rugg et al., 1993); this region may be another mutational hotspot.

A mutation in the helix termination peptide, of keratin 10, has also been detected by analysis of BCIE (Rothnagel et al., 1992a), to add the EBS mutation in K5 in this region (Figure 3). This of course fits well with the earlier truncation, antibody, and peptide experiments which suggested that this highly conserved subdomain was likely to be very important. The EBS mutation site in K14 helix 2B (Bonifas et al., 1991) is still a single mutation, and further mutations need to be isolated in order to know whether this too represents a mutational hotspot. These mutants are likely to be very useful in trying to work out the function of the skip residue feature in intermediate filaments.

XI. WHY ARE THESE MUTATIONS SO DELETERIOUS?

The appearance of restricted locations in a protein molecule at which mutations are identified could indicate critical functional sites at which mutations are especially damaging, or mutational hotspots in the DNA. In this case, it is predicted that many more mutations will come to light, associated with a spectrum of severity of functional consequences, from unremarkable individual variability (these mutations being hard to find, or possibly stable polymorphisms) to the most severe Dowling-Meara EBS. Alternatively, but less likely, they could of course also represent the only sites at which mutations are tolerant, all other mutations being lethal and so never showing.

As more mutations are discovered, it does indeed appear that there are clusters of mutations in a number of regions along the molecule (Figure 3). These appear to be (a) in H1 of the head domain (probably a regulatory region, and mutations can probably affect phosphorylation); (b) in helix 1A of the rod domain, a type-conserved region (probably an important registration or binding site (Magin et al., 1987) which may contain more than one "hotspot"); (c) close to the skip residue; and (d) there is another one in the helix termination peptide. Current ideas about helix packing (Figure 2) show that these sites interact at multiple points—except for the H1 domain, for which we have no information. Figure 2A shows the alignment predicted from Stewart et al. (1989) and Geisler et al. (1992), with a short antiparallel overlap at the helix termination peptide. Figure 2B is a speculation, arising from the variation in packing overlaps observed in the presence or absence of end domains (Heitlinger et al., 1992) of lamins, suggesting that tighter packing would give a dual interaction in helix 2B that would incorporate both the skip residue region and the helix termination peptide. One can imagine that these mutations have their deleterious but non-lethal effect because of the reduced affinity interactions that they introduce in the packing of intermediate filament protein subunits to make the 10 nm filaments seen by electron microscopy. The analogy of a fragile balloon (the cell) containing a more rigid box (the keratin cytoskeleton) inside it, can be useful. If the box is rigid, pressure can be applied to the balloon without harm as the box will take the strain. If however the box is too small, too fragile, too soft, or absent altogether, then pressure on the balloon will eventually burst it. We think this is what is happening to the basal keratinocytes in EBS, and to suprabasal cells in BCIE.

The striking thing about these mutations is that they are far more informative *in vivo* than similar mutations have been when examined as expressed in tissue culture cells *in vitro*. We have observed that immortalized cell lines can look very normal in culture, even though we know they are from patients with a severe skin condition and are still expressing this dominant negative protein. Many of the structural defects will simply not be seen until the whole system is put under the physical and physiological stress involved in normal evolutionary selection pressures. We believe that it is therefore very important to work at the clinical interface, and to see the patients, many of whom have stoically put up with their fragile skin and their mouth ulcers for many years without hope of treatment. They have many more clues to tell us.

Further studies will doubtless lead us before long to an understanding of the processes that lead to the clinical variants of these diseases, and will simultaneously teach us many useful things about how keratins, the cytoskeleton proteins of epithelia, really work *in vivo*.

REFERENCES

Albers, K. & Fuchs, E. (1987). The expression of mutant epidermal keratin cDNAs transfected in simple epithelial and squamous cell carcinoma lines. J. Cell Biol. 105, 791–806.

Albers, K. & Fuchs, E. (1989). Expression of mutant keratin cDNAs in epithelial cells reveals possible mechanisms for initiation and assembly of intermediate filaments. J. Cell Biol. 108, 1477–1493.

Bader, B.L., Magin, T.M., Freudenmann, M., Stumpp, S., & Franke, W.W. (1991). Intermediate filaments formed de novo from tail-less cytokeratins in the cytoplasm and in the nucleus. J. Cell Biol. 115, 1293–1307.

Baribault, H. & Oshima, R.G. (1991). Polarized and functional epithelia can form after the targeted inactivation of both mouse keratin 8 alleles. J. Cell Biol. 115, 1675–1684.

Baribault, H., Price, J., Miyai, K., & Oshima, R.G. (1993). Mid-gestational lethality in mice lacking keratin 8. Genes & Dev. (in press).

Birkenberger, L. & Ip, W. (1990). Properties of the desmin tail domain: Studies using synthetic peptides and antipeptide antibodies. J. Cell Biol. 111, 2063–2075.

Boller, K., Kemler, R., Baribault, H., & Doetschman, T. (1987). Differential distribution of cytokeratins after microinjections of anti-cytokeratin monoclonal antibodies. Eur. J. Cell Biol. 43, 459–468.

Bonifas, J.M., Bare, J.W., Chen, M.A., Niemi, K.M., & Epstein, E.H. (1992). Epidermolytic hyperkeratosis: Linkage to keratin gene regions on chromosomes 12q and 17q in two families. J. Invest. Dermatol. 98, 573a.

Bonifas, J.M., Rothman, A.L., & Epstein, E.H. (1991). Epidermolysis bullosa simplex: Evidence in two families for keratin gene abnormalities. Science 254, 1202–1205.

Cheng, J., Syder, A.J., Yu, Q.-C., Letai, A., Paller, A., & Fuchs, E. (1992). The genetic basis of epidermolytic hyperkeratosis: A disorder of differentiation-specific epidermal keratin genes. Cell 70, 811–819.

Chipev, C.C., Korge, B.P., Markova, N., Bale, S.J., DiGiannova, J.J., Compton, J.G., & Steinert, P.M. (1992). A leucine-proline mutation in the H1 subdomain of keratin 1 causes epidermolytic hyperkeratosis. Cell 70, 821–828.

Cohen, C. & Parry, D.A.D. (1986). α-Helix coiled coils—a widespread motif in proteins. Trends Biochem. Sci. 11, 245–248.

Côté, F., Collard, J.-F., & Julien, J.-P. (1993). Progressive neuronopathy in transgenic mice expressing the human neurofilament heavy gene: A mouse model of amyotrophic lateral sclerosis. Cell 73, 35–46.

Coulombe, P.A., Hutton, M.E., Letai, A., Hebert, A., Paller, A.S., & Fuchs, E. (1991). Point mutations in human keratin 14 genes of epidermolysis bullosa simplex patients: Genetic and functional analysis. Cell 66, 1301–1311.

Crick, F.H.C. (1953). The packing of α-helices: Simple coiled-coils. Acta Chrystallogr. 6, 689–697.
Debus, E., Weber, K., & Osborn, M. (1982). Monoclonal cytokeratin antibodies that distinguish simple from stratified squamous epithelia: Characterization on human tissues. EMBO J. 1, 1641–1647.
Djabali, K., Portier, M.M., Gros, F., Blobel, G., & Georgatos, S.D. (1991). Network antibodies identify nuclear lamin B as a physiological attachment site for peripherin intermediate filaments. Cell 64, 109–121.
Domenjoud, L., Jorcano, J.L., Breuer, B., & Alonso, A. (1988). Synthesis and fate of keratins 8 and 18 in nonepithelial cells transfected with cDNA. Exp. Cell Res. 179, 352–361.
Fuchs, E., Esteves, R.A., & Coulombe, P.A. (1992). Transgenic mice expressing a mutant K10 gene reveal the likely genetic basis for epidermolytic hyperkeratosis. Proc. Natl. Acad. Sci. USA 89, 6906–6910.
Gawlitta, W., Osborn, M., & Weber, K. (1981). Coiling of intermediate filaments induced by microinjection of a vimentin-specific antibody that does not interfere with locomotion and mitosis. Europ. J. Cell Biol. 26, 83–90.
Geisler, N., Kaufmann, E., & Weber, K. (1982). Protein chemical characterization of three structurally distinct domains along the protofilament unit of desmin 10 nm filaments. Cell 30, 277–286.
Geisler, N., Schunemann, J., & Weber, K. (1992). Chemical cross-linking indicates a staggered and antiparallel protofilament of desmin and characterizes one higher-level complex between protofilaments. Eur. J. Biochem. 206, 841–852.
Geisler, N. & Weber, K. (1982). The amino acid sequence of chicken muscle desmin provides a common structural model for intermediate filament proteins. EMBO J. 1, 1649–1656.
Gill, S. R., Wong, P. C., Monteiro, M. J., & Cleveland, D. W. (1990). Assembly properties of dominant and recessive mutations in the small mouse neurofilament (NF-L) subunit. J. Cell Biol. 111, – .
Gown, A. M. & Vogel, A. M. (1982). Monoclonal antibodies to intermediate filament proteins of human cells: Unique and cross-reacting antibodies. J. Cell Biol. 95, 414–424.
Hanukoglu, I. & Fuchs, E. (1982). The cDNA sequence of a human epidermal keratin: Divergence of sequence but conservation of structure among intermediate filament proteins. Cell 31, 243–252.
Hanukoglu, I. & Fuchs, E. (1983). The cDNA sequence of a type II cytoskeletal keratin reveals constant and variable structural domains among keratins. Cell 33, 915–924.
Hatzfeld, M. & Franke, W.W. (1985). Pair formation and promiscuity of cytokeratins: Formation in vitro of heterotypic complexes and intermediate-sized filaments by homologous and heterologous recombinations and purified polypeptides. J. Cell Biol. 101, 1826–1841.
Hatzfeld, M. & Weber, K. (1990). The coiled coil of *in vitro* assembled keratin filaments is a heterodimer of type I and II keratins: Use of site-specific mutagenesis and recombinant protein expression. J. Cell Biol. 110, 1199–1210.
Heitlinger, E., Peter, M., Lustig, A., Villiger, W., Nigg, E. A., & Aebi, U. (1992). The role of the head and tail domain in lamin structure and assembly: Analysis of bacterially expressed chicken lamin A and trucated B2 lamins. J. Struct. Biol. 108, 74–91.
Herrmann, H., Hofmann, I., & Franke, W.W. (1992). Identification of a nonapeptide motif in the vimentin head domain involved in intermediate filament assembly. J. Mol. Biol. 223, 637–650.
Ip, W. (1988). Modulation of desmin intermediate filament assembly by a monoclonal antibody. J. Cell Biol. 106, 735–745.
Ishida-Yamamoto, A., McGrath, J.A., Chapman, S.J., Leigh, I.M., Lane, E.B., & Eady, R.A.J. (1991). Epidermolysis bullosa simplex (Dowling-Meara type) is a genetic disease characterized by an abnormal keratin filament network involving keratins K5 and K14. J. Invest. Dermatol. 97, 959–968.
Ishida-Yamamoto, A., McGrath, J.A., Judge, M.R., Leigh, I.M., Lane, E.B., & Eady, R.A.J. (1992). Selective involvement of keratins K1 and K10 in the cytoskeletal abnormality of epidermolytic hyperkeratosis (bullous congenital ichthyosiform erythroderma). J. Invest. Dermatol. 99, 19–26.

Jackson, B.W., Grund, C., Winter, S., Franke, W.W., & Illmensee, K. (1981). Formation of cytoskeletal elements during mouse embryogenesis II. Epithelial differentiation and intermediate-sized filaments in early postimplantation embryos. Diff. 20, 203–216.

Kartasova, T., Roop, D.R., Holbrook, K.A., & Yuspa, S.H. (1993). Mouse differentiation-specific keratins 1 and 10 require a preexisting keratin scaffold to form a filament network. J. Cell Biol. 120, 1251–1261.

Kaufmann, E., Weber, K., & Geisler, N. (1985). Intermediate filament forming ability of desmin derivatives lacking either the amino-terminal 67 or the carboxy-terminal 27 residues. J. Mol. Biol. 185, 733–742.

Klymkowsky, M.W. (1981). Intermediate filaments in 3T3 cells collapse after intracellular injection of a monoclonal anti-intermediate filament antibody. Nature 291, 249–251.

Klymkowsky, M.W., Miller, R.H., & Lane, E.B. (1983). Morphology, behaviour, and interaction of cultured epithelial cells after the antibody-induced disruption of keratin filament organization. J. Cell Biol. 96, 494–509.

Kouklis, P.D., Papamarcaki, T., Merdes, A., & Georgatos, S.D. (1991). A potential role for the COOH-terminal domain in the lateral packing of type III intermediate filaments. J. Cell Biol. 114, 773–786.

Kulesh, D.A., Ceceña, G., Darmon, Y.M., Vasseur, M., & Oshima, R.G. (1989). Posttranslational regulation of keratins: Degradation of mouse and human keratins 18 and 8. Mol. Cell Biol. 9, 1553–1565.

Lane, E.B. (1982). Monoclonal antibodies provide specific intramolecular markers for the study of epithelial tonofilament organization. J. Cell Biol. 92, 665–673.

Lane, E.B., Rugg, E.L., Navsaria, H., Leigh, I.M., Heagerty, A.H.M., Ishida-Yamamoto, A., & Eady, R.A.J. (1992). A mutation in the conserved helix termination peptide of keratin 5 in hereditary skin blistering. Nature 356, 244–246.

Lendahl, U., Zimmerman, L.B., & McKay, R.D.G. (1990). CNS stem cells express a new class of intermediate filament protein. Cell 60, 585–595.

Lu, X. & Lane, E.B. (1990). Retrovirus-mediated transgenic keratin expression in cultured fibroblasts: Specific domain functions in keratin stabilization and filament formation. Cell 62, 681–696.

Lu, X., Quinlan, R.A., Steel, J.B., & Lane, E.B. (1993). Network incorporation of intermediate filament molecules differs between pre-existing and new filaments. Expl. Cell Res. (in press).

Magin, T.M., Hatzfeld, M., & Franke, W.W. (1987). Analysis of cytokeratin domains by cloning and expression of intact and deleted polypeptides in Escherichia coli. EMBO J. 6, 2607–2615.

McCormick, M.B., Coulombe, P.A., & Fuchs, E. (1991). Sorting out IF networks: Consequences of domain swapping on IF recognition and assembly. J. Cell Biol. 113, 1111–1124.

McGrath, J.A., Ishida-Yamamoto, A., Tidman, M.J., Heagerty, A.H.M., Scofield, O.M.V., & Eady, R.A.J. (1992). Epidermolysis bullosa simplex (Dowling-Meara): A clinicopathological review. Brit. J. Dermatol. 126, 421–430.

Miller, R.K., Vikstrom, K., & Goldman, R.D. (1991). Keratin incorporation into intermediate filament networks is a rapid process. J. Cell Biol. 113, 843–855.

Moll, R., Franke, W.W., Schiller, D.L., Geiger, B., & Krepler, R. (1982). The catalog of human cytokeratins: Patterns of expression in normal epithelia, tumors and cultured cells. Cell 31, 11–24.

Parry, D.A.D., Crewther, W.G., Fraser, R.D.B., & MacRae, T.P. (1977). Structure of α-keratin: Structural implication of the amino acid sequences of the type I and type II chain segments. J. Mol. Biol. 113, 449–454.

Paulin-Levasseur, M., Scherbarth, A., Giese, G., Röser, K., Bohn, W., & Traub, P. (1989). Expression of nuclear lamins in mammalian somatic cells lacking cytoplasmic intermediate filament proteins. J. Cell Sci. 92, 361–370.

Powell, B.C. & Rogers, G.E. (1990). Cyclic hair-loss and regrowth in transgenic mice overexpressing an intermediate filament gene. EMBO J. 9, 1485–1493.

Pruss, R.M., Mirsky, R., Raff, M.C., Thorpe, R., Dowding, A.J., & Anderton, B.H. (1981). All classes of intermediate filaments share a common antigenic determinant defined by a monoclonal antibody. Cell 27, 419–428.

Purkis, P.E., Steel, J.B., Mackenzie, I.C., Nathrath, W.B.J., Leigh, I.M., & Lane, E.B. (1990). Antibody markers of basal cells in complex epithelia. J. Cell Sci. 97, 39–50.

Quinlan, R.A., Cohlberg, J.A., Schiller, D.L., Hatzfeld, M., & Franke, W.W. (1984). Heterotypic tetramer (A2D2) complexes of non-epidermal keratins isolated from cytoskeletons of rat hepatocytes and hepatoma cells. J. Mol. Biol. 178, 365–388.

Roop, D.R., Huitfeldt, H., Kilkenny, A., & Yuspa, S.H. (1987). Regulated expression of differentiation-associated keratins in cultured epidermal cells detected by monospecific antibodies to unique peptides of mouse epidermal keratins. Differentiation 35, 143–150.

Rothnagel, J.A., Dominey, A.M., Dempsey, L.D., Longley, M.A., Greenhalg, D.A., Gagne, T.A., Huber, M., Frenk, E., Hohl, D., & Roop, D.R. (1992a). Mutations in the rod domains of keratins 1 and 10 in epidermolytic hyperkeratosis. Science 257, 1128–1130.

Rothnagel, J.A., Dominey, A.M., Dempsey, L.D., Longley, M.A., Greenhalgh, D.A., Gagne, T.A., Huber, M., Frenk, E., Hohl, D., & Roop, D.R. (1992b). Mutations in the rod domains of keratins 1 and 10 in epidermolytic hyperkeratosis. Science 257, 1128–1130.

Stasiak, P.C., Purkis, P.E., Leigh, I.M., & Lane, E.B. (1989). Keratin 19: Predicted amino acid sequence and broad tissue distribution suggest it evolved from keratinocyte keratins. J. Invest. Dermatol. 92, 707–716.

Steinert, P.M. (1990). The two-chain coiled-coil molecule of native epidermal keratin intermediate filaments is a type I-type II heterodimer. J. Biol. Chem. 265, 8766–8774.

Steinert, P.M. & Parry, D.A.D. (1985). Intermediate filaments: Conformity and diversity of expression and structure. Ann. Rev. Cell Biol. 1, 41–65.

Stewart, M., Quinlan, R.A., & Moir, R.D. (1989). Molecular interactions in paracrystals of a fragment corresponding to the α-helical coiled-coil rod portion of glial fibrillary acidic protein: Evidence for an antiparallel packing of molecules and polymorphism related to intermediate filament structure. J. Cell. Biol. 109, 225–234.

Sun, T.-T., Eichner, R., Schermer, A., Cooper, D., Nelson, W.G., & Weiss, R.A. (1984). Classification, expression and possible mechanisms of evolution of mammalian epithelial keratins: A unifying model. In: The Transformed Phenotype, Vol. 1 (Levine, A.J., Vande Woude, G.F., Topp, W.C., & Watson, J.D. eds.), pp. 169–176, Cold Spring Harbor Laboratory, New York.

Tölle, H.-G., Weber, K., & Osborn, M. (1986). Microinjection of monoclonal antibodies to vimentin, desmin, and GFA in cells which contain more than one IF type. Exp. Cell. Res. 162, 462–474.

Tölle, H.G., Weber, K., & Osborn, M. (1985). Microinjection of monoclonal antibodies specific for one intermediate filament protein in cells containing multiple keratins allow insight into the composition of particular 10 nm filaments. Eur. J. Cell Biol. 38, 234–244.

Traub, P. & Vorgias, C.E. (1983). Involvement of the N-terminal polypeptide of vimentin in the formation of intermediate filaments. J. Cell Sci. 63, 43–67.

Tseng, S.C.H., Jarvinen, M.J., Nelson, W.G., Huang, J.W., Woodcock-Mitchell, J., & Sun, T.T. (1982). Correlation of specific keratins with different types of epithelial differentiation: Monoclonal antibody studies. Cell 30, 361–372.

Vassar, R., Coulombe, P.A., Degenstein, L., Albers, K., & Fuchs, E. (1991). Mutant keratin expression in transgenic mice causes marked abnormalities resembling a human genetic skin disease. Cell 64, 365–380.

Waseem, A., Alexander, C.M., Steel, J.B., & Lane, E.B. (1990). Embryonic simple epithelial keratins 8 and 18: Chromosomal location emphasizes difference from other keratin pairs. New Biol. 2, 464–478.

Weber, K., Plessmann, U., & Ulrich, W. (1989). Cytoplasmic intermediate filament proteins of invertebrates are closer to nuclear lamins than are vertebrate intermediate filament proteins: sequence characterization of two muscle proteins of a nematode. EMBO J. 8, 3221–3227.

Wong, P.C. & Cleveland, D.W. (1990). Characterization of dominant and recessive assembly-defective mutations in mouse neurofilament NF-M. J. Cell Biol. 111, – .

Woodcock-Mitchell, J., Eichner, R., Nelson, W.G., & Sun, T.T. (1982). Immunolocalization of keratin polypeptides in human epidermis using monoclonal antibodies. J. Cell Biol. 95, 580–588.

Xu, Z., Cork, L.C., Griffin, J. W., & Cleveland, D.W. (1993). Increased expression of neurofilament subunit NF-L produces morphological alterations that resemble the pathology of human motor neuron disease. Cell 73, 23–33.

INDEX

www.ingramcontent.com/pod-product-compliance
Ingram Content Group UK Ltd.
Pitfield, Milton Keynes, MK11 3LW, UK
UKHW020806290926
15470UKWH00003B/987

9 781559 388450